TODAY'S TECHNICIAN

CLASSROOM MANUAL FOR
AUTOMOTIVE ENGINE
REPAIR AND REBUILDING

FOURTH EDITION

TODAY'S TECHNICIAN ™

CLASSROOM MANUAL FOR AUTOMOTIVE ENGINE REPAIR AND REBUILDING

FOURTH EDITION

Christopher Hadfield
Hennepin Technical College
Brooklyn Park, MN

DELMAR
CENGAGE Learning™

Australia • Canada • Mexico • Singapore • Spain • United Kingdom • United States

DELMAR
CENGAGE Learning™

Today's Technician™: Automotive Engine Repair & Rebuilding, 4th Edition
Christopher Hadfield

Vice President, Career and Professional Editorial: Dave Garza

Director of Learning Solutions: Sandy Clark

Executive Editor: David Boelio

Managing Editor: Larry Main

Senior Product Manager: Matthew Thouin

Editorial Assistant: Jillian Borden

Vice President, Career and Professional Marketing: Jennifer McAvey

Executive Marketing Manager: Deborah S. Yarnell

Marketing Manager: Jimmy Stephens

Marketing Coordinator: Mark Pierro

Production Director: Wendy Troeger

Production Manager: Mark Bernard

Content Project Manager: Cheri Plasse

Art Director: Benj Gleeksman

For product information and technology assistance, contact us at
Professional Group Cengage Learning Customer & Sales Support, 1-800-354-9706
For permission to use material from this text or product,
submit all requests online at **cengage.com/permissions.**
Further permissions questions can be e-mailed to
permissionrequest@cengage.com.

Library of Congress Control Number: 2009902929

ISBN-13: 978-1435426061
ISBN-10: 1435426061

Delmar
5 Maxwell Drive
Clifton Park, NY 12065-2919
USA

Cengage Learning is a leading provider of customized learning solutions with office locations around the globe, including Singapore, the United Kingdom, Australia, Mexico, Brazil and Japan. Locate your local office at:
international.cengage.com/region

Cengage Learning products are represented in Canada by Nelson Education, Ltd.

For your lifelong learning solutions, visit **delmar.cengage.com**

Visit our corporate website at **cengage.com.**

Notice to the Reader
Publisher does not warrant or guarantee any of the products described herein or perform any independent analysis in connection with any of the product information contained herein. Publisher does not assume, and expressly disclaims, any obligation to obtain and include information other than that provided to it by the manufacturer. The reader is expressly warned to consider and adopt all safety precautions that might be indicated by the activities described herein and to avoid all potential hazards. By following the instructions contained herein, the reader willingly assumes all risks in connection with such instructions. The publisher makes no representations or warranties of any kind, including but not limited to, the warranties of fitness for particular purpose or merchantability, nor are any such representations implied with respect to the material set forth herein, and the publisher takes no responsibility with respect to such material. The publisher shall not be liable for any special, consequential, or exemplary damages resulting, in whole or part, from the readers' use of, or reliance upon, this material.

3 4 5 XX 11

CONTENTS

Preface . . . . **vii**

CHAPTER 1 ***Automotive Engines*** **1**
• Introduction 1 • Basic Engine Operation 2 • Cooling System 5 • Lubrication System 6
• Engine Breathing 7 • Engine Performance 8 • Cylinder Head 8 • Timing Mechanism 10
• Engine Block 10 • Summary 12 • Review Questions 12

CHAPTER 2 ***Theory of Engine Operation*** **14**
• Introduction 14 • Major Engine Components 15
• Engine Operating Principles 18 • Behavior of Liquids and Gases 21
• Pressure and Vacuum 22 • Boyle's Law 22 • Engine Operation 22
• Engine Classifications 24 • Engine Vibration 28 • Engine Displacement 31
• Direction of Crankshaft Rotation 33 • Engine Measurements 33
• Other Engine Designs 39 • Displacement-on-Demand Engines 44
• Engine Identification 45 • Summary 47 • Review Questions 49

CHAPTER 3 ***Engine Repair and Rebuilding Industry*** **51**
• Introduction 51 • Full Service Repair Facilities 51
• Machine Shop and Engine Rebuild Facilities 52
• Engine Repair and Replacement Specialty Facilities 53
• Engine and Component Remanufacturing Facilities 54 • Summary 55
• Review Questions 56

CHAPTER 4 ***Engine Operating Systems*** **58**
• Introduction 58 • The Starting System 59 • Lubrication Systems 64
• Cooling Systems 74 • Cooling Fans 85
• Lubrication and Cooling System Warning Indicators 87 • Fuel System 91
• Automotive Fuels 95 • Summary 99 • Review Questions 100

CHAPTER 5 ***Engine Materials, Fasteners, Gaskets, and Seals*** **102**
• Introduction 102 • Engine Materials 102 • Manufacturing Processes 106
• Fasteners 110 • Gaskets, Seals, Sealants, and Adhesives 115
• Summary 122 • Review Questions 123

CHAPTER 6 ***Intake and Exhaust Systems*** **125**
• Introduction 125 • Air Induction System 125 • Air Intake Ductwork 126
• Air Cleaner/Filter 127 • Intake Manifold 129 • Vacuum Basics 133
• Vacuum Controls 134 • Turbochargers 134 • Superchargers 139
• Exhaust System Components 143 • Mufflers 147 • Summary 150
• Review Questions 150

CHAPTER 7 ***Factors Affecting Engine Performance*** **152**
• Introduction 152 • Spark Plugs 152 • Combustion Chamber Sealing 154
• Fuel and Combustion 158 • Engine Noises 163 • Summary 163
• Review Questions 164

CONTENTS

CHAPTER 8 Engine Configurations, Mounts, and Remanufactured Engines 166
• Introduction 166 • Engine Configurations 166 • Engine Mounts 168
• Remanufactured Engines 171 • Summary 173 • Review Questions 173

CHAPTER 9 Cylinder Heads 175
• Introduction 175 • Cylinder Heads 175 • Cylinder Head Component Relationships 187
• Combustion Chamber Designs 188 • The Combustion Process 192
• Multivalve Engines 195 • Summary 196 • Review Questions 197

CHAPTER 10 Camshafts and Valvetrains 199
• Introduction 199 • Valvetrain Components 199 • Summary 213 • Review Questions 213

CHAPTER 11 Timing Mechanisms 215
• Introduction 215 • Valve Timing System 216 • Chain-Driven Systems 218
• Belt-Driven Systems 219 • Gear-Driven Systems 221
• Variable Valve Timing and Lift Systems 222 • Summary 225 • Review Questions 225

CHAPTER 12 Engine Block Construction 227
• Introduction 227 • Block Construction 229 • Crankshaft 234 • Camshaft 238
• Lifter Bores 238 • Harmonic Balancer 239 • Flywheel 239
• Short Blocks, Long Blocks, and Crate Engines 240 • Summary 241
• Review Questions 241

CHAPTER 13 Pistons, Rings, Connecting Rods, and Bearings 243
• Introduction 243 • Bearings 243 • Camshaft and Balance Shaft Bearings 249
• Balance Shafts 250 • Pistons 250 • Piston Designs and Construction 256
• Piston Rings 258 • Connecting Rods 261 • Summary 262
• Review Questions 263

CHAPTER 14 Alternative Fuel and Advanced Technology Vehicles . . .265
• Introduction 265 • Alternative Fuel Vehicle Use 266 • Propane Vehicles 266
• E85 and Flexible Fuel Vehicles 269 • Compressed Natural Gas Vehicles 272
• The Honda Civic GX CNG Vehicle 277 • Electric Vehicles 278 • Hybrid Electric Vehicles 279
• HEV Operation 280 • Toyota Prius HEV 285 • Fuel Cell Vehicles 287 • Summary 287
• Review Question 288

APPENDIX A 290

APPENDIX B 291

Glossary 294

Index 308

PREFACE

Thanks to the support the Today's Technician™ series has received from those who teach automotive technology. Delmar, a division of Cengage Learning and the leader in automotive related textbooks, is able to live up to its promise to provide new editions regularly. We have listened and responded to our critics and our fans and present this new updated and revised fourth edition. By revising our series regularly, we can and will respond to changes in the industry, changes in the technology, changes in the certification process, and to the ever-changing needs of those who teach automotive technology.

The Today's Technician™ series features textbooks that cover all mechanical and electrical systems of automobiles and light trucks (while the heavy-duty trucks portion of the series does the same for heavy-duty vehicles). Principally, the individual titles correspond to the main areas of ASE (National Institute for Automotive Service Excellence) certification. Additional titles include remedial skills and theories common to all of the certification areas and advanced or specific subject areas that reflect the latest technological trends. Each text is divided into two volumes: a Classroom Manual and a Shop Manual.

Unlike yesterday's mechanic, the technician of today and for the future must know the underlying theory of all automotive systems and be able to service and maintain those systems. Dividing the material into two volumes provides the reader with the information needed to begin a successful career as an automotive technician without interrupting the learning process by mixing cognitive and performance learning objectives into one volume.

The design of Delmar's Today's Technician™ series was based on features that are known to promote improved student learning. The design was further enhanced by a careful study of survey results, in which the respondents were asked to value particular features. Some of these features can be found in other textbooks, while others are unique to this series.

Each Classroom Manual contains the principles of operation for each system and subsystem. The Classroom Manual also contains discussions on design variations of key components used by the different vehicle manufacturers. This volume is organized to build upon basic facts and theories. The primary objective of this volume is to allow the reader to gain an understanding of how each system and subsystem operates. This understanding is necessary to diagnose the complex automobiles of today and tomorrow. Although the basics contained in the Classroom Manual provide the knowledge needed for diagnostics, diagnostic procedures appear only in the Shop Manual. An understanding of the basics is also a requirement for competence in the skill areas covered in the Shop Manual.

A spiral bound Shop Manual covers the "how-tos." This volume includes step-by-step instructions for diagnostic and repair procedures. Photo Sequences are used to illustrate some of the common service procedures. Other common procedures are listed and are accompanied with line drawings and photos that allow the reader to visualize and conceptualize the finest details of the procedure. This volume also contains the reasons for performing the procedures, as well as when that particular service is appropriate.

The two volumes are designed to be used together and are arranged in corresponding chapters. Not only are the chapters in the volumes are linked together, but also the contents of the chapters are linked. This linking of content is evidenced by marginal callouts that refer the reader to the chapter and page that the same topic is addressed in the other volume. This feature is valuable to instructors. Without this feature, users of other two-volume textbooks must search the index or table of contents to locate supporting information in the other volume. This is not only cumbersome, but also creates additional work for an instructor when planning the presentation of material and when making reading assignments. It is also valuable to the students; with page references, they also know exactly where to look for supportive information.

Both volumes contain clear and thoughtfully selected illustrations. Many of which are original drawings or photos specially prepared for inclusion in this series. This means that the art is a vital part of each textbook and not merely inserted to increase the numbers of illustrations.

The page layout, used in the series, is designed to include information that would otherwise break up the flow of information presented to the reader. The main body of the text includes all of the "need-to-know" information and illustrations. In the wide side margins of each page are many of the special features of the series. Items that are truly "nice-to-know" information include simple examples of concepts just introduced in the text, explanations or definitions of terms that will not be defined in the glossary, examples of common trade jargon used to describe a part or operation, and exceptions to the norm explained in the text. This type of infromation is placed in the margin, out of the normal flow of information. Many textbooks attempt to include this type of information and insert it in the main body of text; this tends to interrupt the thought process and cannot be pedagogically justified. By placing this information off to the side of the main text, the reader can select when to refer to it.

Jack Erjavec
Series Advisor

HIGHLIGHTS OF THIS EDITION—CLASSROOM MANUAL

The Classroom Manual for this edition of *Today's Technician: Automotive Engine Repair and Rebuilding* has been updated to include a new chapter on the engine repair and rebuilding industry, all new content on hybrid electric vehicles and alternative fuel-related issues, an updated list of FFVs, and new material on VVT systems. Increased coverage on the basics of engine operation is also provided, as well as a new Engine Specification Chart included in Appendix E. The Classroom Manual also features a thoroughly updated art program, now in full color, and will appeal greatly to visual learners.

HIGHLIGHTS OF THIS EDITION—SHOP MANUAL

The fourth edition of Shop Manual now provides all new content on basic testing, initial inspection, and service writing. Information on scan tool usage has been expanded, along with content on diagnostic strategies from real-world scenarios, as experienced by working technicians. The Shop Manual covers 100 percent of the 2008 NATEF Standards, includes nearly 30 new Job Sheets, and features new full-color photo sequences that detail step-by-step diagnostic and repair procedures. ASE questions for the chapters have been updated and are designed to assess students' chapter comprehension and prepare them for future certification testing. As with the Classroom Manual, all photos and illustrations are now full color and offer students outstanding visual support.

CLASSROOM MANUAL

Features of this manual include the following:

COGNITIVE OBJECTIVES

These objectives define the contents of the chapter and define what the student should have learned upon completion of the chapter.

Each topic is divided into small units to promote easier understanding and learning.

CAUTIONS AND WARNINGS

Throughout the text, warnings are given to alert the reader to potentially hazardous materials or unsafe conditions. Cautions are given to advise the students of things that can go wrong if instructions are not followed or if a nonacceptable part or tool is used.

CROSS-REFERENCES TO THE SHOP MANUAL

Reference to the appropriate page in the Shop Manual is given whenever necessary. Although the chapters of the two manuals are synchronized, material covered in other chapters of the Shop Manual may be fundamental to the topic discussed in the Classroom Manual.

MARGINAL NOTES

These notes add "nice-to-know" information to the discussion. They may include examples or exceptions, or may give the common trade jargon for a component.

Chapter 1

OVERVIEW OF ENGINE PERFORMANCE

UPON COMPLETION AND REVIEW OF THIS CHAPTER, YOU SHOULD BE ABLE TO:

- Describe the relationship of basic engine construction to engine performance.
- Describe the basic function of the cooling, fuel, and air induction systems.
- Explain similarities between distributor ignition systems and distributorless ignition systems.
- Explain the relationship of intake and exhaust systems to engine performance.
- Explain the basic function of the Exhaust Gas Recirculaton (EGR) valve.
- Explain the basic function of the Positive Crankcase Ventilation (PCV) valve.
- Describe the function of the catalytic converter.
- Describe the basic function of the evaporative emissions system.
- Examine basic differences between OBD I and OBD II.

INTRODUCTION

Congratulations! Since you are taking a course in Engine Performance, you are about to combine most of the knowledge you have gained to date and use it to start diagnosis of "drivability" problems. The Engine Performance Specialist is usually one of the most highly talented people in the shop. It takes a good foundation in electricity and electronics. It requires knowledge of the engine and how it works along with basic transmission operation. You will be surprised at how many transmission complaints are actually related to engine performance. The engine performance person also does most of the electrical diagnosis and scan tool work, so you will also need some knowledge of basic engine testing, computer operation, networking, body control modules, and so forth, which is included to help you in your career. I hope you find this text helpf___ your pursuit of the most challenging and rewarding facets of automoti___

classified as an **internal combustion** ___ place inside the engine. Burning ___ mounts of pressure which powers ___ stand the high temperatures and ___

engine and is usually cast from iron ___ holes called **cylinders** which are

1

WARNING: Do not let the used coolant into the shop drains; collect it in a drain pan. The used coolant should be recycled or removed as a hazardous waste.

CAUTION:
Thermostats have a high failure rate and are a very common cause of cooling system concerns. They are typically replaced during cooling system maintenance, repair, and engine overhaul. Do not install a thermostat with a lower operating temperature rating than specified or without a thermostat. Allowing the engine to run cooler than designed accelerates engine wear and increases emissions and fuel consumption.

Many shops now use a coolant flushing machine that automatically expels the used coolant and refills the system with fresh coolant (Figure 4-28). Many of these pieces of equipment also recycle the coolant to eliminate costly removal from the shop. Follow the instructions provided by the equipment manufacturer to flush the cooling system. Do not use recycled coolant if the manufacturer recommends against it; that could nullify the powertrain warranty.

While the system is drained, it is a good time to replace the thermostat and hoses. Following is a typical procedure for thermostat replacement (Figure 4-29).

Drain the coolant to a level below the thermostat housing. Doing this will reduce the potential to spill coolant onto the floor. Remove the radiator hose from the housing. Some hoses use a worm-gear-type hose clamp. These can be removed by using a flat blade screwdriver to turn the worm gear. Some hoses will have a spring wire clamp; use special clamp pliers or large pliers to remove these clamps. Some engines will have a bypass hose connected to the thermostat housing; if this is the case on your engine, it should be removed at this point.

Loosen, then remove, the fasteners that attach the thermostat housing to the engine. It may be necessary to strike the housing with a rubber mallet to break the seal between the engine and the housing. Do not drive a chisel or flat-blade screwdriver between the mating surfaces; doing so may damage the mating surfaces and prevent the system from sealing.

Remove the thermostat and gasket or O-ring seal. Pay attention to the direction the thermostat was facing. Confirm the proper direction in the service manual. If the thermostat was installed backwards, that could be the cause of overheating.

Being careful not to gouge the metal, use the proper abrasive disc to clean both mating surfaces. Apply the correct sealant, if one is recommended, to the mating surface of the housing. Carefully install the thermostat in the engine block, being sure the pellet is facing

FIGURE 4-28 A coolant flushing machine.

Upper radiator hose   Thermostat housing
FIGURE 4-29 The thermostat sits on the top front of this engine.

144

___ ___ the throttle body
___ ___ jector has a moveable armature in the center of the injector, and a pintle with a tapered tip is positioned at the lower end of the armature. A spring pushes the armature and pintle downward so the pintle tip seats in the discharge orifice. The injector coil surrounds the armature, and the two ends of the winding are connected to the terminals on the side of the injector. An integral filter is located inside the top of the injector. When the ignition switch is turned on, voltage is supplied to one injector terminal and the other terminal is connected through the computer. Each time the control unit completes the circuit from the injector winding to ground, current flows through the injector coil and the coil magnetism moves the plunger and pintle upward. Under this condition, the pintle tip is unseated from the injector orifice and fuel sprays out this orifice.

TBI Advantages and Disadvantages. Throttle body systems provide improved fuel metering when compared to carburetors. They are also less expensive and simpler to service. TBI units also have some advantages over port injection. They are less expensive to manufacture, simpler to diagnose and service, and do not have injector balance problems to the extent that port injection systems do when the injectors begin to clog.

However, throttle body units are not as efficient as port systems. The disadvantages are primarily manifold related. Like a carburetor system, fuel is still not distributed equally to all cylinders, and a cold manifold may cause fuel to condense and puddle in the manifold. Like a carburetor, throttle-body injection systems must be mounted above the combustion chamber level, which eliminates the possibility of tuning the manifold design for more efficient operation.

Port Fuel Injection

PFI systems use one injector at each cylinder. They are mounted in the intake manifold near the cylinder head where they can inject a fine, atomized fuel mist as close as possible to the intake valve (Figure 9-25). Fuel lines that run to each cylinder from a fuel manifold are usually referred to as a **fuel rail**. The fuel rail assembly on a PFI system of V6 and V8 engines usually consists of a left- and right-hand rail assembly. The two rails can be connected either by crossover and return fuel tubes or by a mechanical bracket arrangement. A typical fuel rail arrangement is shown in Figure 9-26. Fuel tubes crisscross between the two rails. Figure 9-27 shows a fuel rail on a transverse four-cylinder engine. Since each cylinder has its own injector, fuel distribution is exactly equal. With little or

Shop Manual
Chapter 9,
page 393

The **fuel rail** delivers fuel to the fuel injectors.

FIGURE 1-20 The lower crankshaft gear turns at twice the speed of the larger camshaft gear.

FIGURE 1-21 Valve timing failures can be catastrophic.

A BIT OF HISTORY

This feature gives the student a sense of the evolution of the automobile. This feature not only contains nice-to-know information, but also should spark some interest in the subject matter.

Main **journals** are round machined portions on the centerline of the crankshaft where the crankshaft is held in the main bore.

A BIT OF HISTORY

Jean Joseph Lenoir created the first workable internal combustion engine in 1860. Nicholas Otto was credited with creating the first four-stroke internal combustion engine back in 1876. All previous internal combustion engines did not compress the air-fuel mixture. They attempted to draw the air-fuel mixture in during a downward

TIMING MECHANISM

As we discussed, proper valve timing is necessary to keep the valves opening and closing at the correct times. This is achieved by the timing mechanism. A sprocket or gear on the crankshaft is attached by a gear, belt, or chain to a sprocket or gear on the camshaft (Figure 1-20). Timing belts require periodic maintenance. When that is neglected, the timing belt can snap, and serious engine damage can result. In some cases, the valves knock the top of the pistons and bend, and a new set of valves is required (Figure 1-21). In the worst cases, those valves break the pistons, and the whole engine needs to be overhauled. Timing chains also wear out over time, and replacing them requires skill and attention to detail. Variable valve timing systems increase performance and minimize emissions. More and more vehicles employ these systems to vary the times when the valves can be opened. This makes timing mechanism theory and service a new and more complex challenge.

ENGINE BLOCK

The **engine block** is the main supporting structure of the engine (Figure 1-22). It holds the pistons, connecting rods, crankshaft, and sometimes the camshaft. The cylinder head bolts to the top of the block. The block is bored (drilled) to create cylinders. An eight-cylinder engine has eight-cylinder bores (Figure 1-23). The pistons are pushed up and down in the cylinders by the crankshaft. On the power stroke, the piston is actually pushed down by the force of combustion to turn the crankshaft. Rings on the pistons seal the small clearance between the cylinder walls. This keeps the hot expanding gases from combustion within the combustion chamber, so that all the force acts to push the piston down. The rings also scrape excess oil from the cylinder walls, so the oil doesn't burn inside the combustion chamber. When oil is allowed to burn during combustion, it forms blue smoke that exits the tailpipe. The block also holds the crankshaft in its main bore. The soft main bearings are placed within the main bore to provide a place for the finely machined crankshaft **journals** to ride. The journals are the machined round areas of the crankshaft that allow the bearings and bearing caps to bolt around them and hold the crankshaft in the engine block. The clearance between the journals and the bearings must be just right: enough to allow adequate oil, but not so much that precious oil pressure leaks through excessive clearances.

ing out against the lower pressure outside ... per square inch (psi), but inches ...

Vacuum could be measured in pounds per square inch (psi), but inches ... (in. Hg) are most commonly used for this measurement. Let us assume that a plastic "U" tube is partially filled with mercury, and atmospheric pressure is allowed to enter one end of the tube. If a vacuum is supplied to the other end of the "U" tube, the mercury is forced downward by the atmospheric pressure. When this movement occurs, the mercury also moves upward on the side where the vacuum is supplied. If the mercury moves downward

AUTHOR'S NOTE: As altitude increases, atmospheric pressure decreases. As pressure decreases, the boiling point of a liquid also decreases. For example, water boils at 212°F at sea level but will boil at an incrementally lower temperature as altitude increases. So if you are on a camping trip in the Rocky Mountains (say, 6,000 feet above sea level) and the water boils at a lower temperature (less than 212°F), how long would you have to boil a three-minute egg before it has thoroughly cooked?

AUTHOR'S NOTES

This feature includes simple explanations, stories, or examples of complex topics. These are included to help students understand difficult concepts.

TERMS TO KNOW LIST

A list of new terms appears next to the Summary.

SUMMARIES

Each chapter concludes with a summary of key points from the chapter. These are designed to help the reader review the chapter contents.

SUMMARY

TERMS TO KNOW

Acid
Atmospheric pressure
Atom
Bar
Base
Coefficient of drag (Cd)
Compound
Compressible

- An element contains only one type of atom.
- An atom is the smallest particle of an element.
- Compounds contain two or more types of atoms.
- A molecule is the smallest particle in a compound.
- Protons are positively charged particles at the center, or nucleus, of each atom.
- Neutrons have no electric charge, and they are located in the nucleus of most atoms.
- Electrons are negatively charged particles found in various orbits around the nucleus of an atom.
- Work is the result of applying a force.
- Force is measured in pounds and distance.
- Energy is the ability to do work, and there are six basic types of energy.

30

Direct drive
Electron
Element
Energy
Friction
Gear reduction
Inertia
Kilopascal
Mass
Molecule
Momentum
Negative pressure
Neutron
Noncompressible
Overdrive
Positive pressure
Power
Pressure
Proton
Torque
Vacuum
Valence ring
Venturi
Volume
Weight
Work

REVIEW QUESTIONS

Short answer essay, fill-in-the-blank, and multiple-choice questions are found at the end of each chapter. These questions are designed to accurately assess the student's competence in the stated objectives at the beginning of the chapter.

REVIEW QUESTIONS

Short Answer Essays

1. Define an element and a compound.
2. Name the four factors that are necessary for complete combustion.
3. Define a molecule.
4. Describe three particles found in an atom, including the electrical charge and location of each particle.
5. Explain why a venturi decreases pressure.
6. What is atmospheric pressure?
7. What happens to air pressure as the temperature of the air changes?
8. Describe Newton's three laws of motion.
9. Describe six different forms of energy.
10. Describe four different types of energy conversion.

Fill-in-the-Blanks

1. At sea level, atmospheric pressure is _____ psi.
2. In nature, a _____ pressure always moves toward a _____ pressure.
3. The nucleus of an atom is comprised of a(n) _____ and a(n) _____ .
4. Work is calculated by multiplying _____ times _____ .
5. Energy may be defined as the ability to do _____ .
6. When one object is moved over another object, the resistance to motion is called _____ .
7. Weight is the measurement of the earth's _____ on an object.

SHOP MANUAL

To stress the importance of safe work habits, the Shop Manual also dedicates one full chapter to safety. Other important features of this manual include:

PERFORMANCE-BASED OBJECTIVES

These objectives define the contents of the chapter and define what the student should have learned upon completion of the chapter. These objectives also correspond with the list of required tasks for NATEF certification.

Although this textbook is not designed to simply prepare someone for the certification exams, it is organized around the NATEF task list. These tasks are defined generically when the procedure is commonly followed and specifically when the procedure is unique for specific vehicle models. Imported and domestic model automobiles and light trucks are included in the procedures.

BASIC TOOLS LISTS

Each chapter begins with a list of the basic tools needed to perform the tasks included in the chapter.

MARGINAL NOTES

These notes add "nice-to-know" information to the discussion. They may include examples or exceptions, or may give the common trade jargon for a component.

SPECIAL TOOLS LISTS

Whenever a special tool is required to complete a task, it is listed in the margin next to the procedure.

PHOTO SEQUENCES

Many procedures are illustrated in detailed Photo Sequences. These detailed photographs show the students what to expect when they perform particular procedures. They also can provide the student a familiarity with a system or type of equipment, which the school may not have.

CAUTIONS AND WARNINGS

Throughout the text, warnings are given to alert the reader to potentially hazardous materials or unsafe conditions. Cautions are given to advise the student of things that can go wrong if instructions are not followed or if a nonacceptable part or tool is used.

CUSTOMER CARE

This feature highlights those little things a technician can do or say to enhance customer relations.

SERVICE TIPS

Whenever a short-cut or special procedure is appropriate, it is described in the text. These tips are generally those things commonly done by experienced technicians.

Idle Contact Switch Test

The procedure for diagnosing the idle contact switch varies depending on the vehicle make and model year. The following is a typical idle contact switch test. Refer to the wiring diagram in Figure 10-41 for terminal identification during this procedure. Always refer to a similar diagram for the vehicle being tested.

WARNING: Connecting a pair of jumper wires from the terminals of a 12-volt battery to the IAC motor idle contact switch terminals may result in very high current flow and jumper wire heating, which can burn your hands.

1. Backprobe terminal B on the IAC motor and turn the ignition switch on. Connect a digital voltmeter from terminal B to ground and hold the throttle approximately half open. The voltage supplied from the PCM to the idle contact switch should be 4.5 to 8 volts, depending on the system.
2. If the voltage at terminal B on the IAC motor is not within specifications, turn the ignition switch off and disconnect the IAC motor connector and the PCM connector. Connect the ohmmeter leads from terminal B in the IAC ...

CAUTION: Never connect a 12-volt source across the idle contact switch terminals on the ISC motor. If these contacts are closed, the contacts will be ruined.

FIGURE 6-1 GM's electrostatic discharge (ESD) symbol that warns a technician that the component or circuit is sensitive to static.

2. Avoid touching the electrical terminals of the part unless you are instructed to do so in the written service procedures. It is good practice to keep your fingers off all electrical terminals since the oil from your skin can cause corrosion.
3. When you are using a voltmeter, always connect the negative meter lead first.
4. Do not remove a part from its protective package until it is time to install the part.
5. Before removing the part from its package, ground yourself and the package to a known good ground on the vehicle.
6. When replacing a PROM, ground your body by putting a metal wire around your wrist and connect the wire to a good ground.

CUSTOMER CARE: It is always an excellent customer service practice to return the vehicle to the customer with no evidence or side effects of being repaired. This applies to seat position, mirror adjustment, cleaning of smudges, and perhaps even cleaning the windshield. Refrain from changing radio ... or tone controls as well as eating, drinking ... includes resetting electronic dashb... work is performed ...

437

JOB SHEET

Name _____ Date _____

TESTING A MAP SENSOR

Upon completion of this job sheet, you should be able to test a manifold absolute pressure sensor in a variety of ways.

ASE Correlation

This job sheet is related to the ASE Engine Performance Test's content area: computerized engine controls diagnosis and repair; task: inspect, test, adjust, and replace computerized engine co...

Tools an...
Hand-o...
Lab Sc...
DMM...

Descr...
Year...
Mod...

Pro...

SERVICE TIP: Many technicians use a DSO when testing fuel injectors and fuel injector circuits. If the DSO is equipped with a graphing mode, perform a comparative amperage flow test through the fuel injectors. Discrepancies may be easier to observe and may lead to locating or verifying faulty fuel injectors or related circuit problems. The technician can also record and store various problem situations for future reference.

SPECIAL TOOLS
Lab scope

Injectors that fire individually are SFI injectors.

FIGURE 10-23 Checking for injector leaks with the fuel rail pressurized.

8. Repeat the procedure on each injector. If the volume of fuel discharged from any injector varies more than 0.3 cu. in. (5 cc) from the specifications, the injector should be replaced.
9. Reconnect the negative battery cable and disconnect the 12-volt power supply.

Connect the fuel pressure gauge to the fuel system. While the fuel system is pressurized with the injectors removed from the fuel rail after the flow test, observe each injector for leakage from the injector tip (Figure 10-23). Injector leakage must not exceed the manufacturer's specifications. If the injectors leak into the intake ports on a hot engine, the air-fuel ratio may be too rich when a restart is attempted a short time after the engine is shut off. When the injectors leak, they drain all the fuel out of the rail after the engine is shut off for several hours. This may result in slow starting after the engine has been shut off for a longer period of time.

While checking leakage at the injector tips, observe the fuel pressure in the pressure gauge. If the fuel pressure drops off and the injectors are not leaking, the fuel may be leaking back through the check valve in the fuel pump. Repeat the test with the fuel line plugged. If the fuel pressure no longer drops, the fuel pump check valve is leaking. If the fuel pressure drops off and the injectors are not leaking, the fuel pressure may be leaking through the pressure regulator and the return fuel line. Repeat the test with the return line plugged. If the fuel pressure no longer drops off, the pressure regulator valve is leaking.

Oscilloscope Checks

An oscilloscope can be used to monitor the injector's pulse width and duty cycle when an injector-related problem is suspected. The pulse width is the time in milliseconds that the injector is energized. The duty cycle is the percentage of on-time to total cycle time.

To check the injector's firing voltage on the scope, a typical hookup involves connecting the scope's positive lead to the injector supply wire and the scope's negative lead to an engine ground.

Fuel injection signals vary in frequency and pulse width. The pulse width is controlled by the PCM which varies it to control the air-fuel ratio. The frequency varies with engine speed. The higher the speed, the more pulses per second there are. Most often, the injector's ground circuit is completed by a driver circuit in the PCM. All of these factors are important to remember when setting a lab scope to look at fuel injector activity. Set the scope to read 12 volts, then set the sweep and trigger to allow you to clearly see the on signal on the left and the off signal on the right. Make sure the entire waveform is clearly seen. Also remember that the setting may need to be changed as engine speed increases or decreases.

Fuel injectors are either fired individually or in groups. When the injectors are fired in groups, a driver circuit controls two or more injectors (Figure 10-24). On some V-type engines, one driver fires the injectors on one side of the engine while another fires the other side. Each fuel injector has its own driver transistor in sequential and throttle body injection. It is extremely important while troubleshooting that you recognize how the injectors are fired. When the injectors are fired in groups, there can be a common or noncommon

426

CROSS-REFERENCES TO THE CLASSROOM MANUAL

Classroom Manual
Chapter 6, page 148

Reference to the appropriate page in the Classroom Manual is given whenever necessary. Although the chapters of the two manuals are synchronized, material covered in other chapters of the Classroom Manual may be fundamental to the topic discussed in the Shop Manual.

JOB SHEETS

Located at the end of each chapter, the Job Sheets provide a format for students to perform procedures covered in the chapter. A reference to the NATEF Task addressed by the procedure is referenced on the Job Sheet.

CASE STUDIES

Case Studies concentrate on the ability to properly diagnose the systems. Beginning with Chapter 3, each chapter ends with a case study in which a vehicle has a problem, and the logic used by a technician to solve the problem is explained.

TERMS TO KNOW LIST

Terms in this list can be found in the Glossary at the end of the manual.

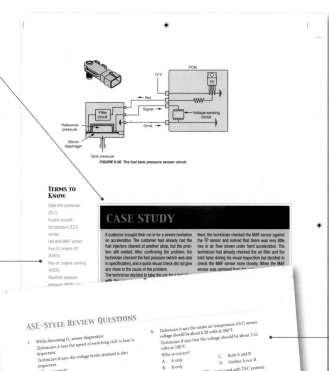

FIGURE 6-85 The fuel tank pressure sensor circuit.

TERMS TO KNOW

Data link connector (DLC)

Engine coolant temperature (ECT) sensor

Hot wire MAF sensor

Key on, engine off (KOEO)

Key on, engine running (KOER)

Manifold absolute pressure (MAP) sensor

CASE STUDY

A customer brought their car in for a severe hesitation on acceleration. The customer had already had the fuel injectors cleaned at another shop, but the problem still existed. After confirming the problem, the technician checked the fuel pressure (which was also in specification), and a quick visual check did not give any clues to the cause of the problem.

The technician decided to take the car for a test drive with the scan tool connected.

Next, the technician checked the MAF sensor against the TP sensor and noticed that there was very little rise in air flow shown under hard acceleration. The technician had already checked the air filter and the inlet hose during his visual inspection but decided to check the MAF sensor more closely. When the MAF sensor was removed from the vehicle, the technician noticed...

ASE-STYLE REVIEW QUESTIONS

Each chapter contains ASE-style review questions that reflect the performance-based objectives listed at the beginning of the chapter. These questions can be used to review the chapter as well as to prepare for the ASE certification exam.

ASE-STYLE REVIEW QUESTIONS

1. While discussing O₂ sensor diagnostics:
   Technician A says the speed of switching rich to lean is important.
   Technician B says the voltage levels attained is also important.
   Who is correct?
   A. A only
   B. B only
   C. Both A and B
   D. Neither A nor B

2. While discussing ECT sensor diagnosis:
   Technician A says a defective ECT sensor may cause hard cold engine starting.
   Technician B says a defective ECT sensor may cause improper operation of emission control devices.

6. *Technician A* says the intake air temperature (IAT) sensor voltage should be about 0.28 volts at 260°F.
   Technician B says that the voltage should be about 2.51 volts at 100°F.
   Who is correct?
   A. A only
   B. B only
   C. Both A and B
   D. Neither A nor B

7. *Technician A* says some TP sensors used with TAC systems are actually two TP sensors in one housing.
   Technician B says some TP sensors are non-contact type Hall effect sensors.
   Who is correct?
   A. A only
   B. B only
   C. Both A and B
   D. Neither A and B

...flow sensor diagnosis:

ASE CHALLENGE QUESTIONS

Each technical chapter ends with five ASE challenge questions. These are not more review questions, rather they test the students' ability to apply general knowledge to the contents of the chapter.

ASE CHALLENGE QUESTIONS

...uld be worn.

...be utilized.

...s are more ...output devices. ...y result in

3. EMI is being discussed.
   Technician A says EMI can cause driveability problems.
   Technician B says that EMI is worse with poor grounds.
   Who is correct?
   A. A only
   B. B only
   C. Both A and B
   D. Neither A nor B

4. The idle tracking switch is being discussed.
   Technician A says a poor idle may be caused by a high voltage signal present when the switch is open.
   Technician B says an infinity ohmmeter reading at part throttle indicates a defective switch.
   Who is correct?
   A. A only
   B. B only
   C. Both A and B
   D. Neither A nor B

5. A low voltage on a sensor's reference wire could be caused by any of the following except:
   A. poor computer ground.
   B. shorted reference wire.
   C. high resistance in the sensor.
   D. excessive voltage drop on sensor ground.

ASE PRACTICE EXAMINATION

A 50 question ASE practice exam, located in the appendix, is included to test students on the contents of the Shop Manual.

APPENDIX A — ASE PRACTICE EXAMINATION

1. *Technician A* says a hydrometer reading of 1.200 at 80°F means the battery must be recharged before performing a capacity test.
   Technician B says a capacity test can be correctly performed with the battery cables connected.
   Who is correct?
   A. A only
   B. B only
   C. Both A and B
   D. Neither A nor B

2. *Technician A* says engine detonation may be caused by low octane fuel.
   Technician B says that detonation can be caused by engine overheating.
   Who is correct?
   A. A only
   B. B only
   C. Both A and B
   D. Neither A nor B

3. The PCV system is being discussed.
   Technician A says oil in the crankcase breather filter will confirm that the PCV valve is clogged.
   Technician B says the PCV valve must be disconnected to be checked with the engine operating.
   Who is correct?
   A. A only
   B. B only
   C. Both A and B
   D. Neither A nor B

4. An EI-equipped vehicle will not start.
   Technician A says a good first step is to make certain that the MIL is illuminated with the key on.
   Technician B says to check for spark at more than one plug wire during diagnosis.
   Who is correct?
   A. A only
   B. B only
   C. Both A and B
   D. Neither A nor B

5. Electronic fuel injection systems are being discussed.
   Technician A says a hard to start engine may have an open electrical fuel pump relay bypass circuit.
   Technician B says a hesitation when accelerating from idle may be caused by dirty injectors.
   Who is correct?
   A. A only
   B. B only
   C. Both A and B
   D. Neither A nor B

6. *Technician A* says that OBDII AIR systems do not supply air to the catalytic converter.
   Technician B says OBD II secondary air systems are all belt driven.
   Who is correct?
   A. A only
   B. B only
   C. Both A and B
   D. Neither A nor B

7. A vehicle has a consistent slow- or no-crank condition. This may be caused by:
   A. A slipping AC generator drive belt
   B. A defective AC generator
   C. A low regulator voltage
   D. Any of the above

8. A fuel-injected vehicle comes in with a miss on acceleration.
   Technician A says that there could be an ignition problem.
   Technician B says that the intake to MAF duct could be opening up at a tear on acceleration.
   Who is correct?
   A. A only
   B. B only
   C. Both A and B
   D. Neither A nor B

9. *Technician A* says static electricity generated by clothing in contact with vehicle upholstery must be discharged before beginning work on electronic devices.
   Technician B says the best control for static electricity is the wearing of a waist or wrist grounding strap.
   Who is correct?
   A. A only
   B. B only
   C. Both A and B
   D. Neither A nor B

10. *Technician A* says a camshaft installed one tooth retarded can be compensated by adjusting the ignition's base timing.
    Technician B says to use the starter to crank the engine with the timing belt removed to determine if the engine is interference type or free-wheeling.
    Who is correct?
    A. A only
    B. B only
    C. Both A and B
    D. Neither A nor B

SUPPLEMENTS

INSTRUCTOR RESOURCES

The Instructor Resources DVD is a robust ancillary that contains all preparation tools to meet any instructor's classroom needs. It includes PowerPoint slides with images, video clips, and animations that coincide with each chapter's content coverage, a computerized test bank with hundreds of test questions, a searchable image library with all pictures from the text, theory-based worksheets in Word that provide homework or in-class assignments, the Job Sheets from the Shop Manual in Word, a NATEF correlation chart, and an Instructor's Guide in electronic format.

WEBTUTOR ADVANTAGE

Newly available for this title and to the Today's Technician™ Series is the *WebTutor Advantage*, for Blackboard and Angel online course management systems. The *WebTutor for Today's Technician: Automotive Engine Repair and Rebuilding*, 5e will include PowerPoint presentations with video clips and animations, end-of-chapter review questions, pre-tests and post-tests, worksheets, discussion springboard topics, Job Sheets, and more. The WebTutor is designed to enhance the classroom and shop experience, engage students, and help them prepare for ASE certification exams.

REVIEWERS

The author and publisher would like to extend a special thanks to the instructors who reviewed this text and offered their invaluable feedback:

Paul Bean
Utah Valley University
Orem, UT

Brad Lyne
North Central Kansas Technical College
Beloit, KS

Wade Esplin
Southwest Applied Technical College
Cedar City, UT

John Fowler
Iowa Central Community College
Fort Dodge, IA

Gary Ham
South Plains College
Levelland, TX

Tim Reaves
Cosumnes River College
Sacramento, CA

Chapter 1

AUTOMOTIVE ENGINES

UPON COMPLETION AND REVIEW OF THIS CHAPTER, YOU SHOULD BE ABLE TO:

- Describe the purpose of the automotive engine.
- Describe the basic operation of the internal combustion four-stroke engine.
- Understand the basic function of the cooling and lubrication systems.
- Understand the basic function of the intake and exhaust systems.

- Be familiar with basics of the cylinder head and its related components.
- Be familiar with the engine block and its related components.
- Identify many of the major components of the engine.

INTRODUCTION

The automotive engine is the power source that drives the automobile. Today's engines are capable of delivering excellent acceleration, smooth running in all weather conditions, and high cruising speeds. They also achieve good fuel economy and low toxic emissions. The basic operation of the four-stroke gasoline **internal combustion engine (ICE)** has not changed in over one hundred years.

There have been, however, many refinements to the base engine and its supporting systems, which have allowed huge gains in performance and in control of fuel consumption and emissions. Lighter materials within the engine have allowed weight reductions. The precision and accuracy of the manufacturing process of modern engines has increased dramatically. A modern engine can produce more power than its counterpart from the 1960s and 1970s. Improved materials and manufacturing processes have improved the durability of engines significantly. Most engines should provide over 150,000 miles of trouble-free service with proper maintenance. Even maintenance requirements have diminished significantly as improved parts, new component design, and electronic controls have been introduced. **Powertrains** of the new millennium are over 95 percent cleaner in terms of toxic emissions than those of the 1960s. While the number of vehicles on the road has increased tremendously, the amount of air pollution generated by automobiles has diminished.

This text will focus on the operation and service of these powerful and efficient new gasoline engines (Figure 1-1). You must thoroughly understand the operation of the engine and the functions of its components to become a skilled automotive repair technician. In order to provide good customer service and ensure proper engine performance, you must accurately diagnose faults and perform precise repairs of the engine and its supporting systems. This chapter will highlight the contents of the classroom and shop manuals. Each chapter in the classroom manual will explain the theory and operation of a system, and the corresponding shop manual will describe the diagnostic and repair strategies for those systems.

> **Internal combustion engines (ICEs)** burn their fuels within the engine.

> A **powertrain** includes the engine and all the components that deliver that force to the road, including the transmission and axles.

FIGURE 1-1 Today's smaller engines can produce some hefty horsepower.

The **combustion chamber** is a sealed area in the engine where burning (combustion) of the air and fuel mixture takes place.

The **piston** is a round-shaped part that is driven up and down in the engine cylinder bore.

An engine **cylinder** is a round hole bored into the cylinder block.

The **powertrain control module (PCM)** is an onboard computer that controls functions related to the powertrain, such as fuel delivery and spark timing.

BASIC ENGINE OPERATION

The gas engine is often described as an air pump because it uses a tremendous amount of air, but to make power it also uses a small percentage of gasoline. The engine pulls air into a sealed **combustion chamber** through the intake system. The combustion chamber is sealed on the bottom by the **piston** and its rings (Figure 1-2) and on the top by the combustion chamber formed in the cylinder head (Figure 1-3).

Air is pulled into the **cylinders** by the vacuum formed as the pistons move downward. Fuel is injected into the airstream near (or sometimes in) the combustion chambers. The timing and quantity of fuel delivery is precisely controlled by the **powertrain control module (PCM)**.

The air and fuel mixture is then compressed to make it more combustible as the piston moves toward the top of its travel. At the optimum instant, the PCM initiates ignition that will deliver a spark to the spark plug. When the spark jumps across the gap of the spark plug, the air and fuel mixture burns rapidly, and the engine harnesses some of the heat energy available in the fuel. This rapid, controlled burning is called **combustion**.

FIGURE 1-2 This piston and rings, with the connecting rod attached, fits closely in the engine cylinder to seal the bottom of the combustion chamber.

FIGURE 1-3 The bottom of this cylinder head (shown) shows the four combustion chambers sealed by the valves.

Connecting rod

FIGURE 1-4 The connecting rod connects the piston to the crankshaft.

FIGURE 1-5 An engine valve.

FIGURE 1-6 The camshaft has large, round journals that fit into bearings in the block and small eccentric lobes that open and close the valves.

The power of the expanding gases pushes down with the force comparable to an elephant standing on top of the piston. The piston is connected to the crankshaft by a connecting rod (Figure 1-4). The crankshaft causes the engine to turn. Then the spent gases are allowed to exit the combustion chamber through the exhaust system. This process is repeated in each of the engine cylinders to keep the crankshaft spinning.

The engine valves allow airflow into and out of the combustion chamber. There are typically one or two intake valves per cylinder and one or two exhaust valves per cylinder (Figure 1-5). Refer to Figure 1-3 to see the heads of the valves within the combustion chamber. The valves are opened by the **camshaft** (Figure 1-6). The camshaft is gear, belt, or chain driven by the crankshaft. The relationship between the camshaft and crankshaft rotation creates the proper **valve timing**. It is carefully timed to ensure that the valves open at the correct time. The camshaft has egg-shaped lobes on it to force the opening of the valves. Valve springs close the valves.

Engine Location

Most engines are installed in the front of the vehicle. They may be installed longitudinally, with the front of the engine toward the bumper and the rear toward the passenger compartment, or transversely, with the engine placed in the engine compartment sideways. Trucks and rear-wheel drive (RWD) vehicles most commonly use a longitudinally mounted engine. Many passenger vehicles and most front-wheel drive (FWD) vehicles use a transversely mounted engine (Figure 1-7). This design allows a shorter hood line, desired in today's vehicles to provide greater passenger space. A few vehicles use a mid-engine design, where the engine is placed between the rear axle and the driver. Often, this well-balanced configuration is associated with modern sports cars.

Engines are held in place using motor mounts. These are attaching pieces with a flexible rubber portion that isolates much of the engine vibration from the automobile frame and passenger compartment. The mounts connect the engine to the vehicle chassis (Figure 1-8).

The crankshaft and connecting rod work together to change the reciprocating motion of the piston into the rotary motion of the crankshaft.

Combustion is the controlled burn created by the spark igniting the hot, compressed air and fuel mixture in the combustion chamber.

The **camshaft** is a shaft with eccentric lobes on it to force the opening of the valves.

Valve timing refers to the relationship between the rotation of the crankshaft and camshaft. That correct relationship ensures that the valves open at the correct time.

FIGURE 1-7 This transversely mounted engine sits sideways in the engine compartment with the transaxle bolted on its end.

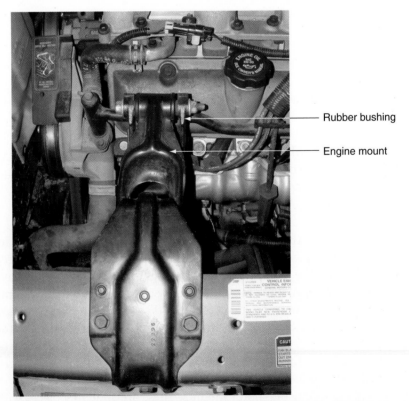

Rubber bushing

Engine mount

FIGURE 1-8 The engine mount has a rubber bushing to isolate the vibration of the engine from the chassis.

FIGURE 1-9 The flywheel bolts to the crankshaft and is attached to the transmission through the clutch on a manual transmission.

Engine Function

The engine is attached to the transmission through the flywheel and clutch on a manual transmission (Figure 1-9) or through the flexplate and torque converter on an automatic transmission. The engine delivers rotational speed and a twisting force to spin the transmission components. The twisting force is called torque and the speed that develops is called horsepower. The transmission uses this torque and horsepower and manipulates them to provide the appropriate level of torque and speed to drive the wheels and propel the vehicle. The transmission multiplies the torque of the engine through gear sets to allow good acceleration. In higher gear(s), torque is reduced and speed is increased to improve fuel economy.

COOLING SYSTEM

Combustion develops a tremendous amount of heat in the engine. The expansion of gases from the heat pushes the piston down, but some of the heat is absorbed in the engine components. A liquid cooling system is used on all current gas automobile engines to remove excess heat from the engine. An engine belt or gear drives a water pump that circulates coolant throughout the engine to keep all areas at an acceptable temperature, around 200°F (110°C). When the coolant reaches a certain temperature, a **thermostat** (Figure 1-10) opens to allow coolant to flow through the radiator to cool it down. Air flowing across the radiator lowers the temperature of the coolant, and then the water pump pulls the coolant back

> A **thermostat** is a mechanical component that blocks or allows coolant flow to the radiator.

FIGURE 1-10 A thermostat.

FIGURE 1-11 The headgasket seals between the block and the cylinder head.

into the engine and recirculates it. The cooling system must retain heat to speed warm-up to enhance combustion after cold starts. It must also maintain an even engine operating temperature for better engine efficiency. To prevent meltdown of components, the cooling system must remove excess heat.

Cooling system problems can cause very serious damage to an engine. Heat expansion of parts can cause the engine to seize. When an engine overheats, it is likely to blow the seal between the cylinder head and the cylinder block. That seal is the headgasket, and replacement can cost several hundred dollars and many hours of work (Figure 1-11).

LUBRICATION SYSTEM

The lubrication system is critical to reduce the tremendous friction created between moving components in the engine. This also reduces engine heat and wear. The oil pump is mounted on the bottom of the engine or on the crankshaft to circulate engine oil under pressure to all friction areas in the engine (Figure 1-12). The oil is then filtered to remove particles of metal or debris that could damage finely machined engine components. The oil travels through oil passages drilled throughout the engine to reach the crankshaft, the camshaft, the cylinder walls, and other key friction areas. Just a few minutes of engine operation without adequate lubrication can turn an engine into irreparable scrap metal. Proper lubrication system maintenance is essential to the longevity of any engine.

FIGURE 1-12 This oil pump is driven off of
the front of the crankshaft.

ENGINE BREATHING

Engine power is dependent on how well the engine can breathe. The intake and exhaust system provide the breathing tubes for the engine. The **intake manifold** connects the air intake ductwork to the cylinder head (Figure 1-13). Ports into the combustion chamber allow fresh air to be pulled in past the intake valves when they are open. The intake manifold distributes air equally to each cylinder. The **exhaust manifold** connects the exhaust ports on the cylinder head to the rest of the exhaust system. To a major degree, the more air an engine can take in, the more power it can put out. The intake and exhaust systems must flow freely to allow good airflow. Turbocharging and supercharging are systems that increase airflow by forcing air under pressure into the intake. These systems dramatically increase engine power (Figure 1-14).

Problems with the engine's ability to breathe seriously affect engine performance. Something as simple as a plugged air filter can cause the engine to barely run or even not start. You will learn simple maintenance procedures for the intake and exhaust systems, as well as more sophisticated tests to diagnose tricky problems.

The **intake manifold** connects the air inlet tubes to the cylinder head ports to distribute air to each cylinder.

FIGURE 1-13 The intake manifold distributes air equally to the cylinders. Note that this is for an eight-cylinder engine.

FIGURE 1-14 This supercharger creates powerful boost and acceleration.

FIGURE 1-15 This badly worn lifter would cause poor engine performance.

FIGURE 1-16 You will learn many different ways to analyze the engine.

ENGINE PERFORMANCE

There are many conditions that an engine needs in order to function as designed and deliver good performance. An engine needs to have the combustion chamber well sealed, in order to develop compression of the air and fuel mixture. The engine and intake system must hold adequate vacuum to allow the engine to pull in enough air. The lubrication and cooling systems must be operating properly. Each engine component must be within its operating tolerance to allow the engine as a whole to hold together and run well (Figure 1-15). Problems in how the engine performs will keep you busy as an engine technician.

You will need to learn how to evaluate the engine's mechanical condition in order to recommend repairs. Vacuum tests and compressions tests as well as listening and smelling will be some of the weapons in your arsenal to correctly diagnose the cause of engine problems (Figure 1-16). You will act like a detective before you recommend major engine repairs, trying to gather as much information as possible. You'll need a detailed diagnosis, to be able to offer the customer a realistic estimate and to be sure you correct the problem on the first try.

CYLINDER HEAD

The **cylinder head** houses the valves and often the camshaft. In this case it is called an **overhead camshaft engine (OHC)** (Figure 1-17). In cases when the camshaft is in the block, the cylinder head still holds many of the **valvetrain** components. The pushrods, lifters, rocker arms, and springs are all parts of the valvetrain because they work to open and close the valves. When the camshaft is in the block, as the high point of the camshaft lobe comes up, it pushes on a lifter that acts like a plunger to move the **pushrod**. The pushrod is a simple, hollow rod that transfers the motion from the lifter to the rocker arm. The rocker arm rides on a pivot. When the pushrod raises one end of the rocker, the other end pushes down on the tip of the valve to open it, much like a typical first-class lever (Figure 1-18).

Problems within the cylinder head and valvetrain can be quite serious. A valve that is not adjusted properly, for example, can cause that cylinder to lose power, and the engine

The **valvetrain** is made up of the components that open and close the valves.

Rocker arms

FIGURE 1-17 This OHC engine uses rocker arms to open the valves.

Fulcrum (Pivot point)

Rocker arm

Valve spring

Valve guide

Valve

Pushrod

Lifter

Camshaft

FIGURE 1-18 The valvetrain components open the valve in response to the camshaft lobe.

FIGURE 1-19 You will learn to use specialized equipment such as this valve grinder to overhaul engines.

will run roughly. During an engine overhaul, it is typical to perform a valve job. This involves refinishing all the valves and their seats and replacing the valve seals. This requires complete disassembly of the cylinder head and some precision machining and repair (Figure 1-19).

FIGURE 1-20 The lower crankshaft gear turns at twice the speed of the larger camshaft gear.

FIGURE 1-21 Valve timing failures can be catastrophic.

Main **journals** are round machined portions on the centerline of the crankshaft where the crankshaft is held in the main bore.

A BIT OF HISTORY

Jean Joseph Lenoir created the first workable internal combustion engine in 1860. Nicholas Otto was credited with creating the first four-stroke internal combustion engine back in 1876. All previous internal combustion engines did not compress the air-fuel mixture. They attempted to draw the air-fuel mixture in during a downward

TIMING MECHANISM

As we discussed, proper valve timing is necessary to keep the valves opening and closing at the correct times. This is achieved by the timing mechanism. A sprocket or gear on the crankshaft is attached by a gear, belt, or chain to a sprocket or gear on the camshaft (Figure 1-20). Timing belts require periodic maintenance. When that is neglected, the timing belt can snap, and serious engine damage can result. In some cases, the valves knock the top of the pistons and bend, and a new set of valves is required (Figure 1-21). In the worst cases, those valves break the pistons, and the whole engine needs to be overhauled. Timing chains also wear out over time, and replacing them requires skill and attention to detail. Variable valve timing systems increase performance and minimize emissions. More and more vehicles employ these systems to vary the times when the valves can be opened. This makes timing mechanism theory and service a new and more complex challenge.

ENGINE BLOCK

The **engine block** is the main supporting structure of the engine (Figure 1-22). It holds the pistons, connecting rods, crankshaft, and sometimes the camshaft. The cylinder head bolts to the top of the block. The block is bored (drilled) to create cylinders. An eight-cylinder engine has eight-cylinder bores (Figure 1-23). The pistons are pushed up and down in the cylinders by the crankshaft. On the power stroke, the piston is actually pushed down by the force of combustion to turn the crankshaft. Rings on the pistons seal the small clearance between the cylinder walls. This keeps the hot expanding gases from combustion within the combustion chamber, so that all the force acts to push the piston down. The rings also scrape excess oil from the cylinder walls, so the oil doesn't burn inside the combustion chamber. When oil is allowed to burn during combustion, it forms blue smoke that exits the tailpipe. The block also holds the crankshaft in its main bore. The soft main bearings are placed within the main bore to provide a place for the finely machined crankshaft **journals** to ride. The journals are the machined round areas of the crankshaft that allow the bearings and bearing caps to bolt around them and hold the crankshaft in the engine block. The clearance between the journals and the bearings must be just right: enough to allow adequate oil, but not so much that precious oil pressure leaks through excessive clearances.

FIGURE 1-22 An eight-cylinder engine block.

FIGURE 1-23 The cylinder bores are finely finished to provide a good sealing surface for the rings.

Crankshaft bearings that are of the friction type are softer than the metal they support. Bearings protect the crankshaft from damage, but they also wear. When they wear excessively, they cause low oil pressure and engine knocking.

An engine overhaul involves refurbishing the engine to nearly new condition. The rings and bearings are always replaced, but additional work is usually required. You will learn to inspect and measure the cylinder walls to see if they need to be refinished. You will evaluate the oil pump, the lifters, the pistons, and literally every engine component to be sure it can withstand another punishing one hundred thousand miles of trouble-free performance. Each step of engine repair requires careful inspection, measurement, and judgment.

A BIT OF HISTORY
(continued)

movement of the piston, and then ignite the mixture. The expansion of gases would force the piston down the rest of its travel. This design was used to make the piston forceful on each downward stroke. The four-stroke engine, however, proved much more efficient and powerful. Gottlieb Daimler received the first patent for the internal combustion engine in 1885. The basic operation of today's engines is still similar to those first four-stroke engines. Figure 1-24 shows an early working model from Daimler's workshop, which is now a museum.

FIGURE 1-24 This carriage is powered by one of Daimler's first engines.

SUMMARY

TERMS TO KNOW

Camshaft

Combustion

Combustion chamber

Cylinder

Cylinder head

Engine block

Exhaust manifold

Intake manifold

Internal combustion

engine (ICE)

Journals

Overhead camshaft

engine (OHC)

Piston

Powertrain

Powertrain control

module (PCM)

Pushrod

Thermostat

Valve timing

Valvetrain

- Today's smaller engines make more power with longer life, fewer emissions, and better fuel economy than their older counterparts.
- The engine ignites a compressed air and fuel mixture in a confined chamber to harness the heat energy of expanding gases.
- The power of combustion pushes down on the piston. The piston is connected to the crankshaft by the connecting rod.
- The intake and exhaust systems allow the engine to take in and expel the great quantity of air that the engine needs to make power.
- The cooling system prevents meltdown of engine components.
- The lubrication system reduces temperature and friction in the engine.
- The cylinder head houses the valvetrain that opens and closes the engine valves.
- The cylinder block holds the pistons, connecting rods, and crankshaft and provides the main structure of the engine.
- All of the engine systems and components must work as designed and in harmony with each other in order to make the engine run smooth, have sufficient power, and produce low emissions.

REVIEW QUESTIONS

Short Answer Essays

1. Explain the benefits of today's modern engines.
2. Explain what the combustion chamber is.
3. How is air pulled into the cylinders?
4. What controls fuel injection and spark timing?
5. What is the function of the intake manifold?
6. Where can an oil pump be mounted?
7. How does a turbocharger create more power?
8. What is valve timing?
9. Name three components located in the engine block.
10. What forces the crankshaft to turn?

Fill-in-the-Blanks

1. The _____ of modern engines have decreased over 90 percent from earlier models.
2. The engine pulls air into the _____ _____ through the intake system.
3. The PCM controls _____ _____ and _____ _____, among other functions.
4. The _____ _____ distributes air equally to each cylinder.
5. The oil pump may be driven by the _____ .
6. When the thermostat opens, coolant flows through the _____ .

7. The _____ _____ connects the cylinder head to the exhaust system.

8. The _____ and _____ are compressed in the combustion chamber.

9. The _____ opens and closes the valves.

10. The _____ _____ holds the crankshaft.

Multiple Choice

1. Today's powertrains require:
   A. More maintenance   C. Hinker jets
   B. Bigger engine sizes   D. Less fuel

2. The engine is often described as a:
   A. Water pump   C. Impeller
   B. Air pump   D. Gear driver

3. *Technician A* says that the more air an engine can take in, the more power it can put out.
   Technician B says that the exhaust contributes to how well an engine can breathe.
   Who is correct?
   A. A only   C. Both A and B
   B. B only   D. Neither A nor B

4. *Technician A* says the piston is connected to the crankshaft by the rocker arm.
   Technician B says the connecting rod converts the reciprocating motion of the piston to the rotary motion of the crankshaft.
   Who is correct?
   A. A only   C. Both A and B
   B. B only   D. Neither A nor B

5. The cylinder head forms the top of the:
   A. Cylinder
   B. Combustion chamber
   C. Intake manifold
   D. Throttle bore

6. When a spark ignites the air and fuel mixture, _____ occurs.
   A. An explosion   C. Conflagration
   B. A blast   D. Combustion

7. The _____ is the main supporting structure of the engine.
   A. Motor   C. Cylinder head
   B. Block   D. Main bore

8. *Technician A* says the main bearings protect the crankshaft journals.
   Technician B says the camshaft lobes lift the pistons.
   Who is correct?
   A. A only   C. Both A and B
   B. B only   D. Neither A nor B

9. Which of the following is true?
   A. The crankshaft rotates at the same speed as the camshaft.
   B. The camshaft rotates twice as fast as the crankshaft.
   C. The camshaft rotates three times as fast as the crankshaft.
   D. The crankshaft rotates twice as fast as the camshaft.

10. Valve timing is:
    A. When the spark occurs
    B. When the pistons rotate
    C. When the valves open and close
    D. How long combustion takes

Chapter 2

THEORY OF ENGINE OPERATION

UPON COMPLETION AND REVIEW OF THIS CHAPTER, YOU SHOULD BE ABLE TO:

- Identify the major engine components.

- Describe basic engine operation.

- Describe the basic laws of physics involved with engine operation.

- Classify engines according to the number of cycles, the number of cylinders, cylinder arrangement, and valvetrain type.

- Define the four-stroke cycle theory.

- Describe the different cylinder arrangements and the advantages of each.

- Describe the different valvetrains used in modern engines.

- Define engine measurement terms such as bore and stroke, displacement, compression ratio, engine efficiency, horsepower and torque, horsepower losses, mechanical efficiency, and thermal efficiency.

- Describe the relationship between compression ratio and engine power output.

- Define mechanical, volumetric, and thermal efficiencies, and describe factors that affect each.

- Describe the basic operation of alternative engine designs, including diesel, Miller-cycle, and Stratified Charge.

INTRODUCTION

Today's engines are designed to meet the demands of the automobile-buying public and many government-mandated emissions and fuel economy regulations. High performance, fuel economy, reduced emissions, and reliability are demanded. To fulfill these needs, manufacturers are producing engines using lightweight blocks and cylinder heads and non-traditional materials such as powdered metals and composites.

While many of the engine's mechanical components are similar to those of over one hundred years ago, refinements in design, materials, and machining help the manufacturers improve engine performance and reliability. Many engine and support system functions are now closely controlled by the PCM. Fuel delivery, spark timing, and often valve timing are managed so precisely that the PCM can make adjustments within milliseconds; a millisecond is one thousandth of a second. Today's technician is called upon to diagnose and service these advanced engines properly.

The **internal combustion engine** used in automotive applications utilizes several laws of physics and chemistry to operate. Although engine sizes, designs, and construction vary greatly, they all operate on the same basic principles. This chapter discusses these basic principles and engine designs.

Internal combustion engines burn their fuels within the engine.

MAJOR ENGINE COMPONENTS

The engine's cylinder block is the structure that the rest of the engine is built upon (Figure 2-1). The block has precisely machined holes or cylinder bores in which the pistons are pushed up and down.

The number of cylinder bores describes the engine; a three-, four-, five-, six-, eight-, ten-, or twelve-cylinder engine. These bores are typically between 2.5 and 4.25 inches on common production vehicles and light duty trucks. Lengthwise across the bottom of the block is another bore that holds the **crankshaft**. The downward movement of the pistons spins the crankshaft. The pistons are connected to the crankshaft by the **connecting rods**.

The crankshaft connects to the transmission and provides the rotary motion to turn the wheels. On top of the block covering the cylinders is the cylinder head. It houses valves that open and close to allow fresh air into the cylinders and burned exhaust gases out. Look at Figure 2-2 to identify these components in the engine.

The cylinder block also has oil and coolant passages drilled throughout it to circulate these vital fluids to key areas in the block and up to the cylinder head. Usually, the water and oil pumps are mounted directly to the block in several different ways.

The crankshaft is a long iron or steel fabrication with round bearing surfaces machined onto it called journals (Figure 2-3). The main journals form the centerline of the crankshaft. Main **engine bearings** fit into the main bores of the engine block, and the crankshaft can spin freely in this mounting. There is just enough clearance between the bearings and the journals to allow a thin film of pressurized oil to keep the journals from actually riding on the bearings. Half of the main bore is integral to the block. Main caps form the other half of the bore; these can be removed to service the crankshaft or bearings.

The crankshaft also has journals that are offset from the centerline. These are called rod journals because the connecting rods, fitted with rod bearings, bolt around these journals to attach to the crankshaft. The rod bearings are very similar to the main bearings. When the piston pushes the connecting rod down on the offset journal of the crankshaft, it forces the

A **crankshaft** is a shaft held in the engine block that converts the reciprocating motion of the pistons into rotary motion used to turn the transmission.

The **connecting rod** forms a link between the piston and the crankshaft.

Engine bearings are two bearing halves that form a circle to fit around the crankshaft journals.

FIGURE 2-1 This engine block is an eight cylinder. It is a vee style engine with four cylinders on each bank of the vee. This photo shows four cylinders on one bank of the vee.

FIGURE 2-2 This engine cutaway shows the major components of the engine.

Piston rings are hard cut rings that fit around the piston to form a seal between the piston and the cylinder wall.

A **wrist pin** is a hardened steel pin that connects the piston to the connecting rod and allows the rod to rock back and forth as it travels with the crankshaft.

The **transmission** is a device that uses the rotary motion and power of the crankshaft to turn the driveshaft and/or the differential. The transmission multiples the power output and changes the speed of the driveshaft using different gear ratios.

A **differential** is a set of gears that multiples the power and changes the speed of the transmission output.

FIGURE 2-3 A crankshaft for an eight-cylinder engine.

crankshaft to rotate. The crankshaft also has counterweights or crankshaft throws that are offset around the shaft. The weight of these throws helps offset the crankshaft speed fluctuations and vibrations as the different cylinders produce power.

The pistons are slightly smaller than the cylinders, to allow them to move up and down with minimal friction. They are fitted with **piston rings** that seal the area between the piston and the cylinder wall. The piston connects to the small end of the connecting rod through a **wrist pin** (sometimes called a piston or gudeon pin) and a bushing. The force of combustion works on the top of the piston and is transferred to the crankshaft through the connecting rod. As the piston moves up and down in the cylinder, it rotates the crankshaft. This process converts the reciprocating motion of the piston into rotary motion (Figure 2-4). This rotary motion is what is needed to turn the tires and wheels after the energy is transferred though the **transmission** and **differential**. A flywheel or flexplate at the rear of the crankshaft, along with a balancer pulley at the front, provides a large, stable mass for smoothing out the

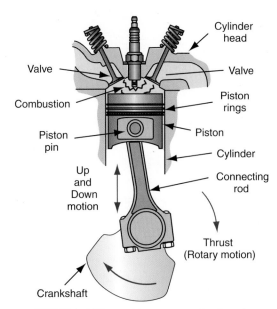

FIGURE 2-4 The engine components at work.

FIGURE 2-5 The cylinder head and valves seal the top of the combustion chamber.

rotation and dampening the pulsation of the power stroke. Automatic transmissions use flexplates, and manual transmissions use flywheels.

The cylinder head forms a tight cover for the top of the cylinders and contains machined chambers into which the air-fuel mixture is forced (Figure 2-5). The area of the

FIGURE 2-6 A camshaft.

cylinder head located above the cylinder bore is called the combustion chamber. It is sealed on the top by the cylinder head and valves and on the bottom by the pistons and rings. The cylinder head also has threaded holes for the spark plugs that screw right into the head so that their tips protrude into the combustion chambers. The valves in each cylinder are opened and closed by the action of the camshaft and related valvetrain. The camshaft is connected to the crankshaft by a gear, chain, or belt, and is driven at one-half crankshaft speed. In other words, for every complete revolution the crankshaft makes, the camshaft makes one-half revolution. The camshaft may be mounted in the engine block itself, or in the cylinder head, depending on design. Some engines use two camshafts in the head; these are called dual overhead cam (DOHC) engines.

The camshaft has lobes that are used to open the valves (Figure 2-6). A camshaft mounted in the block operates the valves remotely through pushrods and rocker arms; others act on followers placed directly on top of the valves. In some overhead camshaft (OHC) engines, the camshaft lobes contact the rocker arms.

Lubricating oil for the engine is normally stored in the oil pan mounted to the bottom of the engine. The oil is force-fed to almost all parts of the engine by the oil pump.

ENGINE OPERATING PRINCIPLES

Thermodynamics is the study of the relationship between heat energy and mechanical energy.

One of the many laws of physics utilized within the automotive engine is **thermodynamics**. The driving force of the engine is the expansion of gases. Gasoline (a liquid fuel) will change state to a gas if it is heated or burned. Gasoline must be mixed with oxygen before it can burn. In addition, the air-fuel mixture must be burned in a confined area in order to produce power. Gasoline that is burned in an open container produces very little power, but if the same amount of fuel is burned in a closed container, it will expand producing usable force. When the air and fuel mixture changes states, it also expands as the molecules of the gas collide with each other and bounce apart. Increasing the temperature of the gasoline molecules increases their speed of travel, causing more collisions and expansion.

Heat is generated by compressing the air-fuel mixture within the combustion chamber. Igniting the compressed mixture causes the heat, pressure, and expansion to multiply. This process releases the energy of the gasoline so it can produce work. The igniting of the mixture is a controlled burn, not an explosion. The controlled combustion releases the fuel energy at a controlled rate in the form of heat energy. The heat, and consequential expansion of molecules, dramatically increases the pressure inside the combustion chamber. Typically, the pressure works on top of a piston that is connected to a crankshaft. The expanding gases push the piston down with tremendous force. As the piston is forced down, it causes the crankshaft to turn.

This twisting force on the crankshaft is called **torque**. Torque is applied to the drive wheels through the transmission. As the engine drives the wheels to move the vehicle, a certain amount of work is done. The rate of work being performed is measured in **horsepower**.

Energy and Work

In engineering terms, in order to have work, there must be motion. Using this definition, work can be measured by combining distance and weight and expressed as foot-pound (ft.-lb.) or Newton-meter (Nm). These terms describe how much weight can be moved a certain distance. A foot-pound is the amount of energy required to lift one pound of weight one foot in distance. The amount of work required to move a 500-pound weight five feet is 2,500 foot-pounds (3,390 Nm). In the metric system, the unit used to measure force is called a newton meter (Nm). A newton is a unit measure of force, similar to a pound. Torque is measured as the amount of force in newtons multiplied by the distance that the force acts in meters. One foot-pound is equal to 1.355 Nm. It is important to understand both systems.

Energy produces work that can be measured in units, such as torque. Basically, energy is anything that is capable of resulting in motion. Common forms of energy include electrical, chemical, heat, radiant, mechanical, and thermal.

The **law of conservation of energy** states that the total amount of energy in an isolated system remains constant. Energy cannot be created or destroyed; however, it can be stored, controlled, and changed into other forms of energy. For example, the vehicle's battery stores chemical energy that is changed to electrical energy when a load is applied. An automotive **engine** converts thermal (heat) energy into mechanical energy or power to produce force and motion. During combustion, when the gasoline burns, thermal energy is released. The engine converts this thermal energy into mechanical energy. Mechanical energy provides movement. The heat, expansion, and pressure during combustion all supply thermal energy within the combustion chamber. This thermal energy is converted to mechanical energy as the piston is pushed downward in the cylinder with great force.

Types of Energy

There are two types of energy: potential and kinetic. **Potential energy** is available to be used but is not being used. An example of this in the automobile is the chemical energy of the battery when the engine and ignition is turned off. Most of the chemical energy is stored as potential energy. (A very small percentage of the available energy is actually being used to keep computer memories alive.) When the ignition key is turned to start and the starter begins to crank the engine over, **kinetic energy** is being used. Kinetic energy is energy that is being used or working. The battery is now providing significant chemical energy to the starter so it can do work.

Energy Conversion

In our example of the battery starting the engine, there are several energy conversions that occur. Energy cannot be destroyed or eliminated, but it can be converted from one state or form to another. The starter converts the chemical energy from the battery into electrical energy to engage the starter. The starter converts this electrical energy into mechanical energy to crank the engine over. At the same time, chemical energy from the battery is being converted to electrical energy to provide power to the fuel and ignition systems so they will have the needed fuel and spark to run. Once the engine is running, combustion occurs. Combustion uses the potential energy in gasoline and converts it into kinetic energy by burning it. The chemical energy of the fuel is converted to thermal energy by the heart produced during combustion. The thermal energy is converted to mechanical energy by the

Torque is a rotating force around a pivot point.

Horsepower is a measure of the rate of work.

The **law of conservation of energy** is a principle law of physics that states that the total energy of an isolated system remains constant despite any internal changes.

An **engine** is defined as a device or machine that converts thermal energy into mechanical energy.

Potential energy is energy that is not being used at a given time, but which can be used.

movement of the pistons and the crankshaft. The following energy conversions are all common automotive applications:

Chemical Energy Conversion to Thermal Energy

As fuel is burned in the combustion chambers, the chemical energy in the fuel is converted to thermal energy.

Thermal Energy Conversion to Mechanical Energy

As the heat, pressure, and expansion of gases develop during combustion, thermal energy increases. This thermal energy is converted to mechanical energy to drive the vehicle, through energy exerted on top of the pistons that turns the crankshaft to drive the transmission and drive wheels.

Chemical Energy Conversion to Electrical Energy

The battery stores chemical energy and converts it to electrical energy as electrical loads such as the starter, the powertrain control module, the fuel injectors, and the radio demand electricity.

Electrical Energy Conversion to Mechanical Energy

The starter, fuel injectors, electric cooling fan, and windshield wiper motor are just a few of the many automotive applications of conversion of electrical energy into mechanical motion. A hybrid electric vehicle (HEV) uses electricity to power an electric motor which then turns the transmission.

Mechanical Energy Conversion to Electrical Energy

The generator provides electrical energy to recharge the battery by using the mechanical energy of rotation through a belt off the spinning crankshaft. Hybrid electric vehicles use regenerative braking during deceleration to recharge the high voltage battery pack.

Mechanical Energy Conversion to Thermal Energy

The braking system on a traditional internal combustion engine vehicle is a fine example of this form of energy conversion. The mechanical energy of the rotating wheels is changed into thermal energy through the **friction** created between the rotating brake discs or drums and the stationary friction surfaces of the brake pads or shoes. The thermal energy is dissipated to the atmosphere.

Newton's Laws of Motion

First Law

A body in motion tends to stay in motion. A body at rest tends to stay at rest. Forces applied to the body can overcome this tendency. Certainly the crankshaft, transmission, and drive wheels will not propel the car forward from a rest position unless significant force is applied to the pistons to rotate the crankshaft.

Second Law

A body's acceleration is directly proportional to the force applied to it. The greater the force applied to the drive wheels, the greater the acceleration of the vehicle.

Third Law

For every action, there is an equal and opposite reaction. A simple application of this law on the automobile involves the suspension system. When the tire travels over a bump, the wheel moves upward with a force that opposes spring tension. When the wheel passes the bump, the spring exerts its opposite reaction to force the wheel back down toward the road.

Inertia

Inertia is the tendency of an object at rest (static inertia) to stay at rest or an object in motion (dynamic inertia) to stay in motion. The heavy flywheel attached to the end of the

A **hybrid electric vehicle (HEV)** uses an electric motor and a gas, diesel, or alternate fuel engine to power the vehicle.

Friction is resistance to motion when one object moves across another object. Friction generates heat and is an excellent method of controlling mechanical action.

crankshaft on manual transmission vehicles uses the inertia of the rotating crankshaft to help it continue to spin smoothly between the combustion events in the cylinders.

Momentum

Momentum is when force overcomes static inertia and causes movement of an object. As the object moves, it gains momentum. Momentum is calculated by multiplying the object's weight times its speed. An opposing force will reduce momentum.

Friction

In the engine, the piston rings generate friction against the cylinder walls. Friction opposes momentum and mechanical energy. Friction creates heat and reduces the mechanical energy of the piston as it is pushed downward to turn the crankshaft. This is an example of how engine efficiency is decreased. Friction may occur in solids, liquids, and gases. The airflow over a vehicle causes friction. The manufacturers carefully engineer their vehicles to be aerodynamic. This means that the resistance to motion from friction between the vehicle and the air is minimized. This is carefully calculated and measured as the coefficient of drag (Cd). Unfortunately, there are conditions and vehicles that can lose up to half of their mechanical energy to aerodynamic drag.

BEHAVIOR OF LIQUIDS AND GASES

Molecular Energy

Electrons are in constant motion around the nucleus of atoms or molecules. Kinetic energy is always occurring. The kinetic energy of liquids, gases, and solids changes in response to temperature. When temperature decreases, so does kinetic energy. As temperature increases, so does kinetic energy. Solids have the slowest reaction to temperature, while gases have the quickest response to changes in temperature. We know this theory already through our discussion of combustion. As temperature increases within the combustion chamber, the gases rapidly expand and the increasing kinetic energy is used to create thermal energy that is converted into mechanical energy.

Temperature

The volume of liquids, gases, and solids increases with temperature. It is the increased volume of gases on top of the pistons that produces the force to push the piston downward. When filling engine fluids, such as tire air pressure and coolant, we must be aware of the effects of temperature on volume. It is easy to overfill a coolant reservoir if the coolant is at ambient temperature and you top the fluid off to the maximum level. As the fluid heats up, it may expand beyond the capacity of the reservoir and develop enough pressure to blow the cap's pressure relief valve.

Pressure and Compressibility

Pressure is always applied equally to all surfaces of a container. If there is a leak in the container, the pressure will leak out and equalize with the atmospheric pressure. Gases and liquids both flow, so they are classified as fluids. But there are distinct and significant differences between the characteristics of gases and liquids. Gases can be compressed; the molecules squeeze tighter together and form greater pressure. You can easily fill a tire with excess pressure. Further compress the gases in a combustion chamber to produce more power. Liquids, on the other hand, are difficult to compress. A liquid fills a chamber, and you cannot add more liquid; you can only increase pressure. Brake and clutch hydraulic systems work with liquids. When you depress the clutch or brake pedals, you move the liquid, under pressure to exert force on a piston. This force produces mechanical work from the brake caliper piston or the clutch hydraulic lever. If there is a leak in the system, air can enter. Air is a gas and is compressible. Air in a hydraulic system will cause a lack of pressure and force.

PRESSURE AND VACUUM

Air is a gas with weight and mass. It is layered high above the surface of the earth and exerts a pressure on the earth's surface. This pressure is called atmospheric pressure. A 1 square inch column of air extending from the earth's surface at sea level to the outer edge of the earth's atmosphere weighs 14.7 pounds. We say that atmospheric pressure is 14.7 pounds at sea level. This pressure changes as we climb in altitude. If we are 14,000 feet above sea level in the mountains, that column of air no longer weighs 14.7 lbs; atmospheric pressure decreases. Also, as temperature increases, the molecules become less dense and therefore weigh less; atmospheric pressure decreases. These changes in atmospheric pressure caused by altitude and temperature are important for the PCM to know about in order to deliver the correct amount of fuel for the actual amount of air in the cylinder.

A **vacuum** is a space in which the pressure is significantly lower than atmospheric pressure.

When air pressure is greater than atmospheric pressure, it is a positive pressure. When air pressure is lower than atmospheric, it is a negative pressure or vacuum. The engine creates a **vacuum** as the engine valves are closing and as the piston moves from the top to the bottom of the cylinder. The volume of the chamber has increased, but no more air has been added to fill the space; pressure is reduced and a vacuum is formed. Vacuum is typically measured in inches of mercury vacuum (in. Hg vacuum). At atmospheric pressure, there are 0 in. Hg vacuum. When there is a total absence of air or 0 absolute pressure (no atmospheric pressure at all), a perfect vacuum can be measured at 29.9 in. Hg vacuum.

An engine develops significant vacuum during its intake strokes as the pistons move downward and the valves begin to close. Liquids, solids, and gases move from an area of high pressure to an area of low pressure. Automotive engines rely on engine vacuum and this principle to fill the combustion chambers with air to support combustion. When the throttle is opened, atmospheric pressure flows rapidly into the low pressure area of the cylinder to fill it with air. An engine that cannot develop adequate vacuum will not pull in enough air to make maximum power. A modern, fine-running engine will typically develop about 16–18 in. Hg vacuum at idle.

It is important to understand that a vacuum does not mean that air is sucked into the space with a lower pressure. When a vacuum is created, air is forced from the area of higher pressure (typically atmospheric pressure outside of the engine and near the intake) to the area of lower pressure (typically the combustion chambers). Vacuum is also dependant on volume and temperature.

Turbocharged and supercharged engines increase this pressure difference before and after the throttle plate to help fill the combustion chambers fully under all conditions. The engine still develops the same amount of vacuum, but the supercharger or turbocharger delivers roughly 8–15 psi of pressure over atmospheric behind the throttle. When the throttle opens, the greater pressure difference increases airflow into the cylinders.

BOYLE'S LAW

Boyle's Law states that if the temperature remains constant, the volume of a given mass of gas is inversely proportional to the absolute pressure. Boyle's Law is a summary of the relationships between pressure, temperature, and volume. The fuel and air mixture enters the combustion chamber because of a vacuum. The combustion chamber still has pressure, but it is less than atmospheric pressure. This lower pressure is a direct result of the increasing volume in the combustion chamber.

The **stroke** is the amount of piston travel from TDC to BDC measured in inches or millimeters.

ENGINE OPERATION

Most automotive and truck engines are four-stroke cycle engines. A **stroke** is the movement of the piston from one end of its travel to the other; for example, if the piston is at the top of its travel and then is moved to the bottom of its travel, one stroke has occurred.

Intake valve
closing

Both valves
closed

Both valves
closed

Exhaust valve
opening

Air/fuel
mixture

Ignition

A

B

C

D

Compression stroke
51°BTDC

Compression stroke
20°BTDC

Power stroke

Power stroke
57°BBDC

FIGURE 2-7 The four strokes of an automotive engine: (A) intake stroke, (B) compression stroke,
(C) power stroke, (D) exhaust stroke.

Another stroke occurs when the piston is moved from the bottom of its travel to the top again. A **cycle** is a sequence that is repeated. In the four-stroke engine, four strokes are required to complete one power-producing cycle.

The internal combustion engine must draw in an air-fuel mixture, compress the mixture, ignite the mixture, then expel the exhaust. This is accomplished in four-piston strokes (Figure 2-7).

The first stroke of the cycle is the intake stroke (see Figure 2-7A). As the piston moves down from **top dead center (TDC)**, the intake valve is opened so the vaporized air-fuel mixture can be pushed into the cylinder by atmospheric pressure. As the piston moves downward in its stroke, a vacuum is created (low pressure). Since high pressure moves toward low pressure, the air-fuel mixture is pushed past the open intake valve and into the cylinder. After the piston reaches **bottom dead center (BDC)**, the intake valve is closed, and the stroke is completed. Closing the intake valve after BDC allows an additional amount of air-fuel mixture to enter the cylinder. Even though the piston is at the end of its stroke and no more vacuum is created, the additional mixture enters the cylinder because it weighs more than air alone. The momentum of air flowing into the cylinder also helps pull more air in after the piston has reached BDC.

The compression stroke begins as the piston starts its travel back to TDC (see Figure 2-7). The intake and exhaust valves are both closed, trapping the air-fuel mixture in the combustion chamber. The movement of the piston toward TDC compresses the mixture. Fuel and air molecules are constantly in motion. As the fuel and air molecules are compressed tightly together, they begin to collide more frequently. This causes an increase in the energy level of the molecules. The energy level is measured by its heat intensity (temperature). When the piston reaches TDC, the mixture is fully compressed and is highly combustible because the mixture is close to its **autoignition temperature**. Compressing the mixture closer to this temperature provides for a more complete burning of the molecules. The ignition system induces a spark at exactly the right time to produce maximum power on top of the piston when it is at about ten degrees of crankshaft rotation past TDC. The actual timing of the spark on a modern engine is controlled by the PCM (powertrain control module), also referred to as ECM (engine control module) by some manufacturers, and it varies based on temperature, speed, fuel content, and load. Some manufactures use different terminology for this computer; such as ECM (Engine Control Module). This ensures that peak pressure will act on the piston when it is just headed downward and before the volume of the chamber has increased significantly.

A **cycle** is a complete sequence of events.

Top dead center (TDC) is a term used to indicate that the piston is at the very top of its stroke.

Bottom dead center (BDC) is a term used to indicate that the piston is at the very bottom of its stroke.

The **autoignition temperature** is the minimum temperature required to ignite a gas in air without a spark or flame to assist.

When the spark occurs in the compressed mixture, the rapid burning causes the molecules to expand, beginning the power stroke (see Figure 2-7C). The expanding molecules create a pressure above the piston and push it downward. The downward movement of the piston in this stroke is the only time the engine is productive concerning power output. During the power stroke, the intake and exhaust valves remain closed.

The exhaust stroke of the cycle begins when the piston reaches BDC of the power stroke (see Figure 2-7D). Just prior to the piston reaching BDC, the exhaust valve is opened. The upward movement of the piston back toward TDC pushes out the exhaust gases from the cylinder past the exhaust valve and into the vehicle's exhaust system. As the piston approaches TDC, the intake valve opens and the exhaust valve is closed a few degrees after TDC. The degrees of crankshaft rotation when both the intake and exhaust valves are open is called **valve overlap**. Valve overlap uses the movement of air to help fully clean out the spent gases and fill the chamber with a fresh charge of air and fuel. The cycle is then repeated again as the piston begins the intake stroke. This cycle is occurring simultaneously in the other cylinders of the engine, though the individual strokes do not occur at the same time. For example, cylinder number one will produce its power stroke, then, while its exhaust stroke occurs, cylinder number three will be on its power stroke. This keeps rotating the crankshaft around with regular impulses. Most engines in use today are referred to as **reciprocating**. Power is produced by the up- and-down movement of the piston in the cylinder. This linear motion is then converted to rotary motion by a crankshaft.

ENGINE CLASSIFICATIONS

Engine classification is usually based on the number of cycles, the number of cylinders, cylinder arrangement, and valvetrain type. Additional descriptors include the location and number of camshafts and the type of fuel and ignition systems. Engine displacement is commonly used to identify engine size.

Sometimes, different methods may be used to identify the same engine. For example, a Chrysler 2.0 liters DOHC is also identified as a 420A and D4FE. In this case, the engine has a displacement of 2.0 liters and uses dual overhead camshafts. The 420A is derived from four valves per cylinder, 2.0 liters displacement, and American built. The D4FE means the engine uses dual overhead camshafts with four valves per cylinder and has front exhaust (the exhaust manifold is mounted facing the front of the vehicle). This section will define many of the terms used by automotive manufacturers to classify their engines.

Number of Cylinders

One cylinder would not be able to produce sufficient power to meet the demands of today's vehicles. Most automotive and truck engines use three, four, five, six, eight, ten, or twelve cylinders. The number of cylinders used by the manufacturer is determined by the amount of work required from the engine. Generally, the greater the number of cylinders, the more power the engine can produce. Also, the more cylinders, the greater number of power pulses in one complete 720-degree cycle. This produces a smoother running engine.

Vehicle manufacturers attempt to achieve a balance among power, economy, weight, and operating characteristics. An engine having more cylinders generally runs and idles smoother than those having three or four cylinders. This is because there is less crankshaft rotation between power strokes. There is less time between the combustion chamber explosions that cause the sudden acceleration of the piston, which is the source of vibrations. However, adding more cylinders to the engine design usually increases the weight of the assembly, the complexity of the design, and the cost of production. Over the years, manufacturers have changed their cylinder number, size, and arrangement designs to meet the ever-changing consumer demands, fuel mileage, and emissions laws.

Valve overlap is the length of time, measured in degrees of crankshaft revolution, that the intake and exhaust valves of the same combustion chamber are open simultaneously.

Reciprocating is an up-and-down or back-and-forth motion.

A BIT OF HISTORY

License plates made their first appearance on vehicles in the United States in 1901. In 1905, California began licensing and registering motor vehicles. By the end of 1905, California had registered 17,015 vehicles. Owners had to conspicuously display the circular or octagonal tags in the vehicle. In order to register the vehicle, it had to have working lamps, satisfactory brakes, and a bell or a horn.

Cylinder Arrangement

Engines are also classified by the arrangement of the cylinders. The cylinder arrangement used is determined by vehicle design and purpose. The most common engine designs are in-line and V-type.

The in-line engine places all of its cylinders in a single row (Figure 2-8). Advantages of this engine design include ease of manufacturing and serviceability. The disadvantage of this engine design is the block height. Since the engine is tall, aerodynamic design of the vehicle is harder to achieve. Many manufacturers overcome this disadvantage by installing a **transverse-mounted engine** in the engine compartment of front-wheel-drive vehicles. A modification of the in-line design is the slant cylinder (Figure 2-9). By tilting the engine slightly, the manufacturer can lower the height of the hood line for improvement in the vehicle's aerodynamics.

The V-type engine has two rows of cylinders set 60 to 90 degrees from each other (Figure 2-10). The V-type design allows for a shorter block height, thus easier vehicle aerodynamics. The length of the block is also shorter than in-line engines with the same number of cylinders.

A common engine design used for rear engine vehicles, as well as the front-wheel drive Subaru, is the opposed cylinder engine (Figure 2-11). This engine design has two rows of cylinders directly across from each other. The main advantage of this engine design is the very small vertical height.

A **transverse-mounted engine** faces from side to side instead of front to back.

The opposed cylinder engine may be called a pancake engine.

FIGURE 2-8 In-line engines are designed for ease of construction and service.

FIGURE 2-9 Slant cylinder design lowers the overall height of the engine.

FIGURE 2-10 The V-type engine design allows for lower and shorter hood lines.

FIGURE 2-11 Opposed cylinder engine design.

Valvetrain Type

An **overhead valve (OHV)** engine has the valves mounted in the cylinder head and the camshaft mounted in the block.

Engines can also be classified by their valvetrain types. The three most commonly used valvetrains are the:

1. **Overhead valve (OHV).** The intake and exhaust valves are located in the cylinder head, while the camshaft and lifters are located in the engine block (Figure 2-12). Valvetrain components of this design include lifters, pushrods, and rocker arms.
2. **Overhead cam (OHC).** The intake and exhaust valves are located in the cylinder head along with the camshaft. The valves are operated directly by the camshaft and a follower, eliminating many of the moving parts required in the OHV engine. If a single camshaft is used in each cylinder head, the engine is classified as a single overhead cam (SOHC) engine. This designation is used even if the engine has two-cylinder heads with one camshaft each.
3. **Dual overhead cam (DOHC).** The DOHC uses separate camshafts for the intake and exhaust valves. A DOHC V-8 engine is equipped with a total of four camshafts (Figure 2-13).

A **dual overhead cam (DOHC)** engine has two camshafts mounted above each cylinder head.

Ignition Types

A **spark-ignition (SI)** engine ignites the air-fuel ratio in the combustion chamber by a spark across the spark plug electrodes.

Most automotive engines use a mixture of gasoline and air, which is compressed and then ignited with a spark plug. This type of engine is referred to as a **spark-ignition (SI) engine**. Diesel engines, on the other hand, do not use spark plugs. The fuel and air mixture is ignited by the heat created during the compression stroke. These engines are referred to as **compression-ignition (CI) engines**.

A **compression-ignition (CI)** engine ignites the air-fuel mixture in the cylinders from the heat of compression.

Spark ignition engines may be further identified by the type of ignition system used. The older, less common design is called a distributor ignition (DI). This form of ignition uses a mechanical device to distribute the spark from one **coil** to the correct cylinder. Newer ignition systems tend to use multiple coils to deliver spark to the cylinders. These systems are termed electronic-ignition (EI) systems. A form of EI is the coil on plug (COP) ignition system, where one coil sits on top of each spark plug to deliver spark directly (Figure 2-14). Another type of EI is called a waste spark-ignition system, where one coil provides spark to two cylinders.

A **coil** is an electrical device that converts low voltage from the battery into a very high voltage that is able to push a spark across the gap of the spark plug.

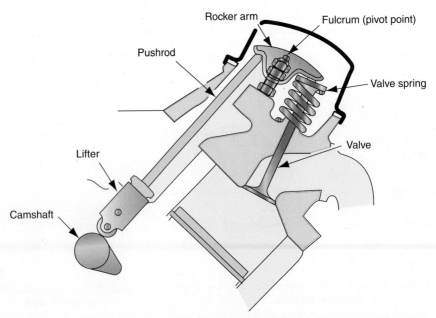

FIGURE 2-12 The overhead valve engine has been one of the most popular valve designs.

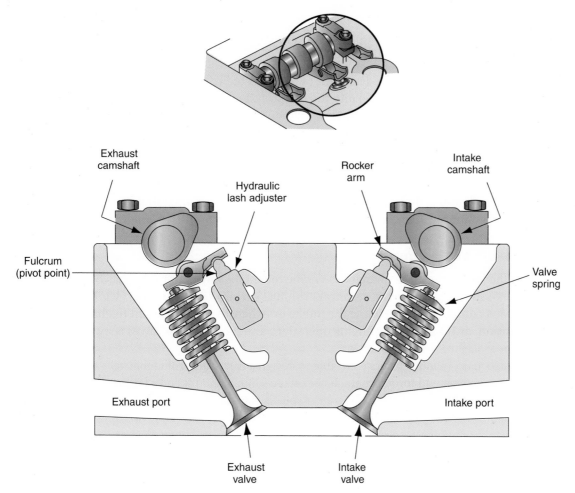

FIGURE 2-13 The dual overhead camshaft engine places two camshafts in the cylinder head.

FIGURE 2-14 This late model V-6 engine has a coil on top of each spark plug (this is the forward cylinder head).

Fuel Injection Types

All new vehicles currently sold in the United States and Canada use a form of **fuel injection** to deliver fuel to the engine. Fuel injectors spray just the right amount of fuel for each combustion event. The required amount of fuel injected is controlled by the PCM. Most fuel injection systems deliver fuel into the intake system very close to the back of the intake valve. These systems use one injector per cylinder and are termed multi-point fuel injection (MPFI). Many manufacturers refine this system to make sure that the injectors each fire on each cylinder's intake stroke for excellent mixing of the air and fuel. These systems are called sequential fuel injection (SFI) systems because the injectors fire in the same sequence as the cylinders. A few newer gas engines use fuel injectors that spray fuel directly into the combustion chamber. These are called gasoline direct injection (GDI) systems. This type of fuel injection improves combustion and reduces fuel consumption and emissions.

Variable Valve Timing

One of the most significant changes to engine technology in recent years has been the addition or modification to the valvetrain timing. **Variable valve timing (VVT) systems** have added components that allow the intake or exhaust valves (or both) to change the lift, duration, or timing during engine operation. The timing of the valves is a critical component of the engine's operation.

In years past, manufacturers were able to design a camshaft profile that gave the engine certain horsepower and torque characteristics at certain engine speeds. Performance and economy at low (or sometimes high) engine speeds was sacrificed because of design limitations and tradeoffs. For example, an engine may run very rough and have low horsepower at idle but then have a lot of power and run very smooth at a higher speed. In an ideal situation, at low engine speeds, the engine should have a lot of torque, good efficiency, and low emissions. At high speeds, the engine should have more power.

At high engine speeds, the engine requires more air to produce more power. A VVT system can keep the intake valves open for a longer duration and thus produce more power. Variable valve timing systems can also lower the cold-start emissions and overall emissions of a vehicle. Some VVT systems control either the exhaust or the intake camshaft, and some control both. Variable valve timing systems are either mechanical, hydraulic, electrical, or a combination. Most modern engines use electrical components to control or monitor the VVT system.

This type of technology allows manufacturers to use smaller displacement engines. Manufacturers are now using VVT systems in many engine designs to get better fuel economy, lower emissions, and more performance. Examples of these designs include Honda's VTEC, BMW's Valvetronic/VANOS, GM's DCVCP, Toyota's VVTL-i, and Nissan's VVL.

ENGINE VIBRATION
Balance Shafts

Many engine designs have inherent vibrations. The in-line four-cylinder engine is an example of this problem. Some engine manufacturers use balance shafts to counteract engine vibration. Four-cylinder engine vibration occurs at two positions of piston travel. The first is when a piston reaches TDC. The inertia of the piston moving upward creates an upward force in the engine. The second vibration occurs when all pistons are level. At this point, the downward moving pistons have accelerated and have traveled more than half their travel distance. The engine forces are downward because the pistons have been accelerating since they have to travel a greater distance than the upward moving pistons to reach the point at which all pistons are level.

Piston travel is further at the top 90 degrees versus the bottom 90 degrees of crankshaft rotation. First, determine the total length. This is rod length, from the center of the big end to the center of the small end, and the crankshaft throw (stroke divided by 2). Next, determine the amount the piston drops in the first 90 degrees of crankshaft rotation (TDC to 90 degrees). Once this is known, the final amount of piston travel can be determined.

The balance shafts rotate at twice the speed of the crankshaft. This creates a force that counteracts crankshaft vibrations. At TDC, the forces of the piston will exert upward; to offset the vibration, the crankshaft and the balance shaft(s) lobes have a downward force (Figure 2-15). At 90 degrees of crankshaft rotation, all pistons are level. Remember, the pistons traveling downward have traveled more than half of their travel distance to reach this point. The engine forces are in a downward direction since these pistons have been accelerated to reach the midpoint at the same time as the upward moving pistons, which need to travel a lesser distance. Since the balance shaft(s) rotate at twice the crankshaft speed, the lobes of the balance shaft(s) are facing up, exerting an upward force to counteract the downward forces of the pistons (Figure 2-16). At 180 degrees of crankshaft rotation, the upward-moving pistons reach TDC, and again exert an upward force. The balance shaft(s) have their lobes facing downward to exert a downward force to counteract the upward force (Figure 2-17). Finally, at 270 degrees of crankshaft rotation, the pistons are level again. At this point, the lobes of the balance shaft(s) are exerting an upward force to counteract the downward force of the pistons (Figure 2-18). Balance shafts may also be used in vee engines (Figure 2-19).

Some balance shafts on OHV engines are mounted in the block V above the camshaft (Figure 2-20). Bearing journals on the balance shaft are mounted on bushings in the block. Some engines use two gears mounted on the front end of the camshaft to drive the balance shaft. The inner camshaft gear is meshed with the gear on the balance shaft, and the timing chain surrounds the outer camshaft gear and the crankshaft gear to drive the camshaft. On some DOHC engines, the balance shaft is mounted in the block above the crankshaft (Figure 2-20).

FIGURE 2-15 Balance shaft operation with piston moving upward.

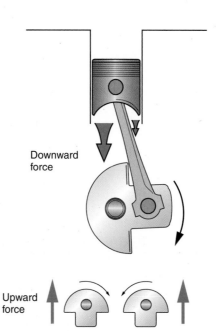

FIGURE 2-16 Balance shaft operation with piston moving downward.

FIGURE 2-17 Balance shaft operation in relation to other pistons moving upward.

FIGURE 2-18 Balance shaft operation with other pistons moving downward.

FIGURE 2-19 A balance shaft used in an OHV engine.

FIGURE 2-20 Balance shaft in a DOHC V-6 engine.

ENGINE DISPLACEMENT

Displacement is the volume of the cylinder between TDC and BDC measured in cubic inches, cubic centimeters, or liters (Figure 2-21). Most engines today are identified by their displacement in liters. Engine displacement is an indicator of engine size and power output. The more fuel that can be burned, the greater amount of energy that can be produced. To burn more fuel, more air must also enter the cylinders. Engine displacement measurements are an indicator of the amount or mass of air that can enter the engine. The mass of air allowed to enter the cylinders is controlled by a throttle plate. If the throttle plate is closed, a smaller mass of air is allowed to enter the cylinder, and the power output is reduced. As the throttle plate is opened, the mass of air allowed to enter the cylinders is increased, resulting in greater power output. The more air that is allowed, the more fuel that can be added to it. As engine speed increases, it will eventually reach a point at which no more air can enter. Any increase in speed beyond this point causes a loss in power output. The major factor in determining the maximum mass of air that can enter the cylinders is engine displacement. The amount of displacement is determined by the number of cylinders, cylinder bore, and length of the piston stroke (Figure 2-22).

Bore and **Stroke**. The **bore** of the cylinder is its diameter measured in inches (in.) or millimeters (mm). The **stroke** is the amount of piston travel from TDC to BDC measured in inches or millimeters. An oversquare engine is used when high rpm (revolutions per minute) output is required. This meets the requirements of many automotive engines. Most truck engines are designed as undersquare since they deliver more low end torque.

> **Displacement** is the measure of engine volume. The larger the displacement, the greater the power output.

> A **bore** is the diameter of a hole.

> If the bore is larger than its stroke, the engine is referred to as oversquare. If the bore is smaller than the stroke, the engine is referred to as undersquare. A square engine has the same bore size as it does stroke.

> **Stroke** is the distance that the piston travels from top dead center (TDC) to bottom dead center (BDC) or vice versa.

> Some engine enthusiasts say "there is no replacement for displacement."

FIGURE 2-21 Displacement is the volume of the cylinder between TDC and BDC.

$$\frac{Bore}{Stroke} = \text{Bore–stroke ratio}$$

FIGURE 2-22 The bore and stroke determine the amount of displacement of the cylinder.

Calculating Engine Displacement

Mathematical formulas can be used to determine the displacement of an engine. As with many mathematical formulas, there is more than one method that can be used. The first is as follows:

Engine displacement $= 0.785 \times B^2 \times S \times N$
where 0.785 is a constant
   $B = $ bore
   $S = $ stroke
   $N = $ number of cylinders
The second formula is:
Engine displacement $= p \times R^2 \times S \times N$
where $p = 3.1416$
   $R = $ bore radius (diameter/2)
   $S = $ stroke length
   $N = $ number of cylinders

Pi times the radius squared is equal to the area of the cylinder cross section. This area is then multiplied by the stroke, and this product is multiplied by the number of cylinders.

The calculation for total cubic inch displacement (CID) of a V-6 engine with a bore size of 3.800 inches (9.65 cm) and stroke length of 3.400 inches (8.64 cm) would be as follows:

Engine displacement $= 0.785 \times 3.800^2 \times 3.400 \times 6 = 231.24$ CID
or
Engine displacement $= 3.1416 \times 1.9^2 \times 3.400 \times 6 = 231.35$ CID

The difference in decimal values is due to rounding pi to four digits. Round the product to the nearest whole number. In this case, the CID would be expressed as 231.

Most engine manufacturers designate engine size by metric displacement. To calculate engine displacement in cubic centimeters or liters, simply use the metric measurements in the displacement formula. In the preceding example, the result would be 3,789.56 cc, which would be rounded off to 3.8 liters. Another way to convert to metric measurements is to use the CID and multiply it by .01639 to find the size of the engine in liters. For example, $231 \times .01639 = 3.79$ or 3.8 liters. Use a service manual to determine the bore and stroke of the engine.

An alternate method of determining the stroke is to measure the amount of crankshaft throw (Figure 2-23). The stroke of the engine is twice the crankshaft throw. You can measure the diameter of the cylinder to find the bore size.

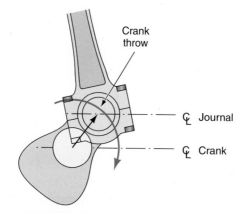

FIGURE 2-23 An alternate method of determining the piston stroke is to double the crankshaft throw.

Crank throw

℄ Journal

℄ Crank

If the engine is machined or modified during the engine rebuild process, the engine displacement will change. For example, during the inspection process, an engine rebuild technician may find that the cylinders need to be bored out 0.060 inch (0.1524 cm). This new cylinder size would have to be included in the new displacement calculation.

DIRECTION OF CRANKSHAFT ROTATION

The Society of Automotive Engineers (SAE) standard for engine rotation is counterclockwise when viewed from the rear of the engine (flywheel side). When viewing the engine from the front, the rotation of the crankshaft is clockwise. Most automotive engines rotate in this direction, but there are a few engines that rotate the opposite direction. Some marine and older Volkswagen and Honda engines are known as reverse rotation engines.

ENGINE MEASUREMENTS

As discussed previously, engine displacement is a measurement used to determine the output of the engine and can be used to identify the engine. Other engine measurements include compression ratio, horsepower, torque, and engine efficiency. Some of these are also used to identify the engine; for example, an engine with 3.8 liters displacement may be available in different horsepower and torque ratings.

Compression Ratio

The **compression ratio** expresses a comparison between the volume the air-fuel mixture is compressed into at TDC and the total volume of the cylinder (piston at BDC). The ratio is the result of dividing the total volume of the cylinder by the volume with the piston at TDC (Figure 2-24).

Compression ratios affect engine efficiency. Higher compression ratios result in increased cylinder pressures, which compress the air-fuel molecules tighter. This is desirable because it causes the flame to travel faster. A higher compression ratio may be desirable because the mechanical efficiency is higher and typically more power will be produced from the same displacement.

There are some drawbacks to just having higher compression ratios, such as a higher octane fuel requirement and increased emissions. These are unwanted outcomes of higher compression ratio engines. However, some modern engines have been able to increase the compression ratio while still allowing for the use of regular octane fuel and having similar emission levels. The use of a VVT system or a variable length manifold, along with

The **compression ratio** is a comparison between the volume above the piston at BDC and the volume above the piston at TDC. Compression ratio is the measure used to indicate the amount the piston compresses each intake charge.

33

Volume before compression: 480 cc

Volume after compression: 60 cc

TDC

BDC

Compression ratio: 8 to 1

FIGURE 2-24 The compression ratio is a measurement of the amount the air-fuel mixture will be compressed.

alternative fuels and advanced computer programs have enabled design engineers more flexibility when designing higher compression ratio engines.

Theoretically, more power can be produced from a higher compression ratio. An increase in the compression ratio results in a higher degree of compression. Combustion occurs at a faster rate because the molecules of the air-fuel mixture are packed tighter. It is possible to change the compression ratio of an engine by performing some machining operations or changing piston designs. Increasing the compression ratio may increase the power output of the engine, but the higher compression ratio increases the compression temperature. This can cause **preignition** and detonation. To counteract this tendency, higher octane gasoline must be used. A higher octane fuel will burn slower than a lower octane fuel. The octane number of a fuel is also known as the anti-knock index; the higher the number, the more resistant to knock, or **detonation**, the fuel is.

Most gas automotive engines have a compression ratio between 8:1 (expressed as eight to one) and 10:1 and develop cylinder pressures during compression of between 125 and 200 psi. Diesel engines typically have compression ratios between 17:1 and 22:1. They can produce up to 600 psi of compression pressure. The usable compression ratio of an engine is determined by the following factors:

- The temperature at which the fuel will ignite.
- The temperature of the air charge entering the engine.
- The density of the air charge.
- Combustion chamber design.

Calculating Compression Ratio

When calculating the total volume above the piston, the combustion chamber in the cylinder head must also be considered. The formula for calculating compression ratio is:

$$CR = (VBDC + VCH)/(VTDC + VCH)$$
where CR = compression ratio
VBDC = volume above the piston at BDC
VCH = combustion chamber volume in the cylinder head
VTDC = volume above the piston at TDC
Example: Volume above the piston at BDC = 56 cubic inches
Combustion chamber volume = 4.5 cubic inches
Volume above the piston at TDC = 1.5 cubic inches

The compression ratio would be:

$$\frac{56 + 4.5}{1.5 + 4.5} = 10{:}1$$

During an engine rebuild, certain machining procedures may affect the compression ratio. Some examples of this would be resurfacing the cylinder head, resizing or boring the cylinders, and installing a different crankshaft. The compression ratio should always be considered when making choices during an engine rebuild.

Horsepower and Torque

Torque is a mathematical expression for rotating or twisting force around a pivot point. As the pistons are forced downward, this pressure is applied to a crankshaft that rotates. The crankshaft transmits this torque to the drivetrain and ultimately to the drive wheels. **Horsepower** is the rate at which an engine produces torque.

To convert terms of force applied in a straight line to force applied rotationally, the formula is: torque = force × radius.

If a 10-pound force is applied to a wrench one foot in length, 10 foot-pounds (ft.-lb.) of torque is produced. If the same 10-pound force is applied to a wrench two feet in length, the torque produced is 20 ft.-lb. (Figure 2-25).

Power is the rate at which work is done. The power an engine produces is rated in **horsepower**. Horsepower is a term that can be used to measure many mechanical actions. When discussing engines, we are measuring mechanical horsepower. One horsepower is equal to 33,000 foot-pounds per minute. The horsepower of an engine can be measured on a dynamometer which places a load on the engine and measures torque. Knowing the torque of an engine and the rpm (engine speed) at which it was measured, you can calculate horsepower by using the following equation.

$$\text{Horsepower} = \frac{\text{torque} \times \text{rpm}}{5{,}252}$$

Torque and horsepower will peak at some point in the rpm range (Figure 2-26). This graph is a mathematical representation of the relationship of torque and horsepower in one engine. The graph indicates that this engine's torque peaks at about 1,700 rpm, while brake horsepower peaks at about 3,500 rpm. The third line in the graph represents the amount of horsepower required to overcome internal resistances or friction in the engine. The amount of **brake horsepower** is less than **indicated horsepower** due to this and other factors.

The Society of Automotive Engineers (SAE) standards for measuring horsepower include ratings for gross and net horsepower. **Gross horsepower** expresses the maximum power developed by the engine without additional accessories operating. These accessories include the air cleaner and filter, cooling fan, charging system, and exhaust system. **Net horsepower** expresses the amount of horsepower the engine develops as installed in the vehicle. Net horsepower is about 20 percent lower than gross horsepower. Most manufacturers express horsepower as SAE net horsepower.

Brake horsepower is the usable power produced by an engine. It is measured on a dynamometer using a brake to load the engine.

Indicated horsepower is the amount of horsepower the engine can theoretically produce.

FIGURE 2-25 Example of how torque can be increased.

Brake horsepower

- - - - Torque

.... Friction horsepower

FIGURE 2-26 Example of torque and horsepower curves.

For example, if the torque of the engine is measured at the following values, a graph can be plotted to show the relationship between torque and horsepower (Figure 2-27):

Torque output @ 1,500 rpm = 105 lb.-ft.
Torque output @ 2,800 rpm = 110 lb.-ft.
Torque output @ 3,500 rpm = 115 lb.-ft.
Torque output @ 5,000 rpm = 130 lb.-ft.
Torque output @ 6,000 rpm = 125 lb.-ft.
Torque output @ 6,500 rpm = 100 lb.-ft.

In this instance, the horsepower is as follows:

Horsepower @ 1,500 rpm = 30
Horsepower @ 2,800 rpm = 59
Horsepower @ 3,500 rpm = 77
Horsepower @ 5,000 rpm = 124
Horsepower @ 6,000 rpm = 143
Horsepower @ 6,500 rpm = 124

Engine Efficiency

There are several different measurements used to describe the **efficiency** of an engine. Efficiency is mathematically expressed as output divided by input. The terms used to define the input and output must be the same. Some of the most common efficiencies of concern are mechanical, volumetric, and thermal efficiencies.

Mechanical efficiency is a comparison between brake horsepower and indicated horsepower; in other words, it is a ratio of the power actually delivered by the crankshaft to

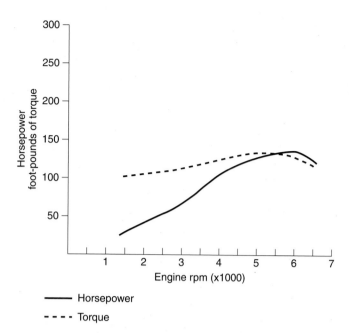

FIGURE 2-27 It is possible to graph the relationship between torque and horsepower. Highest torque output is usually at lower engine speeds, while peak horsepower is at higher engine speeds.

the power developed within the cylinders at the same rpm. The formula used to calculate mechanical efficiency is:

Mechanical efficiency = brake horsepower/indicated horsepower

Ideally, mechanical efficiency would be 100 percent. However, the power delivered will always be less due to power lost in overcoming friction between moving parts in the engine. The mechanical efficiency of an internal combustion four-stroke engine is approximately 90 percent. If the engine's brake horsepower is 120 hp and the indicated horsepower is 140 hp, the mechanical efficiency of the engine is 85.7 percent. Mechanical efficiency is a top concern for engine developers. Increasing the mechanical efficiency can increase the power output and allow for a smaller, more efficient engine to be installed into a larger vehicle. Low friction lubricants, small manufacturing tolerances, and precision manufacturing are just a few examples of the techniques of how engine developers are increasing engine efficiency.

Volumetric efficiency (VE) is a measurement of the amount of air-fuel mixture that actually enters the combustion chamber compared to the amount that could be ingested at that speed. Volumetric efficiency describes how well an engine can breathe. The more air an engine can take in, the more power the engine will put out. As the volumetric efficiency increases, the power developed by the engine increases proportionately. The formula for determining volumetric efficiency is:

$$\frac{\text{Actual volume of air in cylinder}}{\text{Maximum volume of air in cylinder}}$$

An example of a calculation for volumetric efficiency is an engine with a 350 CID displacement that actually has an air-fuel volume of 341 cubic inches at a particular rpm. The actual amount of air taken in is less than the available space; this engine is 97 percent volumetrically efficient at this rpm.

Ideally, 100 percent volumetric efficiency is desired, but actual efficiency is reduced due to pumping losses. It must be realized that a given mass of air-fuel mixture occupies different volumes under different conditions. Atmospheric pressures and temperatures will affect the volume of the mixture. Pumping losses are the result of restrictions to the flow of air-fuel mixture into the engine. These restrictions are the result of intake manifold passages, throttle plate opening, valves, and cylinder head passages. There are some machining techniques used by builders of performance engines that will reduce these restrictions. Pumping losses are influenced by engine speed; for example, an engine may have a volumetric efficiency of 75 percent at 1,000 rpm and increase to an efficiency of 85 percent at 2,000 rpm, then drop to 60 percent at 3,000 rpm. As engine speed increases, volumetric efficiency may drop to 50 percent.

If the airflow is drawn in at a low engine speed, the cylinders can be filled close to capacity. A required amount of time is needed for the airflow to pass through the intake manifold and pass the intake valve to fill the cylinder. As engine speed increases, the amount of time the intake valve is open is not long enough to allow the cylinder to be filled. Because of the effect of engine speed on volumetric efficiency, engine manufacturers use intake manifold design, valve port design, valve size, multiple valves, valve overlap camshaft timing, exhaust tuning, variable length manifolds, and computer control to improve the engine's breathing. Current production models have fast computers that can calculate the volumetric efficiency by using computer data.

The VE of a typical modern spark ignition internal combustion normally aspirated engine is 75–95 percent. There are some examples of naturally aspirated engines having more than 100 percent VE. Older engines were limited in their designs and did not employ the same criteria of their modern counterparts. Today, many engines have four valves per cylinder, whereas an engine made in 1970 typically has only two valves per cylinder. A turbocharger or supercharger will force air into the engine (called forced induction). This process will give the engine a VE over 100 percent also.

Other factors of VE are elevation, temperature, fuel type, and humidity. A higher VE is desired because it means less pumping and parasitic losses and greater power output of the engine. Volumetric efficiency should not be confused with the total engine efficiency, although it does have a great influence on it.

Thermal efficiency is a measurement comparing the amount of energy present in a fuel and the actual energy output of the engine. Thermal efficiency is the measurement of how much energy is actually used to turn the wheels as compared to the total energy available. Heat is a form of energy just like electricity. Heat energy can be measured using two terms: intensity and quantity. The measurement of the intensity of heat energy is measured in the temperature scale (Celsius or Fahrenheit). Temperature is the measurement of an object's internal energy. Something that is hot has more heat intensity than something that is cold; therefore, it has more heat intensity and more heat energy. For example, we may say that a liquid is hot because it is 200 °F. This just means that it has a higher temperature that another familiar object.

Heat quantity (often referred to as heat value or heat's power) is the amount of heat energy a piece of matter has. The most common methods of measuring heat quantity are the Btu, the calorie (c), and the Joule (J). One Btu is the amount of heat required to raise the temperature of 1 pound of pure liquid water by 1 °F. The temperature at which water has its greatest density is 39 °F. Testing for a Btu is done at this temperature. One Btu is equal to 1,055 J or 252 C.

Some of the different fuels used for internal combustion engines and their energy content are listed below. Keep in mind that each of these values will vary with different mixtures, seasons, manufacturers, retailers, and geographic location.

1 gallon of mid-grade gasoline = 108,500 – 117,000 Btus
1 gallon of ethanol fuel = 76,100 Btus
1 gallon of #2 diesel = 139,000 Btus
1 gallon of propane = 91,600 Btus

Thermal efficiency is the difference between potential and actual energy developed in a fuel measured in Btus per pound or gallon.

The formula used to determine a gasoline engine's thermal efficiency has a couple of constants. First, 1 hp equals 42.4 Btus per minute. Second, gasoline has approximately 110,000 Btus per gallon. Knowing this, the formula is as follows:

Thermal efficiency = (bhp – 42.4 Bpmin)/(110,000 Bpg – gpmin)
where bhp = brake horsepower
Bpmin   = Btu per minute
Bpg   = Btu per gallon
gpmin   = gallons used per minute

As a whole, gasoline engines waste about two-thirds of the heat energy available in gasoline. Approximately one-third of the heat energy is carried away by the engine's cooling system. This is required to prevent the engine from overheating. Another third is lost in hot exhaust gases. The remaining third of heat energy is reduced by about 5 percent due to friction inside of the engine. Another 10 percent is lost due to friction in drivetrain components. Due to all of the losses of heat energy, only about 19 percent is actually applied to the driving wheels.

Of the 19 percent applied to the driving wheels, an additional amount is lost due to the rolling resistance of the tires against the road. Resistance to the vehicle moving through the air requires more of this energy. By the time the vehicle is actually moving, the overall vehicle efficiency is about 15 percent.

Manufacturers are continuing to increase the thermal efficiency of the engine in hopes that it will help increase fuel economy. As you begin to work with engines, you will be able to compare older engines to modern engines and identify different parts and computer-management strategies that have been employed to increase thermal efficiency.

OTHER ENGINE DESIGNS

Two-Stroke Engines

Throughout the history of the automobile, the gasoline four-stroke internal combustion engine has been the standard. As technology has improved, other engine designs have been used. This section of the chapter provides an overview of some of these other engine designs.

Two-stroke engines are capable of producing a power stroke every revolution. Valves are not used in most two-stroke engines; instead, the piston movement covers and uncovers intake and exhaust ports. The air-fuel mixture enters the crankcase through a **reed valve** or **rotary valve** (Figure 2-28). Upward movement of the piston causes pressure to increase in the cylinder above the piston, and creates a slight vacuum below the piston. Atmospheric pressure pushes the air-fuel mixture into the low pressure area below the piston. At the same time, a previously ingested air-fuel mixture is being compressed above the piston. This action combines the compression stroke and intake stroke.

When the compressed air-fuel mixture is ignited, the piston moves down the cylinder. This slightly compresses the air-fuel mixture in the crankcase. The air-fuel mixture in the crankcase is directed to the top of the piston through the transfer port to the intake port. The reed valve or rotary valve closes to prevent the mixture from escaping from the crankcase. The downward piston movement opens the exhaust port, and the pressure of the expanding gases pushes out the spent emissions. Continued downward movement of the piston uncovers the intake port and allows the air-fuel mixture to enter above the piston. This action combines the power and exhaust strokes.

Most gasoline two-stroke engines do not use the crankcase as a sump for oil since it is used for air-fuel intake. To provide lubrication to the engine, the oil must be mixed with the fuel or injected into the crankcase. This results in excessive exhaust emissions. Diesel-design two-stroke engines may have a sump. Many automotive engine manufacturers are trying to

A **reed valve** is a one-way check valve. The reed opens to allow the air-fuel mixture to enter from one direction, while closing to prevent movement in the other direction.

A **rotary valve** is a valve that rotates to cover and uncover the intake port.

FIGURE 2-28 Two-stroke cycle engine design.

find ways to reduce the emission levels of the two-stroke engine to allow for its use in automobiles. The advantage of the two-stroke engine is that it produces more power per cubic inch of displacement than the four-stroke engine.

Today, two-stroke engines are commonly used in small applications, such as lawn and garden equipment and recreational/sports applications (snowmobiles, ATVs, watercraft, etc.). There have been a few automobiles and light trucks produced in the past that used two-stroke engines.

Diesel Engines

The first diesel engine was operated and invented by Rudolf Diesel in 1897. Since then, diesel engines have been mainly used in heavy duty applications including light, medium, and heavy duty trucks, vans, tractors, and semis. Diesels used in marine and stand-by generator applications are desirable for their power output, low fuel consumption, and longevity. Diesel engines tend to be built heavier and often cost more to manufacture.

Domestic automobiles had a few diesel engine options for a few years, but the engine option has not sold in very well over the past few decades with American consumers in small cars. Within recent years, the domestic pickup truck market has more widely accepted the diesel engine option. The more powerful and fuel-efficient diesel engine designs are suited well in pickups where the noise, vibration, and harshness level expectations are lower than they are for cars. By contrast, European automobiles are mostly powered by diesel engines and have been for many years. Within the last few years of the publish date of this book, the number of diesel engines in cars and light trucks has been increasing significantly. Examples of such vehicles with popular diesel engine options are Volkswagen cars and domestic light duty trucks.

Diesel and gasoline engines have many similar components; however, the diesel engine does not use an ignition system consisting of spark plugs and coils. Instead of using a spark initiated and developed by the ignition system, the diesel engine uses the heat produced by compressing air in the combustion chamber to raise the temperature of the air (not the fuel) closer to (and sometimes above) the autoignition temperature of the fuel. Once the fuel is injected under high pressure, it instantly increases in temperature because of its contact

| Intake | Compression | Power | Exhaust |

FIGURE 2-29 The four strokes of a diesel engine.

with the high temperature air in the combustion chamber. This begins the combustion process of the diesel fuel in the power stroke (Figure 2-29). This explains why diesel engines have a higher compression ratio when compared to gasoline engines. Diesel engines are also categorized as CI (compression ignition) engines.

Since starting the diesel engine is dependent on heating the intake air to a high enough level to ignite the fuel, a method of preheating the intake air is required to start a cold engine. Some manufacturers use **glow plugs** to accomplish this. Another method includes using a heater grid in the air intake system.

In addition to ignition methods, there are other differences between gasoline and diesel engines (Figure 2-30). The combustion chambers of the diesel engine are designed to accommodate the different burning characteristics of diesel fuel. There are three common combustion chamber designs:

An open-type combustion chamber is located directly inside of the piston. The fuel is injected directly into the center of the chamber. Turbulence is produced by the shape of the chamber.

A precombustion chamber is a smaller, second chamber connected to the main combustion chamber. Fuel is injected into the precombustion chamber, where the combustion process is started. Combustion then spreads to the main chamber. Since the fuel is not injected directly on top of the piston, this type of diesel injection may be called indirect injection.

A turbulence combustion chamber is a chamber designed to create turbulence as the piston compresses the air. When the fuel is injected into the turbulent air, a more efficient burn is achieved.

Glow plugs are threaded into the combustion chamber and use electrical current to heat the intake air on diesel engines

	Gasoline	**Diesel**
Intake	Air-fuel	Air
Compression	8–10 to 1, 130 psi 545°F	13–25 to 1, 400–600 psi 1,000°F
Air-fuel mixing point	Carburetor or before intake valve with fuel injection	Near TDC by injection
Combustion	Spark ignition	Compression ignition
Power	464 psi	1,200 psi
Exhaust	1,300–1,800°F CO=3%	700–900°F CO=0.5%
Efficiency	22–28%	32–38%

FIGURE 2-30 Comparisons between gasoline and diesel engines.

Diesel engines can be either four-stroke or two-stroke designs, though like gas engines, production automotive applications currently use four-stroke engines. Two-stroke engines complete all cycles in two strokes of the piston, much like a gasoline two-stroke engine.

Diesel engines typically have lower fuel consumption rates and a high power output (compared to a same displacement gasoline engine). Many manufacturers are considering diesel engines as options for their line of cars. Diesel engines operate similar to gasoline engines and have extra operating components to support clean and efficient operation.

Atkinson-Cycle Engine

The Atkinson-cycle engine is a common gasoline internal combustion engine design used in some HEVs such as the Toyota Prius, Chevrolet Tahoe, Toyota Camry, and Ford Escape/Mercury Mariner. The Atkinson-cycle engine has a lower power output when compared to similar displacement Otto-cycle engines but has increased efficiency. Originally, the Atkinson-cycle engine had a connecting link attached to the crankshaft that allowed all four cycles of a four-stroke cycle to occur in one crankshaft revolution. Today's Atkinson-cycle engines in HEVs refer to the principles of the cycles and the ratios, rather than having an additional linkage on the crankshaft.

The Atkinson-cycle engine holds the intake valve open for a longer duration than the Otto-cycle engine design. This action allows for more air to be pushed back into the intake manifold and less compression to occur. The power stroke is then forced to perform more work with less air. This event leads to an increase in the heat energy that is converted to mechanical energy. This event makes the volumetric ratio of the compression stroke to the power stroke less than 1:1.

During the power stroke, the volume of air that was in the compression stroke has to expand to a volume that is greater. This makes the gases during power stroke expand more than they would in an Otto-cycle engine. The mechanical efficiency of the engine is now higher, at the expense of the lower power density of the charge during the power stroke.

Miller-Cycle Engine

The Miller-cycle engine is a modification of the four-stroke cycle engine. Some vehicles such as the Mazda Millennia are powered by a Miller-cycle engine. In this type of engine, a supercharger supplies highly compressed air through an intercooler into the combustion cylinders. The intake valves remain open longer compared to a conventional four-stroke cycle engine. This action prevents the upward piston movement from compressing the air-fuel mixture in the cylinder until the piston has moved one-fifth of the way upward on the compression stroke. The later valve closing allows the supercharger to force more air into the cylinder. The shorter compression stroke lowers cylinder temperatures and still gets the same (or more) compression as a standard Otto four-stroke design because the Miller design uses a supercharger to increase compression. Normally, without a supercharger, holding the intake valve open for a longer duration would result in a loss of compression and power. This operation makes the compression stroke less of a consumer of the total engine power.

By delaying the start of the compression stroke, the engine is not wasting energy, pushing the piston up on the compression stroke. As compression builds in the cylinder, the energy needed to turn the crankshaft and move the piston increases. This energy would normally be delivered by the crankshaft, which is powered by the other cylinders in the engine that are on the power stroke. Allowing for this delay to occur also allows for the crankshaft to be in a more mechanically advantageous position (past BDC).

The greater volume of air in the cylinder creates a longer combustion time, which maintains downward pressure on the piston for a longer time on the power stroke. This action increases engine torque and efficiency. The Miller-cycle principle allows an engine to produce a high horsepower and torque rating for the displacement of the engine. The Miller-cycle engine is similar to a modern Atkinson-cycle engine, except the Miller-cycle engines have a supercharger associated with their incoming air charge.

A few manufactures are producing prototype vehicles that are HEV that use a Miller-cycle gas engine. The greater volume of air in the cylinder creates a longer combustion time, which maintains downward pressure on the piston for a longer time on the power stroke. This action increases engine torque and efficiency. The Miller-cycle principle allows an engine to produce a high horsepower and torque for the cubic inch displacement (CID) of the engine.

Stratified Charge Engine

The stratified charge engine uses a precombustion chamber to ignite the main combustion chamber (Figure 2-31). The advantage of this system is increased fuel economy and reduced emission levels.

In this engine type, the air-fuel mixture is stratified to produce a small, rich mixture at the spark plug while providing a lean mixture to the main chamber. During the intake stroke, a very lean air-fuel mixture enters the main combustion chamber. At the same time, a very rich mixture is pushed past the auxiliary intake valve and into the precombustion chamber (Figure 2-32). At the completion of the compression stroke, the spark plug fires to ignite the rich mixture in the precombustion chamber. The burning rich mixture then ignites the lean mixture in the main combustion chamber.

A stratified charge engine is in some ways similar to a diesel engine. The rich air-fuel mixture ignites easily in the precombustion chamber, which allows the excessively lean air-fuel mixture from the intake to ignite without detonation or preignition. This design has not been terribly successful in automobiles over the years, but manufactures are still working toward a solution. One primary example of a stratified charge engine was the Honda CVCC engine produced in the 1970s.

Honda's CVCC engine lead to the development of other successful and efficient engines produced by other manufactures, such as lean burn and direct injection designs.

FIGURE 2-31 Stratified engine design.

A   B

C   D

FIGURE 2-32 The stratified engine design uses a rich mixture in the precombustion chamber to ignite the lean mixture in the main chamber.

DISPLACEMENT-ON-DEMAND ENGINES

Many manufactures are now making eight-cylinder engines that can run on six, or four cylinders when conditions permit, to provide better fuel economy. These systems are known as displacement-on-demand, variable displacement, variable cylinder management, or active fuel management engines. These systems are currently is some GM, Honda, and Chrysler vehicles.

This type of system is not new technology, but rather a new configuration of an older system that didn't work well. The 1981 Cadillac 8-6-4 used a similar system to deactivate cylinders for fuel economy. Unfortunately, the car never succeeded with consumers because the technology was too ahead of its time. Semiconductors in vehicles were still in their infant stage, and there were many consumer complaints of engine stalling and rough running when the computer automatically turned cylinders off. Modern computer speed and complexity make the new versions of displacement-on-demand barely noticeable to the consumer. The system allows for maximum power when the consumer demands acceleration and a "smaller" engine when the driver is cruising. The engines use special hydraulic valve lifters to stop valve opening to selected cylinders during cruising and deceleration. The powertrain control module (PCM) activates a **solenoid** that allows oil pressure to unlock a pin in the valve lifter. Then, rather than opening the valve, it collapses as the camshaft tries to push it up. The PCM also

A **solenoid** is an electromechanical device. It is typically operated by the PCM through an electrical signal and mechanically controls flow of air, oil, fuel, or other substances. A fuel injector is a type of solenoid.

stops delivering fuel to the four cylinders during cylinder deactivation. Drivers say the transition is seamless; they can't even feel when the engine changes from four to eight or back again. Manufacturers claim up to 25 percent fuel economy increases. One V-8 engine tested achieved 30 mpg during highway driving.

ENGINE IDENTIFICATION

Proper engine identification is required for the technician and/or machinist to obtain correct specifications, service procedures, and parts. The vehicle identification number (VIN) is one source for information concerning the engine. The VIN is located on a metal tag that is visible through the vehicle's windshield (Figure 2-33). The VIN provides the technician with a variety of information concerning the vehicle (Figure 2-34). The eighth digit represents the engine code. The service information provides details for interpretation of the engine code. The VIN is also stamped in a number of other locations around the vehicle to make falsification more difficult.

A vehicle code plate is also located on the vehicle body, usually inside the driver's side door or under the hood. This plate provides information concerning paint codes, transmission codes, and so forth (Figure 2-35). The engine sales code is also stamped on the vehicle code plate.

AUTHOR'S NOTE: One day I was working on an older import vehicle and a police officer walked up to the car and explained that he had reason to believe this was a stolen vehicle. In order to check it out, he lifted the rear carpet in the trunk and compared the VIN with the VIN located on the dashboard. Neither appeared to have been tampered with, and the number did not match the VIN of the stolen vehicle. He left me to my work and went on his way to search for the real stolen vehicle.

The engine itself may have different forms of identification tags or stamped numbers (Figure 2-36). In addition, color coding of engine labels may be used to indicate transitional low emission vehicles (TLEVs) and flex fuel vehicles (FFVs). These numbers may provide build date codes required to order proper parts. Use the service information to determine the location of any engine identification tags or stamps and the methods to interpret them; for example: a Jeep engine with the build date code of *401HX23* identifies a 2.5-liter engine with a multi-point fuel-injection system, 9:1 compression ratio, built on January 23, 1994.

FIGURE 2-33 The VIN number can be seen through the windshield.

1G2FS32P8RE100000

ORIGIN	
1	USA
2	CANADA

MANUFACTURER	
G	GENERAL MOTORS

MAKE	
	CHEVROLET
2	PONTIAC

VEHICLE LINE	
F/S	FIREBIRD
N/E	GRAND AM

BODY	
1	2 DR COUPE
3	CONVERTIBLE

RESTRAINT
MANUAL BELTS WITH DRV / PASS INFLATABLE RESTRAINTS

SEQUENCE NO.
LAST 6 DIGITS

PLANT CODE	
E	PONTIAC EAST

YEAR	
4	2004
5	2005

CHECK DIGIT
9 TH DIGIT

ENGINE				
CODE	ENGINE	RPO	FUEL SYS.	DISP.
L	V8	L27	MFI	3.8L
P	V8	LT1	MFI	5.7L

FIGURE 2-34 The VIN number is used to provide the required information about the vehicle and engine.

FIGURE 2-35 Vehicle code plates provide sales code information about the vehicle.

This is determined by the first digit representing the year of manufacture (4 = 1994). The second and third digits represent the month of manufacture (01 = January). The fourth and fifth digits represent the engine type, fuel system, and compression ratio (HX = 2.5-liter engine with a multi-point fuel-injection system, 9:1 compression ratio). The sixth and seventh digits represent the day of manufacture (23).

NOTE: VIN is stamped on the bedplate

Label located on valve cover

5.0	2355		743	
III II		III IIIIII III		III
2235	1234		743	3
DATE	SEQ NUM		B/CODE	PLT

Block foundry ID and date

Engine number

FIGURE 2-36 The engine may have identification numbers stamped on it.

Some engines are manufactured with oversized or undersized components. This occurs due to mass production of the engines. Components that may be different sizes than nominal include:

- Cylinder bores
- Camshaft bearing bores
- Crankshaft main bearing journals
- Connecting rod journals

Engines with oversize or undersize components (from the factory) are usually identified by a letter code stamped on the engine block (Figure 2-37). This information is helpful when ordering parts and measuring for wear.

Distributor

Letter code

Oil filter boss

FIGURE 2-37 Additional markings indicate if components are undersize or oversize.

SUMMARY

- One of the many laws of physics utilized within the automotive engine is thermodynamics. The driving force of the engine is the expansion of gases. Heat is generated by compressing the air-fuel mixture within the engine. Igniting the compressed mixture causes the heat, pressure, and expansion to multiply.
- Engine classification is usually based on the number of cycles, the number of cylinders, cylinder arrangement, and valvetrain type. In addition, engine displacement is used to identify the size of the engine.

TERMS TO KNOW

Atkinson-cycle engine

Bore

Bottom dead center (BDC)

Brake horsepower

Coil

Compression-ignition (CI) engines

Compression ratio

Connecting rods

Crankshaft

Cycle

Displacement

Dual overhead cam (DOHC)

SUMMARY

TERMS TO KNOW
(continued)

Efficiency

Engine bearings

Friction

Fuel injection

Glow plugs

Gross horsepower

Horsepower

Indicated horsepower

Internal combustion
engine

Kinetic energy

Mechanical efficiency

Miller-cycle engine

Net horsepower

Overhead cam (OHC)

Overhead valve (OHV)

Piston rings

Potential energy

Preignition

Reciprocating

Reed valve

Rotary valve

Solenoid

Spark-ignition (SI)
engine

Stroke

Thermal efficiency

Thermodynamics

Top dead center
(TDC)

Torque

Transverse-mounted
engine

Valve overlap

Variable valve timing
(VVT) systems

Volumetric efficiency
(VE)

Wrist pin

- Most automotive and truck engines are four-stroke cycle engines. A stroke is the movement of the piston from BDC to TDC of the cylinder and is usually measured in inches or millimeters.

- The first stroke of the cycle is the intake stroke. The compression stroke begins as the piston starts its travel back to TDC. When the spark is introduced to the compressed mixture, the rapid burning causes the molecules to expand, and this is the beginning of the power stroke. The exhaust stroke of the cycle begins with the upward movement of the piston back toward TDC and pushes out the exhaust gases from the cylinder past the exhaust valve and into the vehicle's exhaust system.

- The three most commonly used valvetrains are the overhead valve (OHV), overhead cam (OHC), and dual overhead cam (DOHC).

- Other engine designs and features include Diesel, Miller-cycle, stratified charge, gasoline direct injection, variable valve timing, and displacement-on-demand.

- Displacement is the total volume of the cylinder between BDC and TDC measured in cubic inches, cubic centimeters, or liters. Engine displacement is usually an indicator of the engine's capability of producing power.

- Torque is a turning or twisting force. As the pistons are forced downward, this pressure is applied to a crankshaft that rotates. The crankshaft transmits this torque to the drivetrain, and ultimately to the drive wheels.

- Horsepower is the rate at which an engine produces torque in a given amount of time.

REVIEW QUESTIONS

Short Answer Essays

1. Describe how the basic laws of thermodynamics are used to operate the automotive engine.

2. List and describe the four strokes of the four-stroke cycle engine.

3. Describe the different cylinder arrangements and the advantages of each.

4. Describe the different valvetrains used in modern engines.

5. Describe the basic operation of the stratified charge engine.

6. Describe how the compression stroke of a gasoline four-stroke engine raises the temperature of the fuel and air mixture and why this step is so important.

7. Define the term stroke.

8. Describe the differences of how the fuel and air are mixed and injected in a four-stroke diesel engine when compared to a four-stroke gasoline engine.

9. Explain the basic operation of the two-stroke engine.

10. Describe the difference between the Miller-cycle engine design and the Otto-cycle engine design.

Fill-in-the-Blanks

1. Most automotive and truck engines are _____ cycle engines.

2. During the intake stroke, the intake valve is _____, while the exhaust valve is _____.

3. The _____ engine places all of the cylinders in a single row.

4. An _____ _____ engine has the valve mechanisms in the cylinder head, while the camshaft is located in the engine block.

5. A _____ is a sequence that is repeated.

6. In a four-stroke engine, four strokes are required to complete one _____.

7. Instead of using a spark produced by the ignition system, the diesel engine uses the _____ produced by compressing air in the combustion chamber to ignite the fuel.

8. A _____ _____ engine has a power stroke every revolution.

9. In a two-cycle engine the air-fuel mixture enters the crankcase through a _____ or _____ _____.

10. The _____ _____ engine uses combustion from the precombustion chamber to ignite the air-fuel mixture in the main combustion chamber.

Multiple Choice

1. Increasing the compression ratio of an engine is being discussed.
   Technician A says increased compression ratios result in a reduction of power produced.
   Technician B says higher compression ratio engines require higher octane gasoline.
   Who is correct?
   A. A only
   B. B only
   C. Both A and B
   D. Neither A nor B

2. Engine measurement is being discussed.
   Technician A says the bore is the distance the piston moves within the cylinder.
   Technician B says the stroke is the diameter of the cylinder.
   Who is correct?
   A. A only
   B. B only
   C. Both A and B
   D. Neither A nor B

3. *Technician A* says to determine horsepower, torque must be known first.
   Technician B says horsepower will usually peak at a lower engine speed than torque.
   Who is correct?
   A. A only
   B. B only
   C. Both A and B
   D. Neither A nor B

4. *Technician A* says mechanical efficiency is a comparison between brake horsepower and indicated horsepower.
   Technician B says volumetric efficiency is a measurement of the amount of air-fuel mixture that actually enters the combustion chamber compared to the amount that could be drawn in.
   Who is correct?
   A. A only
   B. B only
   C. Both A and B
   D. Neither A nor B

5. During the compression stroke in a four-cycle engine:
   A. The intake valve is open, and the exhaust valve is closed.
   B. The exhaust valve is open, and the intake valve is closed.
   C. Both valves are open.
   D. Both valves are closed.

6. All of these statements about diesel engines are true EXCEPT:
   A. The air-fuel mixture is ignited by the heat of compression.
   B. Glow plugs may be used to heat the air-fuel mixture during cold starting.
   C. Spark plugs are located in each combustion chamber.
   D. Diesel engines may two-cycle or four-cycle.

7. While discussing volumetric efficiency:
   Technician A says volumetric efficiency is not affected by intake manifold design.
   Technician B says volumetric efficiency decreases at high engine speeds.
   Who is correct?
   A. A only   C. Both A and B
   B. B only   D. Neither A nor B

8. In a two-stroke cycle engine:
   A. A power stroke is produced every crankshaft revolution.
   B. Intake and exhaust valves are located in the cylinder head.
   C. The lubricating oil is usually contained in the crankcase.
   D. The crankcase is not pressurized.

9. In an engine with a balance shaft:
   A. The balance shaft does not have to be timed to the crankshaft.
   B. The balance shaft turns at twice the crankshaft speed.
   C. The balance shaft may be mounted on the cylinder head.
   D. The balance shaft increases engine torque.

10. While discussing engine identification:
    Technician A says that the tenth character of the VIN defines the vehicle model year.
    Technician B says that an engine identification tag may identify the date on which the engine was built.
    Who is correct?
    A. A only   C. Both A and B
    B. B only   D. Neither A nor B

Chapter 3

ENGINE REPAIR AND REBUILDING INDUSTRY

UPON COMPLETION AND REVIEW OF THIS CHAPTER, YOU SHOULD BE ABLE TO:

- Understand what types of repair are performed at a full service repair facility.

- Describe the reasons for the existence of specialty shops and repair facilities.

- Understand what changes are occurring in the engine repair and rebuilding industry.

- Describe the skills needed for employment in various settings.

- Understand how specialized repair facilities operate.

- Describe the basic rebuilding process at a remanufacturing facility.

INTRODUCTION

The engine repair and rebuilding industry has changed significantly in the past decade. This chapter was designed to give you a brief introduction to the current industry and the changes that have been made to machining equipment, repair facilities, and the daily operations.

In the past, engines and vehicles were not built to the high standards they are now. If an engine and vehicle drove past 100,000 miles without a major repair, you were considered lucky and running on borrowed time. Today, consumers expect their engines and vehicles to give them reliable service for more miles. The engines in the past were not built with high standards and repairs such as valve jobs and rebuilds were common. Today, things have changed significantly. There are still a few shops that will perform a valve job, or completely rebuild an engine in-house, but that number is dwindling.

When you are employed in the field, one of your jobs may be to help the customers choose what option is best for them. The information in this book, your experiences, and the shop you work in will all help you inform the customers as to what action should be taken.

If repairing and rebuilding engines is what you have chosen to do as your career, you will have to know what options you will have and what types of operations the particular places of employment may do.

FULL SERVICE REPAIR FACILITIES

There are a lot of full service repair shops across the nation. The full service repair shops include independently owned, franchised, and new and used dealership facilities. The majority of these shops are willing to work on anything that comes in the door. Full service repair shops will typically perform oil changes and other small repairs at a lower advertised cost to drive customers to their door. This technique is common and allows the customer to become aware of the amount and type of work that is performed there. Once the customer's trust is earned, there is a good chance that they may come back for a different repair and become long-term customers.

FIGURE 3-1 This valve machine has not been
used for the last few years.

The average full service repair facility will have several bays and hoists to work with, along with computer terminals to write repair orders and look-up service information. Rebuilding an engine in this type of repair facility is not as common as it once used to be. Many independent shops used to own their own machine equipment, such as valve grinding machines. Since today's engines need service less often, those same machines collect more dust and the shop owner/manager has to consider whether it is worth still keeping those machines around (and paying to maintain them) or selling them (Figure 3-1). The space that a machine takes up on a floor must be useable if a shop is to remain profitable and stay open.

Today, independent shops perform a significant amount of engine repairs instead of rebuilds. These repairs may include, but are not limited to, valve jobs, head gaskets, and oil leaks. When a major engine repair is needed, they may often choose to either install a crate engine/component or send repair parts out to a separate facility. One example is if an engine were to overheat, blow a cylinder head gasket and crack the cylinder head. If this were to happen, the full service repair shop will estimate the costs of repairing and rebuilding the cylinder head, versus sending it out to a machine shop as is. The machine shops will take the work in and perform a complete service to the head and then charge the shop for their services. The shop in turn will charge the customer. Depending on the vehicle and engine, it may be more cost-effective to purchase crate components, especially if it is a popular component where there is a lot of supply.

Most shops now prefer to install remanufactured components and engines because the company they purchase from is mostly responsible for the warranty of their product. But their preferred installation choice may be limited by the customer choosing a lower cost alternative.

Engine repair and rebuilding work at a full service shop does not usually require a technician to obtain training beyond what this book provides, but some shops prefer to have the lead technician or a specialized technician perform the work. Sometimes, this technician is chosen because of their interest, experience, background, or specialized training.

MACHINE SHOP AND ENGINE REBUILD FACILITIES

One-third of the full service shops surveyed in the past few years indicate that they still own engine machining equipment, but most of them are now planning on selling them. Machine shops and engine rebuild facilities have been around for a long time. They exist as part of another facility, independent, or connected to a parts supply store. Most of these

FIGURE 3-2 This milling machine can be used to resurface cylinder heads, block decks, or flywheels.

FIGURE 3-3 A CNC machine is one of the many pieces of expensive equipment at a machine shop.

repair facilities have not changed their type of business much over the years but have found themselves focusing on the changes in technology and how it affects them.

Machine shops will take in either complete engine repair and rebuilding or component repairs and rebuilding. Most machine shops survive and exist because they have a relationship with the local surrounding independent shops, franchises, and dealers. These supporters will send work out to the machine shop. Often the vehicle will not be housed at the machine shop, but there are a few exceptions.

Some of the most common repairs at a machine shop are cylinder head remanufacturing and repairs, cylinder block remanufacturing, repairs, and resizing, piston pressing, and component assembly. The extent of these repairs will vary. Machine shops have a significant advantage over purchasing crate engines. Both shipping costs and local support are two things that machine shops offer to distinct themselves from large nationwide rebuilders.

Often machine shops will come across a repair for an older, antique, or high performance engine. Each of these cases is handled separately, and the shop foreman/manager usually talks directly with the customer. Some machine shops have started to accept more high performance work because of the instability of other work. Machine shops used to rely on cylinder heads and internal components failing at high enough rates to keep them in business. Modern engines have lower failure rates that keep them from needing extensive service frequently.

Machine shops usually have a very costly inventory of machines and equipment (Figure 3-2). These machines not only have a high initial purchase, but they must be periodically maintained, calibrated, cleaned, and updated. **Computer numerically controlled (CNC)** equipment is sometimes used in a machine shop (Figure 3-3). This type of equipment requires special training, as does many of the machines and equipment in a machine shop. A CNC machine can cut and mill engine components to a smaller tolerance (more accuracy) than most human operated and controlled machines. This is desirable because the smaller tolerances usually mean less comebacks and a higher quality product. Machine shop technicians are required to have a good working knowledge of engine components, precision measuring tools, math, and be able to locate resources quickly (Figure 3-4). There are a few specialized schools across the United States that specialize in teaching automotive machining.

A **CNC** machine uses special computer software to machine close tolerances.

ENGINE REPAIR AND REPLACEMENT SPECIALTY FACILITIES

Engine repair and replacement has almost become a separate industry in itself. In the past, the transmission repair and rebuilding industry had separated itself from the rest of the industry by having shops offer complete transmission work at one facility. Transmissions

FIGURE 3-4 Machinists need to be able to
understand and apply math.

are very complicated and can require extensive testing and highly skilled technicians to work on them. Modern engines require similar skilled technicians and extensive testing due to their growing complexity. In order to replace or rebuild the engine in a modern vehicle, a technician must have a lot of skill and knowledge. They must also be able to transfer that knowledge over to a different vehicle and a different setup. The facility must be well-equipped if a profit is to be made.

The amount and cost of the equipment and technician training is a significant factor to consider when offering engine rebuilding and repair at a facility. It is for these reasons that specialized shops exist. These shops offer engine and related work. They can usually diagnose and repair the cause in-house, saving money for the customer and keeping the profit in one place. Sometimes, a component replacement is all that is needed.

Specialized shops rely on a good source of parts and equipment to operate. If cost is a strong customer consideration, then installing a good, used engine may be an option to consider with the customer.

Some of the specialized shops offer and advertise high performance, race, and restoration work. This may account for a significant amount of their business in some cases. Specialized shops may also perform custom machining, cylinder porting, and engine tuning.

ENGINE AND COMPONENT REMANUFACTURING FACILITIES

As mentioned earlier, today's engines are lasting longer, and repairs are less frequent. Since the bodies and other mechanical components do not rust and wear out as frequently either, many consumers are repairing their high mileage engines that need replacement, because it costs less than purchasing another vehicle.

For many years, manufactures have been placing the same engine in several models. For example, the Chevrolet 5.7L engine was used in many light truck and car applications for decades. Many of the components are interchangeable from one to the next. Because of the volume and number of components out there, the cost of the parts and crate components (and engines) is significantly lower than many other engine families. This is because of the volume of parts and components produced for this engine. Many manufactures beside GM have adapted this strategy.

A modern example of engine production in volume is the Global Engine Manufacturing Alliance LLC (GEMA) World Engine, which has a joint venture with Chrysler. The World Engine has five factories worldwide and produces about 2 million engines per year. There are approximately 20 different vehicles that use the World Engine by three different manufacturers, each being slightly different for their application. The World Engine is used

FIGURE 3-5 Machine shops and engine remanufacturing plants usually have expensive equipment. This crankshaft is being inspected for cracks using a magnet.

in vehicles all over the globe. In the United States, the World Engine is found in smaller Chrysler, Dodge, Plymouth, and Jeep products in the 2.0L and 2.4L engine.

With more manufacturers producing similar products, another segment of engine rebuilding and remanufacturing was created because of the volume. There are only a few engine and component remanufacturing facilities in the United States. Their products can be purchased directly from them, from a parts store, or from an authorized dealer/installer. These facilities are large and have the capability of "assembly line production style" engine remanufacturing because of their volume.

Remanufacturing starts out with a core engine. The **core engine** is an old, used, or discarded engine that was received from a variety of sources. These sources include salvage yards and a customer's old engine that was returned. A core charge is usually placed on a new engine when it is ordered. This is an extra charge that the customer (or repair facility) must pay until the old engine that was removed is sent back in the same shipping crate. The old engine is then inventoried and completely stripped down to the bare components.

The major components, such as the cylinder head, block, and crankshaft are checked for cracks (Figure 3-5) and are machined, repaired, or discarded as needed. They are then inventoried and stored for later use. The major components are then thoroughly cleaned and ready for reassembly. These engine components are also often sold as individual remanufactured components.

During reassembly, new parts are installed, and everything is measured, tested, and torqued to specification. The engines are then started and tested to ensure that the customer receives a high quality product. Some of the main advantages of purchasing an engine from a remanufacturer are the warranty, quality control, and lower cost.

> A **core engine** is a used engine that is disassembled for purposes of rebuilding in large volumes.

SUMMARY

- Automotive and light truck engines last longer today than they did in the past. This has lead to changes in the engine repair and rebuilding industry.
- Machine shops perform the majority of engine machining and repairs today.
- A **CNC** machine uses special computer software to machine close tolerances.
- Specialized shops sometimes offer high performance engine work.
- Engine remanufacturing in large volumes helps lower the price.
- A core engine is a used engine that is disassembled for purposes of rebuilding in large volumes.

TERMS TO KNOW

CNC

Core engine

REVIEW QUESTIONS

Short Answer Essays

1. Describe why an engine remanufacturer receives core engines.

2. Describe what a CNC machine is.

3. Explain why full service repair shops perform more engine repairs and component replacements than engine rebuilding.

4. Mention where engine remanufacturers receive their core engines from.

5. Describe the relationship between a full service repair facility and a machine shop.

6. Explain why engine repair and rebuilding specialty shops exist.

7. Describe the type of repairs an engine repair and rebuilding specialty shop may perform.

8. List where you can purchase a remanufactured engine.

9. Explain why a shop sometimes prefers to install a remanufactured engine or component instead of rebuilding.

10. List the advantages and disadvantages of a full service repair facility owning engine machining equipment.

Fill-in-the-Blanks

1. A new car dealership is considered a _____ service repair facility.

2. Technicians who work at a machine shop have to use _____ measuring tools as part of their daily job.

3. When an installer purchases a remanufactured engine, a _____ _____ must be paid until the old engine is returned.

4. A CNC is a _____ _____ _____ machine.

5. Engine remanufacturing in large volumes helps lower the _____.

6. Rebuilding a modern engine requires a _____ technician due to their growing complexity.

7. A machine shop's inventory or machines and equipment is usually very _____.

8. High performance and restoration engine work is usually performed at a _____ repair facility.

9. A crate remanufactured engine offers a _____ with its purchase.

10. Modern engines have _____ failure rates that keep them from needing extensive service frequently.

Multiple Choice

1. Where would you typically find an "assembly line" for engine rebuilding?
   A. Specialty shop
   B. Full service shop
   C. New car dealership
   D. Remanufacturing facility

2. The World Engine is currently found in what vehicles?
   A. Chrysler
   B. Ford
   C. Chevrolet
   D. Both B and C

3. Specialized engine rebuilding shops usually work on:
   A. High performance engines
   B. Restoration engines
   C. Race engines
   D. All of the above

4. *Technician A* says that a remanufactured engine that comes from an engine remanufacture facility typically has a warranty.
   Technician B says that an engine remanufacture facility will charge you a core charge if you do not send the old engine back to them.
   Who is correct?
   A. A only
   B. B only
   C. Both A and B
   D. Neither A nor B

5. *Technician A* says that a machine shop's inventory is high compared to other automotive repair shops.
   Technician B says that there are very few machines at new car dealerships.
   Who is correct?
   A. A only
   B. B only
   C. Both A and B
   D. Neither A nor B

6. *Technician A* says that a CNC machine can cut and mill engine components.

   Technician B says that a CNC machine is computer controlled.

   Who is correct?

   A. A only   C. Both A and B

   B. B only   D. Neither A nor B

7. *Technician A* says that a CNC machine can mill an engine part with more accuracy than a regular machine.

   Technician B says that using a CNC will usually result in more customer comebacks.

   Who is correct?

   A. A only   C. Both A and B

   B. B only   D. Neither A nor B

8. *Technician A* says that a remanufacturing facility may receive their core engines from salvage yards.

   Technician B says that a remanufacturing facility may receive their core engines from customers who purchased engines from them.

   Who is correct?

   A. A only   C. Both A and B

   B. B only   D. Neither A nor B

9. The failure rate on a modern engine is:

   A. Lower than an older engine

   B. Higher than an older engine

   C. Similar to that of an older engine

   D. None of the above

10. *Technician A* says that an engine rebuilding technician at a machine shop must be able to make precision measurements.

    Technician B says that not much math is involved when you are working as an engine rebuilding technician at a machine shop.

    Who is correct?

    A. A only   C. Both A and B

    B. B only   D. Neither A nor B

Chapter 4

ENGINE OPERATING SYSTEMS

UPON COMPLETION AND REVIEW OF THIS CHAPTER, YOU SHOULD BE ABLE TO:

- Explain the purpose of the starting system.
- List and identify the components of the starting system.
- Explain the purpose of the battery.
- List the purposes of the engine's lubrication system.
- Describe the operation of the lubrication system and its major components.
- Describe the basic types and purposes of additives formulated into engine oil.
- Explain the purpose of the Society of Automotive Engineers' (SAE) classifications of oil.
- Explain the purpose of the American Petroleum Institute's (API) classifications of oil.
- Explain the purpose of the International Lubricant Standardization and Approval Committee (ILSAC) classifications of oil.

- Describe the two basic types of oil pumps: rotary and gear.
- Describe the purpose of the cooling system.
- Explain the operation of the cooling system and its major components.
- Explain the advantages of a reverse flow cooling system.
- Describe the purpose of antifreeze and explain its characteristics.
- Explain the purpose and basic operation of the fuel system.
- Explain the purpose and basic operation of the ignition system.
- Explain the purpose of the emission control systems.

INTRODUCTION

There are five basic requirements for the engine to run:

- The intake and exhaust systems must allow the engine to take air in and exhaust air out.
- The fuel system must supply the correct amount of fuel to mix with the air.
- The ignition system must deliver a spark to the cylinder at the correct time.
- The engine must develop adequate compression to allow ignition of the air and fuel.
- The electrical system must supply adequate electrical energy to start the engine and keep the auxiliary systems operating.

In order to maintain engine operation, the lubrication system must reduce friction in the engine and provide cooling. The cooling system must regulate engine temperature and remove excess heat. The emissions control systems must minimize the toxic pollutants emitted in the exhaust. The proper operation of these supporting systems helps ensure the engine's good performance, efficiency, and longevity.

In this chapter, we will discuss the purpose and operation of these systems and their components. The engine cannot run properly without fully functional supporting systems.

Failures in the lubrication system, as an example, can cause very serious engine problems. As a professional technician, you will need to understand the operation of all the systems and components that keep the engine running properly. You will use this knowledge to help you provide excellent maintenance and perform expert diagnosis and repairs on the engine and its supporting systems.

THE STARTING SYSTEM

The internal combustion engine must be rotated before it will run under its own power. The starting system is a combination of mechanical and electrical parts working together to start the engine. The starting system is designed to change the electrical energy that is being stored in the battery into mechanical energy. For this conversion to be accomplished, a starter or cranking motor is used. The conventional starting system includes the following components (Figure 4-1).

1. Battery
2. Cable and wires
3. Ignition switch
4. Starter solenoid or relay
5. Starter motor
6. Starter drive and flywheel ring gear
7. Starting safety switch

The battery delivers electrical power to the starter solenoid when the driver turns the ignition key to the crank position. The solenoid delivers battery power to the starter motor. The starter motor drive gear is engaged with the **ring gear** on the **flywheel** or **flexplate**; the flywheel (or flexplate) is bolted to the engine's crankshaft. When the starter motor spins, it rotates the engine at about 200–250 rpm (Figure 4-2). This is fast enough to allow the four-stroke cycle to produce combustion and begin running under its own power. Many newer vehicles have anti-theft systems installed on them that can prevent the vehicle from cranking over.

The **flywheel** is a heavy disk mounted at the end of the crankshaft to smooth the engine power pulses. It has a smooth surface in which the clutch assembly of a manual transmission is attached to. A **flexplate** is a lighter version of a flywheel and is bolted to the torque converter of an automatic transmission. The **ring gear** around the circumference of a flywheel or flexplate allows the starter motor drive gear to turn the engine.

FIGURE 4-1 The major components of a typical starting system.

FIGURE 4-2 A typical permanent magnet starter; it is more compact than a starter with pole shoes and field windings.

Battery terminals provide a means of connecting the battery plates to the vehicle's electrical system.

The starting system of a hybrid electric vehicle (HEV) is different from the normal starting system. The high voltage battery pack is typically used to start the gasoline engine. The same set of electric motors that are used for propulsion are used in conjunction with a gear set to start the gasoline engine. A HEVs main computer will start the engine when the conditions are met. These conditions may be low voltage on the HEV battery pack, A/C demand, electrical demand, or acceleration demand. The starter on some modern gas and diesel engines are controlled by the computer, similar to a HEV. Special care must be taken when servicing these vehicles. This is explained in Chapter 14 of the Shop Manual.

The Battery

An automotive battery is an electrochemical device that provides for and stores potential electrical energy. When the battery is connected to an external load such as a starter motor, an energy conversion occurs, causing an electrical current to flow through the circuit. Electrical energy is produced in the battery by the chemical reaction that occurs between two dissimilar plates that are immersed in an electrolyte solution.

When discharging the battery (current flowing from the battery), the battery changes chemical energy into electrical energy. It is through this change that the battery releases stored energy. During charging (current flowing through the battery from the charging system), electrical energy is converted into chemical energy. Because of this, the battery can store energy until it is needed.

The largest demand is placed on the battery when it must supply current to operate the starter motor. The amperage requirements of a starter motor may be over 200 amperes. This requirement is also affected by temperature, engine size, and engine condition.

A typical 12-volt automotive battery is made up of six cells connected in series (Figure 4-3). This means the positive side of a cell element is connected to the negative side of the next cell element. This is repeated throughout all six cells. The six cells produce 2.1 volts each. Wiring the cells in series produces the 12.6 volts required by the automotive electrical system. The cell elements are submerged in a cell case that is filled with electrolyte solution.

All automotive batteries have two terminals. One terminal is a positive connection, the other is a negative connection. The **battery terminals** extend through the cover or the side of the battery case. Following are the most common types of battery terminals (Figure 4-4):

1. *Post or top terminals*: Used on most automotive batteries. The positive post is larger than the negative post to prevent connecting the battery in reverse polarity.

FIGURE 4-3 The 12-volt battery consists of six 2 volt-cells connected together.

Side terminal   Post or top   "L" terminal
                         terminal

FIGURE 4-4 The most common types of automotive battery terminals.

2. *Side terminals*: Positioned in the side of the container near the top. These terminals are threaded and require a special bolt to connect the cables. Polarity identification is by positive and negative symbols.

3. *L-terminals*: Used on specialty batteries and some imports.

The condition of the battery is critical for proper operation of the starting system and all other vehicle electrical systems. The battery is often overlooked during diagnosis of poor starting or no-start problems. Many problems encountered within the starting system may be attributed to poor battery condition.

In addition, the proper battery selection for the vehicle is important. Some of the aspects that determine the battery rating required for a vehicle include engine size, engine type, climatic conditions, vehicle options, and so on. The requirements for electrical energy to crank the engine increase as the temperature decreases. Battery power drops drastically as temperatures drop below freezing (Figure 4-5). The engine also becomes harder to crank due to some oil's tendencies to thicken when cold, resulting in increased friction.

Temperature	% of Cranking Power
80°F (26.7°C)	100
32°F (0°C)	65
0°F (−17.8°C)	40

FIGURE 4-5 As temperature decreases so does the power available from the battery.

The battery's ability to deliver power to crank the engine is described by the cold cranking amperage (CCA). A battery with a CCA rating of 600 is able to deliver 600 amps of electrical current for 30 seconds when the battery is at 0° F. A bigger engine will require a battery with a higher CCA than a small engine. The CCA requirement also increases in the northern regions of the continent, where temperatures regularly fall below freezing.

The battery that is selected must also fit the battery holding fixture and the holddown must be able to be installed. It is also important that the height of the battery not allow the terminals to short across the vehicle hood when it is shut. Battery Council International (BCI) group numbers are used to indicate the physical size and other features of the battery. This group number does not indicate the current capacity of the battery.

All batteries must be secured in the vehicle to prevent damage and the possibility of shorting across the terminals if the battery tips. Normal vibrations cause the plates to shed their active materials. Holddowns reduce the amount of vibration and help increase the life of the battery.

The battery's location may vary with different vehicles. Typically, the 12-volt battery is located under the hood and close to the front of the vehicle, in an area that is easy to get to for testing and replacement. Some vehicles may put the battery farther into the engine compartment, near the transaxle, in the trunk, or under the rear seat. These are known as remote location batteries (see Figure 4-6). There are many reasons for placing the battery in these areas which include weight distribution, temperature and insulation, and hood height and length. The battery's that are located inside the vehicle will have a vent tube that allows for the battery to vent the gas during normal operation. It is important to refer to the manufacturers, service manual if you cannot locate the battery.

Remote battery location

FIGURE 4-6 Some batteries are located remotely.

FIGURE 4-7 Simple electromagnetic motor.

The Starter Motor

Starter motors use the interaction of magnetic fields to convert electrical energy into mechanical energy. Figure 4-7 illustrates a simple electromagnet style of starter motor. The inside windings are called the **armature**. The armature rotates within the stationary outside windings, called the field, which has windings coiled around **pole shoes**. Some starters use stationary permanent magnets rather than the electromagnets created by the pole shoe windings. When current is applied to the field and the armature, both produce magnetic flux lines. The direction of the windings places the left pole at a south polarity and the right pole at a north polarity. The lines of force move from north to south in the field. In the armature, the flux lines circle in one direction on one side of the loop and in the opposite direction on the other side. Current now sets up a magnetic field around the loop of wire, interacting with the north and south fields and causing a turning force on the loop. This force causes the loop to turn in the direction of the weaker field (Figure 4-8).

Problems within the starter or in the wiring to the starter can prevent it from turning the engine over fast enough to start. Engine mechanical problems can also cause this symptom. You must evaluate the starting system before condemning the engine.

The **armature** is the movable component of the motor that consists of a conductor wound around a laminated iron core and is used to create a magnetic field.

Pole shoes are made of high magnetic permeability material to help concentrate and direct the lines of force in the field assembly.

FIGURE 4-8 Rotation of the conductor is in the direction of the weaker field.

LUBRICATION SYSTEMS

The engine's **lubrication system** supplies oil to high friction and high wear locations.

When the engine is operating, the moving parts generate heat due to friction. If this heat and friction were not controlled, the components would suffer severe damage and even weld together. In addition, heat is created from the combustion process. It is the function of the engine's **lubrication system** to supply oil to the high friction and high wear locations and to dissipate heat away from them (Figure 4-9).

It is impossible to eliminate all of the friction within an engine, but a properly operating lubrication system will work to reduce it. Lubrication systems provide an oil film to prevent moving parts from coming in direct contact with each other (Figure 4-10). Oil molecules work as small bearings rolling over each other to eliminate friction (Figure 4-11). Another function of the lubrication system is to act as a shock absorber between the connecting rod and crankshaft.

Besides reducing friction, engine oil absorbs heat and transfers it to another area for cooling. As the oil flows through the engine, it conducts heat from the parts it comes in

FIGURE 4-9 Oil flow diagram indicates the areas where oil is delivered by the lubrication system.

FIGURE 4-10 The oil develops a film to prevent metal-to-metal contact.

FIGURE 4-11 Oil molecules work as small bearings to reduce friction.

contact with. When the oil is returned to the oil pan, it is cooled by the air passing over the pan. Other purposes of the oil include improving the seal between the piston rings and cylinder walls and washing away abrasive metal and dirt. To perform these functions, the engine lubrication system includes the following components:

- Engine oil
- Oil pan or sump
- Oil filter
- Oil pump
- Oil galleries
- Oil coolers (heavy duty applications)

Engine Oil

Engine oil must provide a variety of functions under all of the extreme conditions of engine operation. To perform these tasks, additives are mixed with natural oil. A brief description of the types of additives used follows:

1. Anti-foaming agents—These additives are included to prevent aeration of the oil. Aeration will result in the oil pump not providing sufficient lubrication to the parts and cause low oil pressure.
2. Anti-oxidation agents—Heat and oil agitation result in **oxidation**. These additives work to prevent the buildup of varnish and to prevent the oil from breaking down into harmful substances that can damage engine bearings.
3. Detergents and dispersants—These are added to prevent deposit buildup resulting from carbon, metal particles, and dirt. Detergents break up larger deposits and prevent smaller ones from grouping together. A dispersant is added to prevent carbon particles from grouping together.
4. Viscosity index improver—As the oil increases in temperature, it has a tendency to thin out. These additives prevent oil thinning.
5. Pour point depressants—Prevent oil from becoming too thick to pour at low temperatures.
6. Corrosion and rust inhibitors—Displace water from the metal surfaces and neutralize acid.
7. Cohesion agents—Maintain a film of oil in high pressure points to prevent wear. Cohesion agents are deposited on parts as the oil flows over them and remain on the parts as the oil is pressed out.

Oil has traditionally been rated by two organizations: the Society of Automotive Engineers (SAE) and the American Petroleum Institute (API). The Society of Automotive Engineers has standardized oil viscosity ratings, while the API rates oil to classify its service or quality limits. Engine oil containers have markings on them indicating the qualities defined by these two organizations (Figure 4-12). The API and SAE ratings are written in the "doughnut" on oil containers (Figure 4-13). These rating systems were developed to allow proper selection of engine oil.

The API classifies engine oils by a two-letter system. The prefix letter is either listed as **S-class** or **C-class** to classify the oil's intended usage. S-class oil is designated for use in spark ignition engines; C-class oil should be used in compression ignition diesel engines. The second letter denotes various grades of oil within each classification and denotes the oil's ability to meet the engine manufacturers' warranty requirements. Each category of oil that meets the qualifications has been tested and approved by the API under set conditions. As new oil standards are developed, the next letter in the alphabet is used to designate the latest revision. As vehicles, engines, and driving habits continue to change, the oil formula must change along with it. This system began many years ago and has developed since then. The sequence started with an SA approved oil. An SA oil is a plain mineral oil that doesn't contain any additives. It

Oxidation occurs when hot engine oil combines with oxygen and forms carbon.

S-class oil is used for automotive gasoline engines. **C-class oil** is designed for commercial or diesel engines. The "S" stands for spark ignition, the "C" stands for compression ignition.

FIGURE 4-12 Definition of the ratings shown on an oil container.

The API doughnut

FIGURE 4-13 The engine oil "doughnut" displays oil rating information.

was primarily used in the 1920s. Currently, the SM rating is used for all automotive engines in use. This newest oil development was standardized in 2004. It provides improved oxidation resistance, improved deposit protection, better low temperature performance, and better wear protection over the life of the oil. This means that you can use an SM grade oil in any vehicle, but you cannot use an SL or older oil in a newer vehicle that requires an SM grade oil.

The following is a chart of the current production oil classifications for gasoline and diesel engines. The API intentionally omits the SI and SK from the sequence of categories. SA through SH are now obsolete.

Category	Service Recommendations
SM	Introduced in 2004, the SM oil will work for all automotive gasoline engines currently in use.
SL	For use in 2004 and older automotive engines.
SJ	For use in 2001 and older automotive engines.
CJ-4	Introduced in 2006, CJ-4 was designed to meet the requirements of 2007 model year on-highway exhaust emission standards.
CI-4	For use in high speed, four-stroke diesel engines designed to meet the 2004 exhaust emission standards.
CH-4	For use in high speed, four-stroke diesel engines designed to meet the 1998 exhaust emission standards.
CG-4	For use in severe-duty, high speed, four-stroke diesel engines designed to meet the 1994 exhaust emission standards.
CF-4	For use in high speed, four-stroke diesel engines designed to meet the 1990 exhaust emission standards.
CF-2	For use in severe-duty, two-stroke diesel engines.
CF	For use in off-road indirect-injected diesel engines, including those that use a 0.5% or greater sulfur content fuel.

At present, many manufacturers of gasoline engines demand an SL or SM oil classification for warranty requirements. Most gasoline engine manufacturers also demand an engine oil with the API Starburst symbol displayed on the oil container (Figure 4-14). In addition to the SAE and API ratings, a third organization, the International Lubricant Standardization and Approval Committee (ILSAC) was formed to approve motor oils for use across domestic and import lines and to help speed up the certifying process. SAE, API, the Japanese Automobile Standards Organization (JASO) and Daimler-Chrysler, General Motors, and Ford have all been involved in the development of the latest ILSAC standard oil GF-4 introduced early in 2004. Some SL rated oils will still carry the GF-3 rating, but the new SM oil will meet the GF-4 rating. Only oils with an SAE viscosity rating of 0W, 5W, or 10W may carry the GF-4 rating. It is expected that many vehicles will continue to use SL, GF-3 oils.

SAE ratings provide a numeric system of determining the oil's **viscosity**. The higher the viscosity number, the thicker or heavier the weight of the oil. For example, oil classified as SAE 50 is thicker than SAE 20 oil. The thicker the oil, the slower it will pour. Thicker oils should be used when engine temperatures are high, but they can cause lubrication problems in cooler climates. Using an engine oil with a thicker viscosity than is recommended can cause premature engine wear. The thicker oil will not reach high friction areas fast enough during the critical startup and warmup periods. Viscosity recommendations have become

Viscosity is the measure of the resistance of the oil when placed under shear and extensional stress (the same stresses placed on oil by the crankshaft and connecting rod bearings). It is used to describe an oil's internal resistance to flow. Technicians often use viscosity to describe an oil's thickness or weight.

The **viscosity index** of an oil is a measure of the change in viscosity an oil will have with changes in temperature.

FIGURE 4-14 The API Starburst symbol is displayed on oil containers.

lighter over the years, as engine machining improvements produce engines with closer clearances. The thinner oils are needed to circulate between the close-fitting components. Thinner oils will flow through the engine faster and easier at colder temperatures, but may **break down** under higher temperatures or heavy engine loads. Oil also has to be thick enough to seal piston rings, yet if it is too thin, the engine will not develop enough oil pressure. The oil viscosity will also change with temperature. The higher the temperature, the lower the viscosity, and vice versa.

A compromise is needed. Multigrade oils were developed for this reason. Multigrade oils are designated by their label. For example, an SAE 5W30 oil is a multigrade oil. The "W" stands for winter. SAE has established a standard for testing these oils. The lower number (5 in this case) is the viscosity of the oil at 0°C, and the higher number (30 in this case) is the viscosity of the oil at 100°C (32° F and 212° F, respectively).

The manufacturing of multigrade oil starts out with a base oil. This base oil is typically the lower number (5 in our case). Viscosity-index modifiers are then added to change the viscosity of the oil by changing its behavior at higher temperatures. A single grade oil is different than a multigrade oil because it cannot pass the low temperature SAE standard.

Oils have been developed that may be labeled "energy conserving." These oils use friction modifiers to reduce friction and increase fuel economy. The energy-conserving designation is displayed in the lower portion of the "doughnut."

Selection of oil is based upon the type of engine and the conditions it will be running in. The API ratings must meet the requirements of the engine and provide protection under the normal expected running conditions. One consideration in selecting SAE ratings is ambient temperatures. Some manufacturers will recommend an oil viscosity rating for normal driving and suggest thicker oil while driving at high sustained speeds in a hot climate. A motorist may request that you refill the engine with higher viscosity oil before a summer vacation while pulling a camper, for example. Many consumers run the standard oil in the vehicle year round. Always select oil that meets or exceeds the manufacturer's recommendations.

To a great degree it is the improvement in the quality of oils that have allowed the tighter clearances and improvements in piston fit and ring selection. The incredible gains in oil technology are a significant part of the reason that engines now often last 150,000 miles or longer before needing major service. Use a high quality oil with the viscosity rating specified by the manufacturer and change it regularly to get the most life out of an engine. SL, GF-4 oil is able to meet the extreme demands of late-model, precisely machined engines.

Synthetic Oils

Synthetic oils are oils that are primarily composed of chemical compounds that are synthetically made in a laboratory. Synthetic oils are manufactured as a substitute for and generally provide superior properties when compared to conventional petroleum oils. Most synthetic oil ratings not only meet, but exceed the API and ILSAC ratings for their given SAE viscosity.

Synthetic engine oils have been in use for a long time, but it hasn't been until recently that their use as an automotive engine oil has been widely accepted. Synthetic oils are chemically more stable, have extended oil change intervals, have better viscosity performance at extreme temperatures, increase fuel mileage, provide better lubrication on cold starts, and are very resistive to oil breakdown, sludge, and oxidation. In some cases, synthetic engine oils are a part of the engine design to help lower emissions.

Today's modern engines are built to provide more power from a smaller engine. Engine compartments are typically smaller and have less wind movement and heat dissipation than the previous older, larger vehicles. Combustion chamber temperatures are typically higher as well. For these reasons, all modern engines are under more stress to perform. Overall, engine temperatures are warmer, and if the cooling system is neglected, they are more susceptible to overheating and damage. In a modern engine, one of the major problems with oil is sludge buildup.

Oil sludge buildup is when the oil gels or solidifies after it has been broken down due to high stresses in the engine. Oil sludge is a preventable disaster. When oil additives breakdown the oil, it oxidizes (mixes with oxygen) and begins to gel. When the oil gels, it flows slower and sticks to the walls of the oil galleries. Often, this gel will stick to oil return galleries and cause low oil levels in the oil sump. This, of course, leads to oil starvation and engine damage. Oil sludge problems, emissions standards, and fuel mileage standards, and customer demand have led some manufacturers to install synthetic oils in brand new vehicles. Below is a list of the 2008 vehicles that have installed synthetic oils in their engines. Some of these vehicles have been using synthetic oils from the factory before 2008.

Oil sludge buildup occurs when the oil gels and thickens due to oxidizing.

- Acura RDX
- Aston Martin
- All Bentley Vehicles
- All Cadillac Vehicles
- Chevrolet Corvette C6 and Z06
- Chevrolet TrailBlazer SS
- Chrysler 300C SRT-8
- Cobalt SS S/C Coupe
- Dodge Caliber SRT-4
- Dodge Charger SRT-8
- Dodge Magnum SRT-8
- Dodge Viper SRT-10
- Jeep Cherokee SRT-8
- All Mercedes-Benz AMG Vehicles
- Mercedes SLR
- Mitsubishi Evolution
- Pontiac Solstice GXP
- All Porsche Vehicles
- Saturn Ion Red Line
- Saturn Sky Red Line

The manufacturers of these vehicles are also using synthetic engine oils because of their superior qualities to regular oil. If a customer already has synthetic engine oil in their engine from the factory, it is recommended to continue to use synthetic oil.

Synthetic oils do not gel as much as traditional oils do in colder temperatures. An engine may start easier and fully lubricate faster in extreme cold temperatures with synthetic oils. Synthetic oils are mainly used for their superior engine protection, especially during cold starts when most engine damage occurs. Some customers claim increased fuel mileage, but the increase in mileage by just changing the type of oil alone is very small.

If a customer wants to switch to synthetic oil during a routine oil change, try to advise them with the knowledge you have and the written material the synthetic oil manufacturer provides. Also check with the manufacturers service literature and technical service bulletins for information about the warranty. Most warranties are not voided when using synthetic oils at recommended intervals, but you should still check. Synthetic oils can be used in gas or diesel engines.

Some engine rebuilders do not recommend using synthetic oils during the engine break-in period (typically the first 500 miles). Other sources may point to the fact that some new vehicles are using synthetic oils as a factory fill. New engines from the manufacturer are machined to a lower tolerance than most rebuilt engines and are manufactured in very

FIGURE 4-15 Typical rotor-type oil pump.

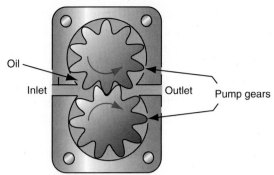

FIGURE 4-16 Typical gear-type oil pump.

controlled environments. The decision to use a synthetic oil during engine break-in should be made after considering the extent of the rebuild, the type, and the quality of any machining performed, and the type and quality of the parts that were installed.

Oil Pumps

The oil pump is the heart of the lubrication system. It develops oil pressure to pump oil throughout the engine. There are two basic types of **oil pumps**: rotor and gear. Both types are **positive displacement pumps**. The rotary pump generally has a four-lobe inner rotor with a five-lobe outer rotor (Figure 4-15). The outer rotor is driven by the inner rotor. As the lobes come out of mesh, a vacuum is created, and atmospheric pressure on the oil pushes it into the pickup tube. The oil is trapped between the lobes as it is directed to the outlet. As the lobes come back into mesh, the oil is expelled under pressure from the pump.

Gear pumps can use two gears riding in mesh with each other or use two gears and a crescent design (Figure 4-16). Both types operate in the same manner as the rotor-type pump. The advantage of the rotor-type pump is its capability to deliver a greater volume of oil since the cavities are larger.

Many of today's engines drive the oil pump by the front of the crankshaft (Figure 4-17). In the past, most oil pumps were driven off of the camshaft by a drive shaft fitting into the bottom of the distributor shaft but not many engines still use an ignition distributor (Figure 4-18).

Most lubrication systems use oil pressures between 10 and 80 psi, depending on engine speed and temperature. A general rule of thumb is that an engine should develop a minimum of 10 psi of oil pressure for each 1,000 rpm of engine speed once the engine is at operating temperature. Since oil pumps are positive displacement types, output pressures must be regulated to prevent excessive pressure buildup. A pressure relief valve opens to return oil to the sump or pump inlet if the specified pressure is exceeded (Figure 4-19). A calibrated spring holds the valve closed, allowing pressure to increase. Once the oil pressure is great enough to overcome the spring pressure, the valve opens and returns the oil to the sump or pump inlet.

Oil Filter

As the oil flows through the engine, it works to clean dirt and deposits from the internal parts. These contaminants are deposited into the oil pan or sump with the oil. Since the oil pump pickup tube syphons the oil from the sump, it can also pick up these contaminants. The pickup tube has a screen mesh to prevent larger contaminants from being picked up and sent back into the engine. The finer contaminants must be filtered from the oil to prevent them from being sent with the oil through the engine. Oil filter elements have a pleated paper or fibrous material designed to filter out particles between 20 and 30 **microns** (Figure 4-20). Most oil filters use a **full-filtration system**. Under pressure from the pump, oil enters the filter on the outer areas of the element and works its way toward the center.

The **oil pump** creates a vacuum so atmospheric pressures can push oil from the sump. The pump then pressurizes the oil and delivers it throughout the engine by use of galleries.

Positive displacement pumps deliver the same amount of oil with every revolution, regardless of speed.

A **micron** is a thousandth of a millimeter or about 0.0008 inch.

A **full-filtration system** means all of the oil is filtered before it enters the oil galleries.

FIGURE 4-17 This crankshaft driven oil pump sits under the timing cover.

FIGURE 4-18 Some oil pumps are driven by a gear on the camshaft.

FIGURE 4-19 The oil pressure relief valve opens to return the oil to the sump if oil pressures are excessive.

In the event the filter becomes plugged, a bypass valve will open and allow the oil to enter the **oil galleries**. If this occurs, the oil will no longer be filtered (Figure 4-21).

An **inlet check valve** is used to prevent oil drainback from the oil pump when the engine is shut off. The check valve is a rubber flap covering the inside of the inlet holes. When the engine is started and the pump begins to operate, the valve opens and allows oil to flow to the filter.

Oil galleries are drilled oil passages in the block and head.

FIGURE 4-20 Oil flow through the oil filter.

FIGURE 4-21 If the oil filter becomes plugged, the bypass valve opens to protect the engine.

Oil Coolers

Turbocharged and some high performance engines or engines in vehicles designed to haul or tow frequently use an **oil cooler** to help keep oil temperatures at a safe level. Oil temperature should generally run between 200°F and 250°F. Excess heat will prematurely break down the oil and allow premature engine wear. Oil coolers look like small radiators and are usually mounted near the radiator. Some designs exchange their heat with cooler incoming air, others use engine coolant to remove the heat.

Oil Flow

The oil pump sends pressurized oil to the oil filter. After the oil leaves the filter, it is directed through galleries to various parts of the engine. A main oil gallery is usually drilled the length of the block. From the main gallery, pressurized oil branches off to upper and lower portions of the engine (Figure 4-22). Oil is directed to the crankshaft main bearings through the main bearing **saddles**. Passages drilled in the crankshaft then direct the oil to the rod bearings (Figure 4-23). Some manufacturers drill a small spit or squirt hole in the connecting rod to spray pressurized oil delivered to the bearings out and onto the cylinder walls (Figure 4-24). When the spit hole aligns with the oil passage in the rod journal, it squirts the oil onto the cylinder wall. Some engines use jet valves to spray oil into the underside of the piston to cool it.

Each camshaft bearing (on OHV engines) also receives oil from passages drilled in the block (Figure 4-25). The valvetrain can receive oil through passages drilled in the block and then through a hollow rocker-arm shaft. Engines using pushrods usually pump oil from the lifters through the pushrods to the valvetrain (Figure 4-26). Oil sent to the cylinder head is returned to the sump through drain passages cast into the head and block.

Not all areas of the engine are lubricated by pressurized oil. The crankshaft rotates in the oil sump, creating some splash-and-throw effects. Oil that is picked up in this manner is thrown throughout the crankcase. This oil splashes onto the cylinder walls below the

The **inlet check valve** keeps the oil filter filled at all times so when the engine is started, an instantaneous supply of oil is available.

An **oil cooler** is a heat exchanger that looks much like a small radiator.

The **saddle** is the portion of the crankcase bore that holds the bearing half in place.

The connecting rod journal is also referred to as the crank pin.

FIGURE 4-22 The lubrication system delivers oil to the high friction and high wear areas of the engine.

FIGURE 4-23 Oil passages drilled into the crankshaft provide lubrication to the main bearings.

FIGURE 4-24 A squirt hole is used to lubricate the cylinder walls and remove excess heat from the piston head.

FIGURE 4-25 OHV engines have branches from the main oil gallery to direct pressurized oil to the camshaft journals.

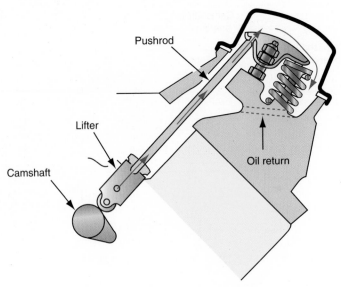

FIGURE 4-26 OHV engines often use hollow pushrods to deliver oil pressure to the cylinder head and valvetrain.

pistons. This provides lubrication and cooling to the piston pin and piston rings. In addition, oil that is forced out of the connecting rod bearings is thrown to lubricate parts that are not fed pressurized oil. Valve timing chains are often lubricated by oil draining from the cylinder head being dumped onto the chain and gears.

AUTHOR'S NOTE: During my experience as an automotive technician and instructor, I encountered a significant number of vehicle owners who attempted to save money on car maintenance by extending engine oil and filter changes beyond the manufacturer's recommended interval. Unfortunately, they spent a lot more on premature, expensive engine repairs than they saved on maintenance costs. Proper lubrication system maintenance is extremely important to provide long engine life; explain this to your customers.

COOLING SYSTEMS

The heat created during the combustion process could quickly increase to a point where engine damage would occur. The **cooling system** is used to disperse the heat to the atmosphere. The cooling system also allows the engine to warm up quickly and provides an even operating temperature. This is important for minimizing emissions and providing efficient engine operation. The cooling system also provides heat for the passenger compartment. The components of the cooling system include:

- Coolant
- Radiator
- Radiator cap
- Recovery system
- Coolant jackets
- Hoses
- Water pump
- Thermostat
- Cooling fan

- Heater system
- Transmission cooler
- Temperature warning system

Coolant

Water by itself does not provide the proper characteristics needed to protect an engine. It works quite well to transfer heat, but does not protect the engine in colder climates. The boiling point of water is 212°F (100°C), and it freezes at 32°F (0°C). Temperatures around the cylinders and in the cylinder head can reach above 500°F (250°C), and the engine will often be exposed to ambient temperatures below 32°F (0°C). In addition, water reacts with metals to produce rust, corrosion, and **electrolysis**. Coolant must protect against freezing, reduce rust and corrosion in the cooling system, and lubricate the water pump and seals. Most manufacturers use a glycol-based engine coolant. In older engines, there was only one type of coolant. Today there are several types of coolant and many different applications for each.

Traditional green ethylene glycol (EG) coolant has been used for many years. It uses inorganic borate salts, silicate, and phosphate to minimize rust and corrosion. These additives break down over time and the coolant becomes acidic. The acidity accelerates corrosion. The newer extended-life, EG-based coolants have a different additive package designed to lengthen the effective life of the coolant. These coolants use a new type of corrosion inhibitor called Organic Additive Technology (OAT) corrosion inhibitors. Many manufacturers using extended-life coolant now specify flushing the cooling system every 5 years or 150,000 miles rather than the traditional coolant flush recommendation of 2 years or 30,000 miles. Many European manufacturers specify coolants with no phosphates to reduce sediment and scale buildup when mixed with hard water. Asian manufacturers may specify a coolant with little or no silicates. These different coolants may be blue, pink, yellow, orange, or red.

Propylene glycol coolant is manufactured because it is less toxic than EG coolant. Less than half a cup of EG coolant is potentially fatal to pets and humans. Animals are attracted to coolant because it is sweet. Some consumers may choose to fill their cooling systems with these environmentally friendlier coolants. If EG and propylene glycol (PG) are mixed, it is more difficult to detect the anti-freeze protection level.

Each manufacture installs their own type of coolant from the factory for specific reasons. Different types and brands of coolant protect engine parts in different ways. Since each manufacturer uses different types and blends of metals, runs different temperatures, and uses different computer programs, it is important to recognize that you should be using the type of coolant recommended by the manufacturer.

Not surprisingly, cooling system failures are the leading cause of other mechanical breakdowns. One of the most common problems related to coolant degradation is **electrolysis**. When two or more dissimilar metals are used in the parts of the engine that touch the coolant, a small electrical charge will occur. This electrical charge is similar to the electrical charge produced by the vehicle's lead acid battery. A battery uses two different metal plates to create and hold its charge. An engine is similar in that some parts are aluminum, iron, copper, or brass. Electrolysis can damage the cooling system walls and make them thinner. Having the proper coolant in the engine protects the metals in the engine and cooling-heating system from electrolysis damage. Excessive electrolysis can lead to overheating, leaking, seal damage, and gasket damage.

Air should not be trapped in the coolant. If air is trapped in the coolant, the engine may not cool properly, and the heater system may not heat properly. Air may get into the system for several reasons; component replacement, coolant leakage, or improper refill. Air pockets can explode in the coolant and cause metal cavitation. Trapped air can also cause the

The temperatures used for boiling points are at sea level.

Electrolysis is the chemical and electrical process of decomposition that occurs when two dissimilar metals are joined with the presence of moisture. The lesser of the two metals is eaten away.

coolant to become acidic. It is important to keep the system bled properly. Some vehicles have a coolant bleed screw to remove the air from the system, while others may need special tools or procedures to remove the air. Removing the air from the cooling system is covered in the Shop Manual.

Use the proper coolant and water mixture to protect the engine under most conditions. Water expands as it freezes. If the coolant in the engine block freezes, the expansion could cause the block or cylinder head to crack. If the coolant boils, liquid is no longer in contact with the cylinder walls and combustion chamber. The vapors are not capable of removing the heat, and the pistons may collapse, or the head may warp or crack from the excessive heat.

Ethylene glycol by itself has a boiling point of 330°F (165°C), but it does not transfer heat very well. The freezing point of ethylene glycol is –8°F (–20°C). To improve the transfer of heat and lower the freezing point, water is added in a mix of about 50/50. At a 50/50 mix, the boiling point is 226°F (108°C), and the freezing point is –34°F (–37°C). In some very cold climates, coolant is mixed with water in a 60/40 ratio to prevent freezing to at least –40°F. These characteristics can be altered by changing the mixture (Figure 4-27).

It may appear that lowering the boiling point by adding water is not desirable. In fact, this is required for proper engine cooling. Under pressure, the boiling point of the coolant mix is about 263°F (128°C). As was stated earlier, the temperature next to the cylinder or cylinder head may be in excess of 500°F (250°C). As coolant droplets touch the metal walls, they boil and are turned into a gas. The superheated gas bubbles are quickly carried away

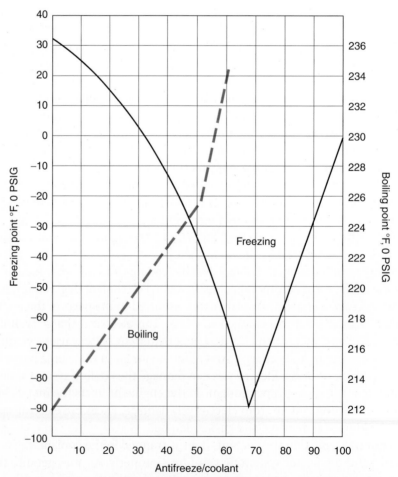

FIGURE 4-27 Changing the strength of the coolant mixture effects its boiling and freezing points.

into the middle of the coolant flow, where they cool and condense back into a liquid. If this **nucleate boiling** did not occur, the coolant would not be capable of removing heat fast enough to protect the engine.

Over time, all coolant can become slightly acidic because of the metals and minerals in the cooling system. All vehicle manufacturers provide a maintenance schedule recommending cooling system flushing and refill based on time and mileage.

Radiator

The **radiator** contains a series of tubes and fins that transfer the heat in the coolant to the air. As the coolant is circulated throughout the engine, it attracts and absorbs the heat within the engine, then flows into the radiator intake tank. The coolant then flows through the tubes to the outlet tank. As the heated coolant flows through the radiator tubes, the heat is dissipated into the air through the fins.

There are two basic radiator designs: downflow and crossflow. Downflow radiators have vertical fins that direct coolant flow from the top inlet tank to the bottom outlet tank (Figure 4-28). Crossflow radiators use horizontal fins to direct coolant flow across the core (Figure 4-29). The core of the radiator can be constructed using a tube-and-fin or a cellular-type design (Figure 4-30). The material used for the core is usually copper, brass, or aluminum. Aluminum-core radiators generally use nylon-constructed tanks.

For heat to be transferred to the air effectively, there must be a difference in temperature between the coolant and the air. The greater the temperature difference, the more effectively heat is transferred. Manufacturers increase the temperature of the coolant by pressurizing the radiator. The radiator cap allows for an increase of pressure within the cooling system, increasing the boiling point of the coolant; for example, if the pressure is increased to 15 psi (103 kPa), the boiling point of the 50/50 mix would be increased to 266°F (134°C). Since the boiling point is increased, the operating temperature of the engine can also be increased.

Nucleate boiling is the process of maintaining the overall temperature of a coolant to a level below its boiling point, but allowing the portions of the coolant actually contacting the surfaces (the nuclei) to boil into a gas.

The **radiator** is a heat exchanger used to transfer heat from the engine to the air passing through it.

For every pound of pressure increase, boiling point of water is raised about $3\frac{1}{4}$°F.

Many vehicles that experience difficulty with air entering the cooling system during service add a bleeder valve at the high point of the cooling system to allow the technician to bleed the system of air.

FIGURE 4-28 A typical downflow radiator.

Top tank

Inlet from engine

Filler cap

Cooling fins (horizontal)

COOLANT FLOW

AIR

AIR

AIR

Core tubes

Transmission oil cooler

Bottom tank

The **coolant recovery system** prevents loss of coolant if the engine overheats, and it keeps air from entering the system.

FIGURE 4-29 A typical crossflow radiator.

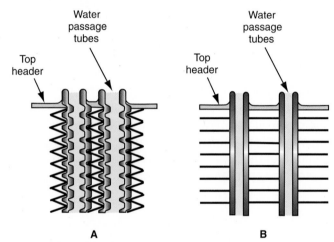

FIGURE 4-30 (A) Construction of a cellular radiator core.
(B) Construction of a tube and fin radiator.

The radiator cap uses a pressure valve to pressurize the radiator to between 14 and 18 pounds (Figure 4-31). If the pressure increases over the setting of the cap, the cap's seal will lift and release the pressure into a recovery tank (Figure 4-32). The cap also has a vacuum valve that will release any vacuum from the system as the coolant temperature decreases after the engine is turned off. If a vacuum is developed in the cooling system, the radiator could collapse.

The **coolant recovery system** contains a reservoir that is connected to the radiator by a small hose. The coolant expelled from the radiator during a high pressure condition is sent to this reservoir. When the engine cools, a void is created in the radiator, and the vacuum valve in the radiator cap opens (Figure 4-33). Under this condition, atmospheric pressure on the coolant in the overflow reservoir pushes it back into the radiator. Due to lower hood lines, many vehicles now use a remote pressure tank. There is no cap on the radiator, the system is filled and checked at the remote pressure tank, sometimes called the reservoir. These are used because the top of the radiator is lower than the highest point on the cooling system. During filling, the system would collect air pockets in the higher spots of the engine cooling systems. These reservoirs are not the same thing as coolant recovery tanks; they contain coolant under pressure, and the top should never be removed while the coolant is hot.

Whenever an engine is rebuilt, the radiator should be thoroughly cleaned and then inspected by pressure testing it. The radiator cap should be replaced with a new one.

FIGURE 4-31 The radiator seals the radiator or reservoir and regulates system pressure.

FIGURE 4-32 A typical coolant recovery system holds coolant released from the radiator.

FIGURE 4-33 When the radiator cap vacuum valve opens, coolant from the recovery reservoir enters the radiator.

Heater Core

The **heater core** is similar to a small version of the radiator. The heater core is located in a housing, usually in the passenger compartment of the vehicle (Figure 4-34). Some of the hot engine coolant is routed to the heater core by hoses. The heat is dispersed to the air inside the vehicle, thus warming the passenger compartment. To aid in quicker heating of the compartment, a heater fan blows the radiated heat into the compartment.

The **heater core** dissipates heat like a radiator. The radiated heat is used to warm the passenger compartment.

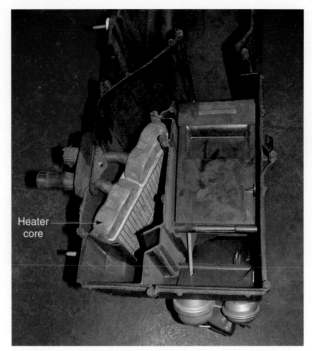

FIGURE 4-34 **The heater core is shown in this cutaway of a heater distribution box.**

Hoses

Hoses are used to direct the coolant from the engine into the radiator, and back to the engine. In addition, hoses are used to direct coolant from the engine into the heater core, and (on some engines) to bypass the thermostat.

Since radiator and heater hoses deteriorate from the inside out, most manufacturers recommend periodic replacement of the hoses as preventive maintenance. Whenever the engine is rebuilt, the hoses should always be replaced.

Water Pump

The water pump is the heart of the engine's cooling system. It forces the coolant through the engine block and into the radiator and heater core (Figure 4-35). The water pump is driven by an accessory belt, the timing belt, or directly from the camshaft. Most water pumps are centrifugal design, using a rotating impeller to move the coolant (Figure 4-36). When the engine is running, the impeller rotates, forcing coolant from the inside of the cavity outward toward the tips by centrifugal force. Once inside the block, the coolant flows around the cylinders and into the cylinder heads, absorbing the heat from these components. If the thermostat is open, the coolant will then enter the radiator. The void created by the empty impeller cavity allows the pressurized coolant to be pushed from the radiator to fill the cavity and repeat the cycle. When the thermostat is closed because coolant temperatures are too cold, the coolant is circulated through a bypass. This keeps the coolant circling through the engine block until it gets warm enough to open the thermostat.

The **thermostat** allows the engine to quickly reach normal operating temperatures and maintains the desired temperature.

Thermostat

Control of engine temperature is the function of the **thermostat** (Figure 4-37). The thermostat is usually located at the outlet passage from the engine block to the radiator. When the coolant is below normal operating temperatures, the thermostat is closed, preventing coolant from entering the radiator. In this case, the coolant flows through a bypass passage and returns directly to the water pump (Figure 4-38).

FIGURE 4-35 Coolant flow through the engine.

FIGURE 4-36 Most water pumps use an impeller to move coolant through the system.

FIGURE 4-37 The thermostat controls the temperature of the coolant.

FIGURE 4-38 The bypass hose lets coolant return to the water pump.

The thermostat is rated at the temperature it opens in degrees Fahrenheit. If the rating is 195°, this is the temperature at which the thermostat *begins* to open. When the temperature exceeds the rating of the thermostat by 20°F, the thermostat will be totally open. Once the thermostat is open, it allows the coolant to enter the radiator to be cooled. The thermostat cycles open and closed to maintain proper engine temperature.

Operation of the thermostat is performed by a specially formulated wax and powdered metal pellet located in a heat-conducting copper cup (Figure 4-39). When the wax pellet is exposed to heat, it begins to expand. This causes the piston to move outward, opening the valve (Figure 4-40).

FIGURE 4-39 The temperature-sensitive wax pellet controls the valve in the thermostat.

FIGURE 4-40 When the pellet expands the thermostat opens and allows coolant to flow through the radiator.

Thermostats have a high failure rate and are a very common cause of cooling system concerns. Do not install a thermostat with a lower opening temperature or run the engine without a thermostat. Allowing the engine to run cooler than designed accelerates engine wear and increases emissions and fuel consumption.

Coolant Flow

Coolant flow through the engine can be one of two ways: parallel or series. In a parallel flow system, the coolant flows into the engine block and into the cylinder head through passages beside each cylinder (Figure 4-41). In a series flow system, the coolant enters the engine block and flows around each cylinder. It then flows to the rear of the block and through passages to the cylinder head. The coolant then flows through the head to the front of the engine (usually the highest point of the cooling system) and out to the radiator (Figure 4-42).

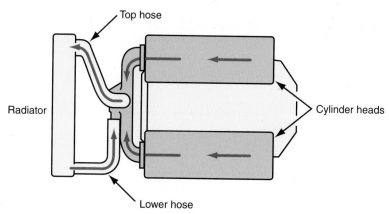

FIGURE 4-41 The parallel flow sends the coolant around each cylinder and up to the cylinder head. There is more than one path for coolant to flow to the head.

FIGURE 4-42 A series flow directs the coolant around all of the cylinders before sending it up through the cylinder head.

The cooling system passages are more than hollow portions in a casting. They are designed and located so no steam pockets can develop. Any steam will go to the top of the radiator. This is accomplished by bleed holes among the cylinder block, **head gasket**, and cylinder head.

Corrosion or contaminants can plug the coolant passages. This results in reduced heat transfer from the cylinder to the coolant. Deposits can build up in small corners around the valve seats, causing hot spots in the cylinder head. This can allow the cylinder head to crack or the valves to burn.

Reverse Flow Cooling Systems

A **reverse flow cooling system** was used in some engines, such as the General Motors LT1 5.7L V8 engine in some model years of Camaro, Firebird, and Corvette. In these engines, the water pump forces coolant through the cylinder heads first and then through the block passages (Figure 4-43). The water pump is driven from a gear and shaft mounted on the front of the engine. The water pump drive gear is rotated by gear teeth on the back of the camshaft sprocket (Figure 4-44). Coolant flows from the block through other water pump passages and through the upper radiator hose to the radiator. Coolant returns through the lower radiator hose to the inlet side of the water pump. On a conventional flow cooling system, the coolant

FIGURE 4-43 Reverse flow cooling system.

FIGURE 4-44 Water pump drive.

flows through the block first and then through the cylinder heads. When the engine is running, the hottest part is the exhaust valve seat. Circulating coolant through the cylinder head first means that the exhaust valve seats (and the entire cylinder head) will operate at cooler temperatures because they have cooler coolant flowing through them, versus warmer coolant that picked up the heat from the cylinder block. This allows the combustion chamber to operate at a higher temperature. The higher temperature is a result of a higher compression ratio. The compression ratio in the LT1 engine is 10.25:1. Improved combustion chamber cooling also allows the engine to operate with increased spark advance without encountering detonation problems. This action improves fuel economy and engine performance. Given these advantages, it is possible that manufacturers will use reverse flow cooling systems again in the future.

COOLING FANS

To increase the efficiency of the cooling system, a fan is mounted to direct flow of air through the radiator core or past the radiator tubes. In the past, at high vehicle speeds air flow through the grill and past the radiator core was sufficient to remove the heat. The cooling fan was really only needed for low speed conditions when air flow was reduced. As modern cars have become more aerodynamic, air flow through the grill has declined; thus proper operation of the cooling fan has become more critical.

The cooling fan can be either driven mechanically by the engine or by an electric motor. Electric drive fans are common on today's vehicles because they only operate when needed, thus reducing engine loads. Some of the earlier designs of electric cooling fans use a temperature switch in the radiator or engine block that closes when the temperature of the coolant reaches a predetermined value. With the switch closed, the electrical circuit is completed for the fan motor relay control circuit and the fan turns on (Figure 4-45). The power feed to the relay is supplied through the ignition switch or directly from the battery. If it is supplied directly from the battery, the cooling fans can come on, even though the engine is not running.

Now, most cooling fans are controlled by the powertrain control module (PCM). The PCM receives engine coolant temperature input signals from the thermistor-type engine coolant temperature (ECT) sensor. When the ECT sensor indicates the temperature is hot enough to turn on the fans, usually around 200–230°F, the PCM activates the fan control relay (Figure 4-46). In this diagram, the relay has power from the fuse waiting at pins number 2 and number 3. When the PCM decides to turn the fan on, it provides a ground to pin number 1 of the relay. This allows current to flow from the 10 amp fuse, through pin number 2 and the relay coil, and to ground through the PCM. Current flowing through the coil inside the relay creates a magnetic field that pulls the contacts between pins number 3 and number 4 closed. Now power can flow from the 40 amp fuse, to pin number 3 and through the closed relay contact points, out through pin number 4 and to the cooling fan. The cooling fan already has a ground supply, so it begins to spin. On some vehicles, the PCM also turns on the fans whenever air-conditioning is turned on, regardless of the engine

FIGURE 4-45 Simplified electric fan circuit.

FIGURE 4-46 The PCM looks at the ECT and A/C inputs to decide when to turn the fans on.

temperature. In some cases, the electric cooling fan may turn on because the engine is hot, even if the ignition is off and the key is out.

Belt-driven fans usually attach to the water pump pulley (Figure 4-47). Some belt-driven fans use a viscous-drive fan clutch to enhance performance (Figure 4-48). The clutch operates

FIGURE 4-47 Belt-driven cooling fans are usually mounted to the water pump pulley.

FIGURE 4-48 Most mechanical fans use a clutch to minimize noise and load on the engine when the engine is cooler.

FIGURE 4-49 The viscous-drive clutch uses silicone oil to lock the fan to the hub.

FIGURE 4-50 As temperature increases, the bimetallic spring expands to operate the clutch and lock the fan on.

FIGURE 4-51 This clutch fan is electronically controlled by the PCM.

the fan in relation to engine temperature by a silicone oil (Figure 4-49). If the engine is hot, the silicone oil in the clutch expands and locks the fan to the pump hub. The fan now rotates at the same speed as the water pump. When the engine is cold, the silicone oil contracts and the fan rotates at a reduced speed. Some fan clutches use a thermostatic coil that winds and unwinds in response to the engine temperatures (Figure 4-50). The coil operates a valve that prevents or allows silicone to leave the reservoir and flow into the working chamber. Some performance engines use a clutch fan that is electronically operated (Figure 4-51).

LUBRICATION AND COOLING SYSTEM WARNING INDICATORS

Vehicle manufacturers provide some method of indicating problems with the lubrication or cooling system to the driver. This is done by the use of an indicating gauge or by the illumination of a light. Some manufacturers control the operation of the gauge or light by computers. Regardless of the type of control, sensors or switches provide the needed input.

FIGURE 4-52 A thermistor used to sense engine temperature.

FIGURE 4-53 A piezoresistive sensor used to sense oil pressure.

Gauge Sending Units

The most common sensor type used for monitoring the cooling system is a **thermistor**. The lubrication system uses piezoresistive sensors to measure oil pressure.

In a simple coolant temperature-sensing circuit, current is sent from the gauge unit into the top terminal of the sending unit, through the variable resistor (thermistor), and to the engine block (ground). The resistance value of the thermistor changes in proportion to coolant temperature (Figure 4-52). As the temperature rises, the resistance decreases, and the current flow through the gauge increases. As the coolant temperature lowers, the resistance value increases, and the current flow decreases. Now, many instrument panels are operated by an internal computer or by a **body control module (BCM)**. The coolant temperature sensor is still a thermistor, but its signal serves as an input to the instrument panel (IP) or to the BCM. The computer then drives an electronic gauge assembly to display coolant temperature.

The **piezoresistive sensor** sending unit is threaded into the oil delivery passage of the engine and the pressure that is exerted by the oil causes the flexible diaphragm to move (Figure 4-53). The diaphragm movement is transferred to a contact arm that slides along the resistor. The position of the sliding contacts on the arm in relation to the resistance coil determines the resistance value and the amount of current flow through the gauge to ground. Again, the oil pressure sending unit may not be wired directly to the gauge; it may deliver its signal to the instrument panel cluster or BCM. See the diagram showing this circuit (Figure 4-54).

Warning Lamps

A **warning light** may be used to warn of low oil pressure or high coolant temperature. Unlike gauge sending units, the sending unit for a warning light is nothing more than a simple switch. The style of switch can be either normally open or normally closed, depending on the monitored system.

Most oil pressure warning circuits use a normally closed switch (Figure 4-55). A diaphragm in the sending unit is exposed to the oil pressure. The switch contacts are controlled by the movement of the diaphragm. When the ignition switch is turned to the on position with the engine not running, the oil warning light turns on. Since there is no pressure to the diaphragm, the contacts remain closed, and the circuit is complete to ground. When the engine is started, oil pressure builds and the diaphragm moves the contacts apart. This opens the circuit, and the warning light goes off. The amount of oil pressure required to move the diaphragm is about 3 psi. If the oil warning light comes on while the engine is running, it indicates that the oil pressure has dropped below the 3 psi limit.

Most coolant temperature warning light circuits use a normally open switch (Figure 4-56). The temperature sending unit consists of a fixed contact and a contact on a bimetallic strip. As the coolant temperature increases, the bimetallic strip bends. As the strip bends, the contacts move closer to each other. Once a predetermined temperature level

The **thermistor** is a resistor whose resistance changes in relation to changes in temperature. It is often used as a coolant temperature sensor.

A **body control module (BCM)** is a computer used to control many systems such as the instrument panel, radio, interior lighting, climate control, and driver information.

A **piezoresistive sensor** is sensitive to pressure changes. The most common use of this type of sensor is to measure the engine oil pressure.

A **warning light** is a lamp that is illuminated to warn the driver of a possible problem or hazardous condition.

FIGURE 4-54 An electronic instrument panel with an internal integrated circuit.

(A)   (B)

FIGURE 4-55 (A) Oil pressure sending unit. (B) Oil pressure warning light circuit.

FIGURE 4-56 Temperature warning light circuit.

FIGURE 4-57 A prove-out circuit included in the normally open (NO) coolant warning light circuit.

FIGURE 4-58 One warning light used with two sensors.

A **prove-out circuit** completes the warning light circuit to ground through the ignition switch when it is in the start position. The warning light will be on during engine cranking to indicate to the driver that the bulb is working properly.

Refer to *Today's Technician: Automotive Engine Performance* for a detailed explanation of fuel system operation.

has been exceeded, the contacts are closed, and the circuit to ground is closed. When this happens, the warning light is turned on.

With normally open switches, the contacts are not closed when the ignition switch is turned to the on position. In order to perform a bulb check on normally open switches, a **prove-out circuit** is included (Figure 4-57).

It is possible to have more than one sender unit connected to a single bulb (Figure 4-58). The light will come on whenever oil pressure is low or coolant temperature is too high.

Message Center

Some vehicles have a message center that displays a number of warning messages to alert the driver regarding dangerous vehicle operating conditions. Warning messages related to the lubrication and cooling system include: LOW COOLANT, CHECK COOLANT TEMPERATURE, ENGINE OVERHEATED, CHECK ENGINE OIL PRESSURE, CHECK ENGINE OIL LEVEL, and CHANGE ENGINE OIL. The messages are displayed in the message center by the PCM. The PCM receives input signals from the ECT sensor. If the engine coolant reaches a specific temperature above the normal operating temperature, the PCM illuminates the CHECK COOLANT TEMPERATURE. If the engine coolant temperature increases a specific number of degrees, the PCM turns on the ENGINE OVERHEATED message. Fluid level sensors are required to send a low coolant signal or a low engine oil signal to the PCM. The PCM looks at several different engine operating parameters to illuminate the CHANGE ENGINE OIL message. If the engine operating parameters such as engine temperature and rpm are continually in the normal operating range, the PCM will probably turn on the CHANGE ENGINE OIL message at the approximate interval recommended by the vehicle manufacturer for oil change intervals. However, under severe vehicle operating conditions such as trailer towing, the engine temperature may be above normal, and the PCM may illuminate the CHANGE ENGINE OIL message at a lower mileage interval. As an example, on some vehicles, the CHANGE ENGINE OIL message comes on for 10 to 25 seconds each time the engine is started. After this time period, the message is turned off. After the oil is changed, the CHANGE ENGINE OIL message must be reset. As an example, on some vehicles, this is done by pushing the accelerator pedal wide open three times in a 5-second interval with the ignition switch on and the engine not running. Always consult the vehicle manufacturer's service manual for the proper procedure.

FUEL SYSTEM

The **fuel system** is designed to deliver the correct amount of fuel to each of the cylinders for any engine operating condition. In this way, the engine can perform well while minimizing toxic emissions and increasing fuel economy. The PCM precisely controls the fuel delivery through its operation of fuel injectors. Air is brought into the engine through the intake system. A fuel injector is placed in the intake manifold near the opening to the cylinder. Most modern vehicles use one fuel injector for each cylinder; this is called **multi-point fuel injection (MPFI)**. A few new vehicles inject the fuel directly into the cylinder; that is gasoline direct injection (GDI). Another system, central port fuel injection (CPFI), uses poppet valves underneath the intake plenum. The fuel delivery system includes the fuel tank to store fuel, an electric fuel pump mounted in the tank to deliver fuel under pressure to the injectors, a fuel filter, a fuel pressure regulator, and fuel injectors (Figure 4-59). The fuel control system includes the PCM and several inputs used to provide the PCM with enough information to determine how much fuel is required. The primary input for fuel delivery calculations is either a mass airflow (MAF) sensor or a manifold absolute pressure (MAP) sensor. The MAF sensor measures the mass of air flowing into the engine so the computer can calculate the proper amount of fuel to mix with it (Figure 4-60). A vehicle may use a MAP sensor instead to provide input for the PCM. The MAP sensor measures the pressure in the intake system to infer the amount of load on the engine. When the driver has the accelerator pedal all the way to the floor, the throttle opens the passage into the intake manifold and the pressure increases to atmospheric pressure. The PCM interprets this as a very heavy engine load and delivers more fuel. The PCM also receives information about engine temperature, intake air temperature, engine speed, throttle position, the level of oxygen in the exhaust and more depending on the vehicle. When the PCM looks at all the data, it decides how much fuel should be injected and it opens the injector for exactly the right

The **fuel system** includes the intake system that is used to bring air into the engine and the components used to deliver the fuel to the engine.

Multi-point fuel **injection** systems uses one injector for each cylinder to deliver fuel.

FIGURE 4-59 Fuel delivery system components.

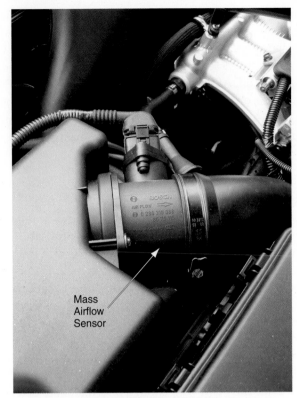

FIGURE 4-60 The MAF sensor sits in the incoming airstream and senses the mass of airflow.

amount of time to deliver that quantity. The theoretically perfect air-to-fuel ratio is 14.7 parts of air to one part of fuel. During acceleration, the mixture must be "richer" (have more fuel). During deceleration, the PCM will "lean" out the mixture (give the engine less fuel to improve fuel economy). The injector has power applied all the time, and the PCM provides a ground pulse that times the injector opening; while the engine is idling, the injector on time may be 1 to 2 milliseconds. The injector has very small openings at its base to atomize the fuel. This means that it breaks the fuel into very small particles that can easily mix with the air to help support complete burning and powerful combustion (Figure 4-61).

FIGURE 4-61 The fuel injector has a small pinhole-sized opening for fuel to be sprayed out under pressure.

Ignition System

The compressed air-fuel mixture must be ignited at the correct time. The delivery of the spark is the function of the **ignition system**. If the spark is not delivered at the correct time, poor engine performance and fuel economy will result. **Burn time** of the fuel plays into the calculation of spark delivery. Combustion should be completed by the time the piston is at 10 degrees after top dead center (ATDC) during the power stroke. If the spark occurs too early, when the piston is moving up during the compression stroke, the piston will have to overcome the combustion pressures. If the spark occurs too late, the combustion pressures are not as effective in pushing the piston down during the power stroke. In either case, power output is reduced.

Engine speed is another consideration in spark delivery. As engine speed increases, the amount of piston movement also increases for the same amount of time. Burn time of the gasoline remains constant, but the piston will travel more as engine speed increases. To compensate for this, the spark must be delivered at an earlier time in the piston movement.

The PCM controls the spark timing. It uses a crankshaft position sensor to determine the precise position of each piston. It may also use a camshaft position sensor to help identify each cylinder. The PCM also looks at many of the same inputs it uses for fuel delivery and then sends an output that allows the ignition coils to fire. The majority of modern ignition systems use multiple coils to fire the spark plugs. These systems are called **electronic-ignition (EI)** systems. Some EI systems use one coil to provide spark for two cylinders; these are called waste spark systems (Figure 4-62). Each coil fires two spark plugs at the same instant. One of the spark plugs is in the cylinder that is near TDC on the compression stroke; this spark will begin combustion on the power stroke. The other spark plug's cylinder is on the exhaust stroke. The spark has no effect on engine operation, hence the term "waste spark." Other EI systems place a coil on top of each spark plug; this eliminates sensitive high voltage wiring. These are called coil-on-plug (COP) systems (Figure 4-63). Other ignition systems use one coil to fire all the spark plugs. These systems use a mechanical distributor to deliver the spark to the correct cylinder; these are termed **distributor-ignition (DI)** systems.

EI systems use coils to deliver high voltage to the spark plug. The high voltage forms a spark as it jumps across the gap of the spark plug electrodes. The coils have power available to them whenever the engine is running. The PCM provides a ground to turn the coils on and off. In some cases, the PCM provides the ground directly to the coils; in other cases,

Refer to *Today's Technician: Automotive Engine Performance* for a detailed explanation of ignition system operation.

The delivery of the spark used to ignite the compressed air-fuel mixture is the function of the **ignition system**.

Burn time is the amount of time from the instant the mixture is ignited until the combustion is complete.

An **electronic-ignition** system uses an electronic module to turn the coils on and off.

A **distributor-ignition (DI)** system has a distributor that distributes spark to each spark plug.

FIGURE 4-62 This coil pack is for a four-cylinder engine; it has two coils, each with two outputs.

FIGURE 4-63 These coils sit directly on top of the spark plugs.

FIGURE 4-64 This EI system uses an ignition module to provide a ground for the coils.

Refer to *Today's Technician: Automotive Engine Performance* for a detailed explanation of emission system operation.

The **emission control system** includes various devices connected to the engine or exhaust system to reduce harmful emissions of hydrocarbons (HC), carbon monoxide (CO), and oxides of nitrogen (NO$_x$).

the PCM provides inputs to an ignition module and it provides the ground to the coils (Figure 4-64).

Emission Control System

The **emission control system** helps to reduce the harmful emissions resulting from the combustion process. The powertrain control module and its engine management system are integral parts of controlling emissions. Spark timing, fuel injection, and several other systems are controlled by the PCM to minimize the output of toxic emissions. Modern engine management systems do an excellent job of cleaning up tailpipe and evaporative emissions. One study showed that driving a two-stroke snowmobile for 7 hours produced more emissions than a 2000 model year vehicle driven 100,000 miles.

While modern engines release less than 10% of the toxic pollutants that they did in the 1970s, the automobile is still a primary source of air pollution in the United States, Canada, and other industrialized countries. Hydrocarbons (HC), carbon monoxide (CO), and oxides of nitrogen (NO_x) are all formed during the combustion process. When an engine is running poorly, the emissions greatly increase. Emission control devices minimize the pollutants by affecting combustion, by cleaning up the post-combustion emissions, and by controlling the evaporative emissions that occur when a vehicle is not being driven or when it is being refueled. The actual controls on an engine depend upon the design of the engine. Many vehicles use these emission control systems:

- Catalytic converter
- Secondary air injection system
- Positive crankcase ventilation valve (PCV) system
- Exhaust gas recirculation (EGR) system
- Evaporative emissions system

AUTOMOTIVE FUELS

Gasoline

Gasoline is a mixture of many ingredients, such as hydrocarbons and various other chemicals and additives. Hydrocarbon compounds are refined from crude oil that is removed from the Earth. Crude oil is refined into many products, including gasoline, diesel fuel, plastics, soaps, fertilizers, and asphalts. The refining process separates the light liquids from the heavy liquids. The lighter liquids are typically used for producing gasoline, and the heavier liquids are used for producing waxes and asphalts.

Octane

The anti-knock quality of gasoline is rated with the term **octane**. The advertised octane number on the gas station pump is the average of two methods of measuring a fuel's octane. Both **RON (research octane number)** and **MON (motor octane number)** are laboratory test methods of measuring a fuel's octane (Figure 4-65). The octane rating of a fuel is used by manufacturers to determine which fuel is appropriate to use for a specific engine.

Gasoline has many ingredients and compounds. It is refined from crude oil and is the most common fuel used in today's engines.

The **octane** number is the ratio of two test fuels that match the knock characteristic of the test fuel. The advertised octane number at a gas station is an average from RON and MON and is an indicator of the fuel's anti-knock quality. Lower octane fuels will burn faster. Higher octane fuels will burn slower.

RON and **MON** are octane testing numbers that are averaged together to get the advertised octane number. Each uses a different test procedure that represents different real-world driving situations.

FIGURE 4-65 These are the most common octane ratings available. Note the (R + M/2) method.

1. Spark occurs 2. Combustion begins

3. Continues 4. Detonation

FIGURE 4-66 Detonation can cause severe damage.

Higher performance engines with higher compression ratios typically use a premium gas that has a higher octane number. More normal engines with normal compression ratios will most likely use a regular or mid-grade fuel. The compression ratio is the main determining factor in the octane choice of a fuel, but other factors such as fuel delivery method, intake manifold design, and combustion chamber design are also considered.

The minimum octane rating must always be observed and used for several reasons. The main reason is to protect the engine from detonation (also known as spark knock). **Detonation** occurs when two flame fronts collide in the combustion chamber (Figure 4-66). The first flame front is normal and created by the spark plug. The second flame front is created because of high temperatures. The collision could cause serious damage to the piston if it is not corrected immediately. Fuels with a lower octane number will burn faster than those with a higher octane number. A higher octane fuel will slow down the flame speed and give less of a chance for detonation to occur. Modern engines and computer systems use a **knock sensor** to detect if there is an abnormal explosion in the combustion chamber (Figure 4-67). If an abnormal explosion is detected, the PCM will usually retard the ignition timing, which lowers the temperature of the combustion chamber and usually stops the detonation.

The octane number itself does not have any affect on fuel efficiency or economy; however, if a lower than recommended octane number is used, the PCM may change the spark timing to compensate and the fuel economy may suffer. Using a higher than recommended octane will not directly affect the engine's performance or the fuel economy. After an engine rebuild, the octane requirement may change due to the excessive machining and possible exchange of parts, such as pistons, crankshaft, or cylinder heads. During an engine rebuild, you should always remain aware of how machining and upgrading components will affect the compression ratio. The octane requirement can also be affected by the air-fuel mixture, valve timing, coolant temperature, carbon deposits, and turbo-/supercharging.

Volatility

Gasoline has to be very volatile so that it can easily evaporate and readily mix with air to support combustion. Vaporized fuel and air will support combustion, and **volatility** describes how easily the gasoline evaporates.

Detonation occurs when two flame fronts are developed and collide in the combustion chamber.

A knock sensor is used to detect abnormal combustion. Its data is used by the PCM to change the ignition timing.

A fuels **volatility** describe how easily it evaporates in the fuel line and system components.

FIGURE 4-67 Knock Sensor

Gasoline blended in the summertime (or year-round in very warm climates) is less volatile than winter blends. The more volatile a fuel is the more vapor pressure it will create in the fuel line and system components at a given temperature. Different pressures are required because engines need to operate in a variety of conditions.

During hot temperatures, a fuel that vaporizes too quickly can develop vapor bubbles in the fuel line and block the flow of liquid to the engine. This is called vapor lock. During cold temperatures, the fuel needs to vaporize quickly so that it can mix with the cold incoming air in the combustion chamber.

Ethanol

Ethanol is an alcohol that can be used as an automotive fuel. Ethanol (ethyl alcohol) behaves very similar to gasoline and is commonly blended with gasoline as an additive. Ethanol is a noncorrosive, low toxic level alcohol that is made from renewable biological sources. In the United States, ethanol is typically made from corn. The starch and sugars from the corn are processed in a fermenting tank at an ethanol producing plant. Distillation towers are then used to separate the water from the alcohol.

> **Ethanol** is an alcohol that can be used as an automotive fuel.

Ethanol is typically blended in with gasoline and is used as an additive and octane improver. Ethanol is an oxygenated fuel that helps reduce the toxicity of burning regular gasoline. Ethanol also helps keep the fuel system cleaner and less subject to corrosion. Ethanol, like other alcohols, absorbs water readily. Water is formed in the fuel system by condensation. If it is not absorbed, it will rust out system components and possibly "freeze" the fuel line in the winter. Currently, ethanol blends of up to 10 percent are approved by all manufacturers and do not void their warranty. Ethanol's volatility is significantly lower to gasoline, the only condition that makes it unsuitable for cold weather use.

Ethanol burns cooler than gasoline does. This reduces the risk of valve burning and detonation, two common side effects of excessive heat. Ethanol also has an octane rating of over 100, which makes it a suitable fuel for advancing the ignition timing and running a higher compression ratio.

Ethanol has a lower energy rating than gasoline does. One hundred percent ethanol has about 76,000 Btus/gallon and 100 percent gasoline has about 125,000 Btus/gallon. The traditional gasoline that you would find at a gas station is mixed with ethanol, which reduces the amount of energy in the fuel because of the mixture rate.

Ethanol is also blended at a rate of 85 percent ethanol and 15 percent gasoline. This blend is called E85. Flex fuel vehicles are specifically set up to operate on E85, gasoline, or any blend in between. More about ethanol and flex fuel vehicles is found in Chapter 14.

Additives

Gasoline by itself does not properly serve the needs of the modern engine. Gasoline additives have changed over the years to meet the demands of different engine types and driving patterns.

Tetraethyl lead (TEL) was added to gasoline in the 1920s. Originally, lead was added to gasoline to improve the anti-knock quality. After lead was added to gasoline, engines started to have higher compression ratios, and vehicles went faster and farther on a tank of fuel. Years later, it was found that the lead was contaminating the drinking water and causing a few cases of lead poisoning. The practice of using lead as an additive was stopped. Unleaded fuels were commonly used as early as the mid-1970's. Leaded fuel is no longer used or available for on-road automotive use.

Oxidation inhibitors prevent the formation of gum and varnish on components. When gasoline is not used for a prolonged period of time, its evaporates can mix with oxygen and form a residue on system components.

Detergents are designed to clean to fuel systems and components. Gasoline can sometimes contain microscopic particles of dirt that can, with time, cause fuel filter or fuel injector clogging. The amount of detergents varies greatly between different manufacturers. The amount of detergent necessary may also be dependant on the design of the fuel system.

Diesel Fuel

Diesel fuel acts very different than gasoline or ethanol and has very different characteristics. A diesel engine does not use spark plugs for the ignition of the fuel and air mixture. It uses the heat developed by the compression stroke. It is very important that diesel fuel is clean and able to flow at low temperatures for good engine longevity and performance.

Diesel is typically a product of crude oil at a distillation station, similar to gasoline. But there are other ways to produce diesel fuel. Some examples are biodiesel, which can be produced from crop and renewable products. Another example is biomass-to-liquid (BTL) which uses waste, garbage, sewage, and used cooking oils. Recently, gas-to-liquid (GTL) technology has proven successful. GTL is a diesel fuel that is produced from natural gas reserves found in the Earth. Diesel fuel is not rated by the octane scale, but rather it uses a different scale called the cetane scale. The **cetane number** is a measure of how easily the fuel can be ignited. The cetane number is an indicator of the fuel's ability to start the engine in low temperatures and smoothly warm up the engine. Typically, the cetane rating for automotive and light truck diesel fuel is between 40 and 50.

Diesel fuel is denser than regular gasoline. Diesel fuel weighs about 7.09 lbs/gallon, while gasoline weighs about 6.01 lbs/gallon.

Diesel fuel contains significantly more energy than gasoline or ethanol. At approximately 138,000 Btus-gallon, it gives diesel engines that extra energy to make more power and get better fuel economy. Diesel engines are also more efficient because of the higher compression ratio, average lower operating speeds, and a slower, more complete burn inside the cylinder. Diesel fuel inherently has high sulfur content. The sulfur was traditionally used for lubrication in the cylinder. Recent ultra low sulfur diesel (ULSD) laws have forced the sulfur level to decrease, and additives are now being used. Biodiesel is becoming a common additive because of its good lubricity properties. Sulfur in the exhaust and the atmosphere are major components of acid rain. Ultra low sulfur diesel standards have been in place in the United States since 2007. These new fuels allow for the use of new technology emissions devices that significantly clean up the exhaust.

Diesel fuel quickly increases its viscosity as the temperature decreases. Two grades of diesel fuel are commonly available. Diesel #1 is a less dense fuel with lower heat content.

TEL is not used as an additive in gasoline anymore. Unleaded gasoline is readily available and is less toxic.

The formation of gum and varnish is prevented by **oxidation inhibitors**.

Detergents are gasoline additives that clean the fuel system.

Diesel #2 is more popular and is more common during summer months and on-road vehicles. It is common to blend a little bit of #1 into #2 in colder temperatures to keep the fuel from gelling. Off-road diesel is also available in the United States. This fuel is untaxed and does not meet the emissions levels for on-road use. It is primarily used for agriculture applications and is dyed red in color.

Diesel fuel is also immiscible with water—meaning it does not mix with water. Small microbes can grow where the water and fuel come in contact. These microbes feed off the fuel and cause filter clogging, injection problems, and engine problems. Diesel engines have filters and water separators that need to be maintained.

SUMMARY

- The starting system is a combination of mechanical and electrical parts that work together to start the engine.
- The starting system components include the battery, cable and wires, the ignition switch, the starter solenoid or relay, the starter motor, the starter drive and flywheel ring gear, and the starting safety switch.
- An automotive battery is an electrochemical device that provides for and stores electrical energy.
- Electrical energy is produced in the battery by the chemical reaction that occurs between two dissimilar plates that are immersed in an electrolyte solution.
- The lubrication system provides an oil film to prevent moving parts from coming in direct contact with each other. Oil molecules work as small bearings rolling over each other to eliminate friction. Another function is to act as a shock absorber between the connecting rod and crankshaft.
- Many different types of additives are used in engine oils to formulate a lubricant that will meet all of the demands of today's engines.
- Synthetic oils are superior to conventional oils because they provide better protection from engine wear.
- Oil is rated by two organizations: the Society of Automotive Engineers (SAE) and the American Petroleum Institute (API). The Society of Automotive Engineers has standardized oil viscosity ratings, while the API rates oil to classify its service or quality limits.
- Oil filter elements have a pleated paper or fibrous material designed to filter out particles between 20 and 30 microns.
- There are two basic types of oil pumps: rotor and gear. Both types are positive displacement pumps.
- The radiator is a series of tubes and fins that transfer heat from the coolant to the air.
- The radiator cap allows for an increase of pressure within the cooling system, increasing the boiling point of the coolant.
- The water pump is the heart of the engine's cooling system. It forces the coolant through the engine block and into the radiator and heater core.
- Control of engine temperatures is the function of the thermostat. When the coolant is below normal operating temperature, the thermostat is closed, preventing coolant from entering the radiator. When normal operating temperature is obtained, the thermostat opens, allowing the coolant to enter the radiator to be cooled.
- In a reverse flow cooling system, coolant flows from the water pump through the cylinder heads and then through the engine block.
- In some instrument panels, a message center displays specific warning messages to alert the driver regarding dangerous vehicle operating conditions.

TERMS TO KNOW

Armature
Battery terminals
Body control module
Burn time
C-class oil
Cetane number
Coolant recovery system
Cooling system
Detergents
Detonation
Distributor ignition (DI)
Electrolysis
Electronic ignition (EI)
Emission control system
Ethanol
Flywheel
Fuel system
Full-filtration system
Head gasket
Heater core
Ignition system
Inlet check valve
Knock sensor
Lubrication system
Micron
Multi-point fuel injection
Nucleate boiling
Octane
Oil breakdown

SUMMARY

TERMS TO KNOW
(continued)

Oil cooler

Oil pump

Oil galleries

Oil sludge

Oxidation

Oxidation inhibitors

Piezoresistive sensor

Pole shoes

Positive displacement
pump

Prove-out circuit

RON & MON

Radiator

Reverse flow cooling
system

Ring gear

Saddle

S-class oil

Synthetic oil

TEL

Thermistor

Thermostat

Viscosity

Volatility

Warning light

- Modern vehicles use multi-point fuel injection systems that provide precise control of fuel delivery.
- Some engines have electronic distributor-ignition (DI) systems, while most engines manufactured at present have electronic-ignition (EI) systems which do not have a distributor.
- Emission control systems minimize emission of toxic pollutants NO, CO, and NO_x.
- Gasoline is a mixture of many ingredients, such as hydrocarbons and various other chemicals and additives.
- Gasoline has to be very volatile so that it can easily evaporate and readily mix with air to support combustion.
- Ethanol (ethyl alcohol) is a noncorrosive, low toxic level alcohol that is made from renewable biological sources. It behaves similar to gasoline and can be mixed with 15 percent gasoline to form E85, which is used in flex fuel vehicles.
- Detergents are designed to clean fuel systems and components
- Diesel fuel contains significantly more energy than gasoline or ethanol and is rated by its cetane number.

REVIEW QUESTIONS

Short Answer Essays

1. What are the purposes of the engine's lubrication system?

2. Explain the purpose of the Society of Automotive Engineers' (SAE) classifications of oil.

3. Explain the purpose of the American Petroleum Institute's (API) classifications of oil.

4. Describe the two basic types of oil pumps: rotary and gear.

5. What is the purpose of the starting system?

6. Explain the operation of the thermostat.

7. Describe the function of the radiator.

8. Explain the function of the water pump.

9. Explain the purpose of the pressure and vacuum valves used in the radiator cap.

10. Describe the purpose of antifreeze and explain its characteristics.

Fill-in-the-Blanks

1. It is the function of the engine's lubrication system to supply oil to the _____ and _____ locations and to remove heat from them.

2. The Society of Automotive Engineers has standardized oil _____ ratings, while the American Petroleum Institute rates oil to classify its _____ or quality limits.

3. Oil pumps are _____ displacement pumps.

4. A pressure relief valve opens to return oil to the sump if the pressure is _____.

5. If the oil filter becomes plugged, a _____ _____ will open and allow the oil to enter the galleries.

6. Coolant can become slightly acidic because of the minerals and metals in the cooling system. A small _____ may flow between metals through the acid and have a corrosive effect on the metals.

7. For heat to be effectively transferred to the air, there must be a _____ in temperature between the coolant and the air.

8. E85 is composed of _____ percent ethanol and _____ percent gasoline.

9. Diesel contains significantly _____ energy than gasoline or ethanol.

10. The fuel system uses a _____ sensor or a _____ sensor as its primary input for fuel calculations.

Multiple Choice

1. Each of the following is a function of the lubrication system, except:
    A. Provides oil pressure to the crankshaft journals
    B. Cushions the shock between the connecting rod and the crankshaft
    C. Splashes lubricate the cylinder walls
    D. Provides oil pressure to the piston pins

2. *Technician A* says that oil pressure increases as temperature increases.
    Technician B says that oil pressure decreases as engine speed increases.
    Who is correct?
    A. A only
    B. B only
    C. Both A and B
    D. Neither A nor B

3. A likely cause of low oil pressure is:
    A. The oil pressure relief valve sticking open
    B. A defective inlet check valve
    C. The oil filter bypass valve opening
    D. Dirty oil galleries

4. Coolant and water should be mixed in a _____ ratio.
    A. 70/30
    B. 50/50
    C. 40/60
    D. 100/1

5. Each of the following is a function of the radiator cap, except:
    A. Holds pressure on the system to raise the boiling point
    B. Releases excess pressure to prevent damage
    C. Seals the radiator to prevent toxic leaks
    D. Holds a vacuum on the system to maintain a proper seal

6. *Technician A* says that when the thermostat opens coolant flows through the radiator.
    Technician B says that thermostats usually last the life of the engine.
    Who is correct?
    A. A only
    B. B only
    C. Both A and B
    D. Neither A nor B

7. *Technician A* says that electric cooling fans reduce the load on the engine, saving horsepower.
    Technician B says that a clutch-type fan slips on the pump hub when the engine is hot.
    Who is correct?
    A. A only
    B. B only
    C. Both A and B
    D. Neither A nor B

8. Each of the following is an input to the PCM for fuel delivery, except:
    A. Engine coolant temperature
    B. Engine speed
    C. Exhaust oxygen content
    D. Thermostat position

9. A _____ ignition system uses one coil to fire two cylinders.
    A. DI
    B. Coil-on-plug
    C. Distributor
    D. Waste spark

10. Each of the following is a toxic pollutant created by the automobile, except:
    A. NO_x
    B. CO
    C. N_2
    D. HC

Chapter 5

ENGINE MATERIALS, FASTENERS, GASKETS, AND SEALS

UPON COMPLETION AND REVIEW OF THIS CHAPTER, YOU SHOULD BE ABLE TO:

- Describe and identify the various types of materials used in engine construction, including iron, steel, aluminum, plastics, ceramics, and composites.

- Explain common usages of aluminum alloys in engine construction.

- Describe the different manufacturing processes, including casting, forging, machining, stamping, and powdered metal.

- Describe various treatment methods, including heat treating, tempering, annealing, case hardening, and shot peening.

- Describe the methods used to locate cracks in castings.

- Properly identify, inspect, and select the correct fasteners required to assemble engine components to the block.

- Explain the purposes of engine gaskets.

- Explain the purpose of engine seals.

- Explain the requirements and construction of a head gasket.

- Describe different gasket materials.

- Explain the proper application of room-temperature vulcanizing (RTV) sealant.

- Explain the proper application of anaerobic sealant.

INTRODUCTION

Gaskets are used to prevent engine fluids or pressure from leaking between two stationary parts.

This chapter covers the materials and machining processes used to construct the components of the engine. Today's technician should be able to identify these characteristics properly in order to perform repairs or machining operations on the components.

The mating surfaces of the engine require proper sealing to prevent leakage and for proper engine performance. **Gaskets** are used to prevent gas, coolant, oil, or pressure from escaping between two stationary parts. In addition, gaskets are used as spacers, shims, wear indicators, and vibration dampers. In recent years, many gaskets have been replaced with the use of sealants. The sealant is less expensive and easy to apply. **Seals** are used to prevent leakage of fluids around a rotating part.

Seals are used to prevent leakage of fluids around a rotating part.

ENGINE MATERIALS

Today's engines are constructed with the use of several different types of materials. The same engine can have iron, steel, aluminum, plastic, ceramic, and composite components. Although a complete understanding of **metallurgy** is not a prerequisite for performing engine repairs, today's technician is challenged to perform repairs requiring the proper handling, machining, and service of these materials.

Metallurgy is the science of extracting metals from their ores and refining them for various uses.

Metals are divided into two basic groups: **ferrous** and **nonferrous**. All metals have a grain structure that can be seen if the metal is fractured (Figure 5-1). Grain size and position determine the strength and other characteristics of the metal.

FIGURE 5-1 The properties and strength of a metal can be seen in its grain.

Iron

Iron is the most common metal used to produce engine components. It comes from iron ore, which is retrieved from the earth. The ore is heated in blast furnaces to burn off impurities. **Coke** is used to fuel the furnaces, resulting in some of the carbon from the coke being deposited into the iron. When the liquid iron is poured from the coke furnace, the resultant billet (called pig iron) will have about a 5 percent carbon content. This makes the iron brittle.

Cast Iron

Once the iron cools, it can be shipped for several uses. When shipped to an engine manufacturing plant, the pig iron is remelted and poured into a cast or mold. The resultant form, when the iron cools, is referred to as cast iron. During the process of remelting the iron, the amount of carbon content can be controlled, **alloys** can be added, or special heat treatments can be performed to achieve the desired strength and characteristics of the cast iron. The most common type of cast iron used in automotive engine construction is **gray cast**. Gray cast is easy to cast and machine. It also absorbs vibrations and resists corrosion.

Some engine applications require additional strength above that provided by gray cast. **Nodular iron** is used in some engines for the construction of crankshafts, camshafts, and flywheels. Nodular iron contains between 2 and 2.65 percent carbon, with small amounts of magnesium and other additives. It is also heat treated, causing the carbon to form as small balls or nodules. The result is a cast iron with reduced brittleness.

Steel

Steel is produced by heating the iron at a controlled temperature to burn off most of the carbon, phosphorous, sulfur, silicon, and manganese. As the process continues, the correct amount of carbon is then readded. This type of **steel** is referred to as **carbon steel**.

Low carbon steel (carbon content between 0.05 and 0.30 percent) can be easily bent and formed. As the carbon content of the steel increases, so does its resistance to denting and penetration. However, the higher the carbon content, the more brittle the steel. Medium carbon steel (carbon content between 0.30 and 0.60 percent) is used for connecting rods, crankshafts, and camshafts in some engines. High carbon steel (carbon content between 0.70 and 1.70 percent) has limited use in engine applications but is used to make drill bits, files, hammers, and other tools.

Ferrous metals contain iron. Cast iron and steel are examples of ferrous metals. These metals will attract a magnet.

Nonferrous metals contain no iron. Aluminum, magnesium, and titanium are examples of nonferrous metals. These metals will not attract a magnet.

Coke is a very hot-burning fuel formed when coal is heated in the absence of air. Coal becomes coke above 1,022°F (550°C). It is produced in special coke furnaces.

Alloys are mixtures of two or more metals. For example, brass is an alloy of copper and zinc.

Iron containing very low carbon (between 0.05 and 1.7 percent) is called **steel**.

Carbon steel has a specific carbon content.

The quality of steel can be improved through the addition of alloys. Some of the most common alloys include nickel, molybdenum, tungsten, vanadium, silicon, manganese, and chromium. Steel with 11 to 26 percent content of chromium is referred to as **stainless steel**. Nickel is often added to make the steel able to withstand sudden shock loads. If increased **tensile strength** is required, vanadium can be added.

The addition of chromium and molybdenum (sometimes called chrome-moly) produces a very strong steel. This alloy is often used to make crankshafts in high performance engines. Chrome-moly is also used in the construction of tube frames used in racing cars.

Magnesium

Magnesium is a very lightweight metal, about 2/3 the weight of aluminum. Pure magnesium is expensive and has a low tensile strength; however, it can be alloyed with other metals (such as aluminum). Magnesium alloy is used in some engine valve, covers, or in the case of the Porsche 911, the engine block.

Aluminum

Aluminum is a silver-white ductile metallic element. It is the most abundant metal in the earth's crust. There is no pure aluminum found in the earth though. It is always mixed with other elements. The aluminum we use comes from an ore called bauxite, containing 50 percent alumina (aluminum oxide). The bauxite is crushed and then ground into a fine powder. Then the impurities are separated from the powdered ore by mixing it with a hot caustic soda solution. It is then pumped into large pressure tanks called digesters. The temperature in the tanks, which is maintained at 300°F (149°C), dissolves the alumina, but the impurities remain solid. The impurities can then be filtered out. The remaining liquid is pumped into precipitator tanks, where it is allowed to cool slowly. The alumina comes out of the liquid as crystals when it cools. The crystals are then heated until white hot to remove any water. The result is a dry, white powder. To complete the process of turning alumina into aluminum, it is dissolved in a substance called cryolite. An electric current is passed through the pot containing the mixture from carbon anodes that hang in the liquid from overhead bars. The current flows through the liquid and causes the liquid to break up. Pure aluminum then falls to the bottom of the pot.

Although aluminum is not as strong as steel, it weighs about one-third less than steel. Pure aluminum is not strong enough to be used for most engine applications, but the attractiveness of weight reduction has led to the development of many types of aluminum alloys that are proving acceptable. Today, most pistons used in automotive engines are constructed of aluminum alloys.

The light weight of aluminum pistons allows for higher engine speed and increased engine responsiveness. The lighter pistons also reduce the amount of load on the connecting rods and crankshaft (Figure 5-2). A few racing applications use aluminum connecting rods to reduce the amount of reciprocating weight. This allows the engine to accelerate faster, operate smoother, last longer, and have an increased maximum speed (also known as the red line on the tachometer).

Many manufacturers are now using aluminum alloys for engine block construction. Aluminum is too soft to be used by itself and is unable to withstand the wear caused by the piston rings traveling in the cylinder. Most aluminum blocks of today's engines are either fitted with special cylinder liners (made of iron, steel, or composite) or special casting.

Most manufacturers use an aluminum alloy to construct cylinder heads, engine mounts, water pump housings, air-conditioning generator housings, intake manifolds, and valve covers.

FIGURE 5-2 This is a lightweight piston used in a racing engine.

Titanium

Titanium alloys offer the benefits of light weight and high strength. Titanium alloys are almost as strong as steel, at about half the weight. The most common use for titanium is in racing applications for the construction of connecting rods and valves. The high cost of titanium alloys, coupled with their difficulty in being welded and machined, have limited titanium use in production vehicles. However, the NSX from Acura was introduced in 1990 with titanium connecting rods.

Plastics

Plastic is a substance made from petroleum by a special method that joins atoms together to make long chains of atoms. The word "plastic" means something that can be pressed into a new shape. After pure plastic is made, it can be improved by adding other materials, to give it additional strength, stiffness, or density.

Plastics were first introduced in the automotive engine when American Motors used plastic to make valve covers on their 258 CID 6-cylinder engine. In subsequent years, other manufacturers have used plastics to help reduce the overall weight of the engine. Today some manufacturers are using forms of plastics to make intake manifolds, pulleys, and valve timing train covers. Plastic intake manifolds transfer less heat to the intake air compared to cast iron or aluminum intake manifolds (Figure 5-3). This action reduces intake air temperature and engine detonation.

Composites

Manufacturers are experimenting with increased use of **composite** materials. These man-made materials are showing great promise in engine applications. Graphite-reinforced fiber or nylon materials have been successfully used as connecting rods, pushrods, rocker arms, intake manifolds, and cylinder liners.

Ceramics

Ceramics also show good promise in automotive engine uses. They are lightweight, provide good frictional reduction, are heat resistant, isolate sound and vibrations, and are brittle. Engine components constructed from ceramics require special handling due to

Plastics are made from petroleum and are used in several engine components because of their light weight and heat transfer properties.

Composites are man-made materials using two or more different components tightly bound together. The result is a material that has characteristics that neither component possesses on its own.

Ceramic is a combination of nonmetallic powdered materials fired in special kilns. The end product is a new product.

FIGURE 5-3 A plastic intake manifold used on a late-model engine.

being so brittle. Ceramic components that are in use today include the compressor and turbine wheels of some General Motors' turbochargers and the rocker arm pads on some Mitsubishi engines. In addition, the aftermarket parts suppliers are making ceramic valves, valve seats, valve spring retainers, and wrist pins available.

Typical uses of ceramics today also include the rotor of some turbochargers, the liner for exhaust ports, intake and exhaust valves, valve seats, and piston pins.

MANUFACTURING PROCESSES

Not only do different types of irons, steels, and aluminum have different qualities, the manufacturing process also determines their strength and properties. The most common processes are casting, forging, machining, and stamping. In addition to the manufacturing process, the material may be specially treated to increase strength.

Casting

As discussed earlier, casting requires heating the metal to a liquid, then pouring it into a mold. When the metal cools, it returns to a solid state. The molds can be made of foundry sand. After the metal cools, the sand is broken away to expose the part.

The mold is made by packing the foundry sand around a wooden pattern. The pattern is removed prior to the liquid metal being poured in. This type of casting usually leaves a rough, grainy appearance.

The lost foam casting method incorporates the use of a polystyrene foam pattern (Figure 5-4). This pattern is left in the mold when the molten metal is poured in. The heat of the molten metal vaporizes the foam. This method leaves the casting with a surface texture similar to styrofoam.

Cast iron is commonly used for engine block, camshaft, connecting rod, and crankshaft construction.

Aluminum is usually cast into permanent, reusable molds. The molten aluminum alloys are forced into the mold under pressure or through the use of centrifugal force. The pressure is used to help eliminate any air pockets that may affect the machining process.

Cast aluminum is commonly used for cylinder head, bell housing, piston, and intake manifold construction.

FIGURE 5-4 This engine is formed by lost foam casting and has a surface appearance similar to styrofoam.

Forging

Forging heats the metal to a state in which it can be worked and reshaped. It is not heated until it becomes a liquid, as it is when casting. When the metal is hot enough to be worked, a forging die is forced onto the metal under great pressure. The metal then assumes the shape of the forging die. If the metal must be worked into complex shapes, several forging dies may be used. Each die will alter the shape until the desired results are obtained.

When the forging die is closed around the hot metal, some of the metal is forced into the parting lines of the die. This excess metal is called flash and is usually removed after the part is removed from the die.

Forged parts are very strong because the high pressure used to force the die onto the component causes the grain structure to follow the shape of the part. The pressure also causes the molecules of the metal to become very compact and tightly bound.

Most engines equipped with turbochargers use forged pistons and some use forged connecting rods and crankshafts. High performance engines often use a forged crankshaft and camshaft to increase the endurance of the engine.

Machining

To construct a component from a piece of steel or iron billet is very time-consuming and very expensive. Despite this, many engine components are constructed in this manner. Machined components are stronger than cast, but not as strong as forged. Machined components used in today's engines include rocker arms, piston pins, lifters, and followers. High performance engines use machined steel crankshafts and camshafts.

Stamping

Some engine components of simple design, and not requiring much thickness, can be stamped out of a sheet of metal. The metal is not heated during this process. Depending on the final design of the component, it can be stamped on a press punch. This process is

Powder metallurgy is the manufacture of metal parts by compacting and sintering metal powders. **Sintering** is done by heating the metal to a temperature below its melting point in a controlled atmosphere. The metal is then pressed to increase its density.

In **tempering**, metal is heated to a specific temperature to reduce the brittleness of hardened carbon steel.

Annealing is a heat-treatment process to reduce metal hardness or brittleness, relieve stresses, improve machinability, or facilitate cold working of the metal.

Case hardening is a heating and cooling process for hardening metal. Because the case hardening is only on the surface of the component, it is usually removed during machining procedures performed at the time the engine is being rebuilt. It is not necessary to replace the case hardening in most applications.

used if the piece is to remain flat. The punch is in the shape of the piece to be cut. The press forces the punch through the sheet of metal, much like a cookie cutter through dough. If the finished product must have some bends to it, it is bent by the stamping process, then trimmed to final specifications. Common components made from stamped steel include oil pans, valve covers, timing train covers, and heat shields.

Powder Metallurgy

Metal powder may be derived by cooling melted metal very quickly. Other methods include reducing the metal oxide, electrolysis, and crushing. The powder can then be blended with other metals to produce an alloy. The powder is then poured into a die and compacted by a cold press. Next it is **sintered** to make the powder bond together.

This process is currently being used to construct some connecting rods and valve seats. After the connecting rod is removed from the die, it is forged into final shape, then shot peened to increase its strength. This process provides a strong component at a light weight. It also eliminates many of the machining processes forged or machined connecting rods must undergo.

Treatment Methods

To obtain the desired result of the component, engine manufacturers may have the component treated by heat, chemical, or shot-peening processes. Heating the metal will change its grain structure. By heat treating a metal component, its properties can be altered. To heat treat a metal properly, it is heated to a desired temperature, depending on the metal used, and the desired results. Once the temperature is reached, it is maintained for a specific amount of time. The last step is to cool the metal at a controlled rate.

To harden a carbon steel, it is heated to 1,400°F (760°C), then quickly cooled. Heating the steel causes its grain structure to become finer. When it is cooled very fast, the grain structure does not have time to change and remains very fine. The harder the steel becomes, the more brittle it becomes.

To counteract the effects of hardening steel, **tempering** of the metal may be the next step. The steel is heated to a temperature between 300 and 1,100°F (150 to 600°C), then cooled slowly. The higher the temperature used to temper the steel, the more hardening is lost, yet the toughness of the metal is increased.

Hard metals must be softened in order to be machined. This process is called **annealing**. Annealing is much like tempering except the temperatures are increased to above 1,000°F (550°C). The metal is then allowed to cool at a slower rate than used for tempering. This makes the grain structure of the metal coarse. If desired, the metal can be hardened and tempered again after the machining process is completed.

To help protect the shell or outer surfaces of a component, the manufacturer may require it to be **case hardened**. This process does not alter the core structure of the metal. Case hardening is performed after all machining processes are completed. Case hardening of crankshafts protects the journals from wear and fractures. Most manufacturers now use a process called ion nitriding to case harden their components. The component is placed into a pressurized chamber filled with hydrogen and nitrogen gases. An electrical current is then applied through the component. This changes the molecular structure and allows the induction of the gases onto the surface area of the component.

Another form of metal treatment is **shot peening**. Shot peening is performed by blasting the component's surface with steel or glass shot. When the balls hit the component, a small dent is made. Thousands of these dents are applied to the component's surface. The dents overlap each other and compress the component's surface. This process of prestressing means any tension forces applied to the part must overcome the compression forces before the part will crack.

Cracks

Cracks are the result of stress in the casting. A crack can cause fluid or compression leakage. When performing engine service requiring the removal of the intake manifold, exhaust manifold, cylinder heads, and so on, check the component for cracks. The following is a list of some of the most common causes for this stress:

1. Fatigue
2. Excessive flexing
3. Impact damage
4. Extreme temperature changes in a very short time
5. Freezing of the engine coolant
6. Excessive overheating
7. Detonation
8. Defects during the casting process

> **AUTHOR'S NOTE:** It has been my experience that one of the common causes of cracked cylinder heads is adding cold water to the cooling system on an overheated engine. This action subjects the cylinder head to a very sudden change in temperature, resulting in a cracked head. If the engine overheats, do not loosen the radiator cap, and do not add water and/or antifreeze to the cooling system until the engine cools down.

Detecting Cracks

Many cracks are detected by a thorough visual inspection. However, very small stress cracks may not be detected in this manner. There are several different methods of crack detection. Following is a sample of the four most common:

1. *Magnetic particle inspection (MPI).* Uses an electromagnet to create a magnetic field in the casting (Figure 5-5). Because all magnets have north and south poles, the magnetic field runs from one pole to the other through the casting. A crack in the casting causes a break in the field, creating opposite poles (Figure 5-6). The magnetic powder is attracted to this area. Magnetic particle inspection cannot be used to detect cracks in aluminum parts because they cannot be magnetized.

Magnaflux is another name for magnetic particle inspection.

FIGURE 5-5 This damaged racing crankshaft is being magnafluxed to check for cracks.

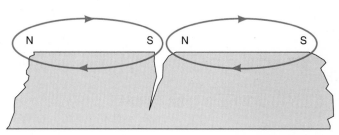

FIGURE 5-6 A crack causes a break in the magnetic field, creating opposite magnetic poles.

2. *Magnetic fluorescent.* This method uses a fluorescent paste dripped onto the casting. The casting is then placed in a magnetic field and observed under a black light. Cracks are visible by white, gray, or yellow streaks.

3. *Penetrant dye.* Using dye penetrant is a three-step process. First, the special dye is sprayed onto the casting surface and allowed to dry. The excess dye is then wiped away. Next, a special remover is sprayed over the surface and the casting is rinsed with water. Third, the developer is sprayed onto the casting. As it dries, any dye left in the cracks seeps through the developer. The crack shows as a red line against a white background.

4. *Pressurizing the block or head.* The casting can be pressurized with air or water. When air is used, the casting is attached to a special fixture and 40 to 60 psi of air pressure is applied. A soapy solution is then sprayed onto the casting so the technician can look for bubbles, indicating a leak. Water pressure testing uses hot water to cause casting expansion. If the casting is leaking, water will be found on the outside of it.

FASTENERS

There are many different types of fasteners used throughout the engine. The most common are threaded fasteners, including bolts, studs, screws, and nuts. These fasteners must be inspected for thread damage, fillet damage, and stretch before they can be reused (Figure 5-7). Some bolts stretch in different areas than that shown in Figure 5-7. Some manufacturers require that you measure the diameter of the bolt's shank to determine if it can be reused. Other manufacturers may call for a measurement of the bolt length to inspect for excessive bolt stretch. In addition, many threaded fasteners are not designed for reuse; the service manual should be referenced for the manufacturer's recommendations. If a threaded fastener requires replacement, there are some concerns the technician must be aware of:

- Select a fastener of the same diameter, thread pitch, strength, and length as the original.
- All bolts in the same connection must be of the same grade.
- Use nut grades that match their respective bolts.
- Use the correct washers and pins as originally equipped.
- Torque the fasteners to the specified value.
- Use torque-to-yield bolts where specified.

Standard threaded fasteners used in automotive applications are classified by the Unified National Series using four basic categories:

1. Unified National Coarse (UNC or NC)
2. Unified National Fine (UNF or NF)
3. Unified National Extrafine (UNEF or NEF)
4. Unified National Pipe Thread (UNPT or NPT)

In recent years, the automotive industry has switched to the use of metric fasteners. Metric threads are classified as coarse or fine, as denoted by an SI or ISO lettering.

The most common type of threaded fastener used on the engine is the bolt (Figure 5-8). To understand proper selection of a fastener, terminology must be defined (Figure 5-9). The head of the bolt is used to torque the fastener. Several head designs are used, including hex, torx, slot, and spline. **Bolt diameter** is the measure across the threaded area or shank. The **pitch** (used in the English system) is the number of threads per inch. In the metric system, thread pitch is a measure of the distance (in millimeters) between two adjacent threads. **Bolt length** is the distance from the bottom of the head to the end of the bolt. The **grade** of the bolt denotes its strength and is used to designate the amount of stress the bolt can withstand. The grade of the bolt depends upon the material it is constructed from, bolt diameter,

Unified National Fine-thread or Unified National Course-thread bolts may be referred to as fine-thread or coarse-thread bolts.

Bolt diameter indicates the diameter of an imaginary line running through the center of the bolt threaded area.

Pitch is the number of threads per inch on a bolt in the USC system.

Bolt length indicates the distance from the underside of the bolt head to the tip of the bolt threaded end.

Unstretched bolt

Threads are straight on line

Stretched bolt

Threads are not straight on line

FIGURE 5-7 Check all bolts for stretch and other damage before reusing them.

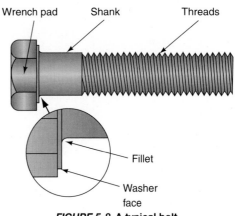

Wrench pad · Shank · Threads

Fillet

Washer face

FIGURE 5-8 A typical bolt.

H = HEAD
G = GRADE MARKETING (BOLT STRENGTH)
L = LENGTH (INCHES)
T = THREAD PITCH (THREAD/INCH)
D = NORMAL DIAMETER (INCHES)

A

H = HEAD
G = PROPERTY CLASS (BOLT STRENGTH)
L = LENGTH (MILLIMETERS)
T = THREAD PITCH (THREAD/MILLIMETER)
D = NORMAL DIAMETER (MILLIMETER)

B

FIGURE 5-9 Bolt terminology.

> **Thread depth** is the height of the thread from its base to the top of its peak.

> **Grade marks** are radial lines on the bolt head.

and **thread depth. Grade marks** are placed on the top of the head (in the English system) to identify the bolt's strength (Figure 5-10). In the metric system, the strength of the bolt is identified by a property class number on the head (Figure 5-11). The larger the number, the greater the tensile strength.

SAE grade markings					
Definition	No lines: unmarked indeterminate quality SAE grades 0-1-2	3 lines: common commercial quality Automotive and AN bolts SAE grade 5	4 lines: medium commercial quality Automotive and AN bolts SAE grade 6	5 lines: rarely used SAE grade 7	6 lines: best commercial quality NAS and aircraft screws SAE grade 8
Material	Low carbon steel	Med. carbon steel tempered	Med. carbon steel quenched and tempered	Med. carbon alloy steel	Med. carbon alloy steel quenched and tempered
Tensile strength	65,000 psi	120,000 psi	140,000 psi	140,000 psi	150,000 psi

FIGURE 5-10 Bolt grade identification marks.

FIGURE 5-11 Property class numbers.

Like bolts, nuts are graded according to their tensile strength (Figure 5-12). As discussed earlier, the nut grade must be matched to the bolt grade. The strength of the connection is only as strong as the lowest grade used; for example, if a grade 8 bolt is used with a grade 5 nut, the connection is only a grade 5.

Proper fastener torque is important to prevent thread damage and to provide the correct clamping forces. The service manual provides the manufacturer's recommended torque value and tightening sequence for most fasteners used in the engine. The amount of torque a fastener can withstand is based on its tensile strength (Figure 5-13). In order to obtain proper torque, the fastener's threads must be cleaned and may require light lubrication.

Torque-to-Yield

Modern engines are designed with very close tolerances. These tolerances require an equal amount of clamping forces at mating surfaces. Normal head bolt torque values have a calculated 25 percent safety factor; that is, they are torqued to only 75 percent of the bolt's maximum proof load (Figure 5-14). Using the chart, it can be seen that a small difference between torque values at the bolt head can result in a large difference in clamping forces. Because torque is actually force used to turn a fastener against friction,

Inch system		Metric system	
Grade	Identification	Class	Identification
Hex nut grade 5	3 dots	Hex nut property grade 9	Arabic 9
Hex nut grade 8	6 dots	Hex nut property grade 10	Arabic 10
Increasing dots represent increasing strength.		Can also have blue finish or paint dab on hex flat. Increasing numbers represent increasing strength.	

FIGURE 5-12 Nut grade markings.

STANDARD BOLT AND NUT TORQUE SPECIFICATIONS					
Size Nut or Bolt	Torque (foot-pounds)	Size Nut or Bolt	Torque (foot-pounds)	Size Nut or Bolt	Torque (foot-pounds)
1/4–20	7–9	7/16–20	57–61	3/4–10	240–250
1/4–28	8–10	1/2–13	71–75	3/4–16	290–300
5/16–18	13–17	1/2–20	83–93	7/8–9	410–420
5/16–24	15–19	9/16–12	90–100	7/8–14	475–485
3/8–16	30–35	9/16–18	107–117	1–8	580–590
3/8–24	35–39	5/8–11	137–147	1–14	685–695
7/16–14	46–50	5/8–18	168–178		

FIGURE 5-13 Standard bolt and nut torque specifications.

the actual clamping forces can vary even at the same torque value. Up to about 25 ft-lb (35 Nm) of torque, the clamping force is fairly constant; however, above this point, variation of actual clamping forces at the same torque value can be as high as 200 percent. This is due to variations in thread conditions or dirt and oil in some threads. Up to 90 percent of the torque is used up by friction, leaving 10 percent for the actual clamping. The result could be that some bolts have to provide more clamping force than others, distorting the cylinder bores.

To compensate and correct for these factors, many manufacturers use torque-to-yield bolts. The yield point of identical bolts does not vary much. A bolt that has been torqued to its yield point can be rotated an additional amount without any increase in clamping force. When a set of **torque-to-yield** fasteners is used, the torque is actually set to a point above the yield point of the bolt. This assures the set of fasteners will have an even clamping force.

Manufacturers vary on specifications and procedures for securing torque-to-yield bolts. Always refer to the service manual for exact procedures. In most instances, a torque wrench

Torque-to-yield is a stretch-type bolt that must be tightened to a specific torque and then rotated a certain number of degrees.

Grade of Bolt	SD	BC	10K	12K		
Minimum Tensile Strength	71,160 psi	113,800 psi	142,000 psi	170,674 psi		
Grade Marking on Head	SD	BC	10K	12K	Socket or Wrench Size	
Metric	TORQUE (In Foot Pounds)				Metric	
Bolt Diameter					Bolt Head	Nut
6mm	5	6	8	10	10mm	10mm
8mm	10	16	22	27	14mm	14mm
10mm	19	31	40	49	17mm	17mm
12mm	34	54	70	86	19mm	19mm
14mm	55	89	117	137	22mm	22mm
16mm	83	132	175	208	24mm	24mm
18mm	111	182	236	283	27mm	27mm
22mm	182	284	394	464	32mm	32mm
24mm	261	419	570	689	36mm	36mm

FIGURE 5-14 Metric bolt and nut torque specifications.

is first used to tighten the bolts to their yield point. Next, the bolt is turned an additional amount as specified in the service manual.

The graph in Figure 5-15 indicates that a bolt can be elongated considerably at its yield point before it reaches its failure point. Also notice, the clamp load is consistent between the proof load and the failure point of the bolt. Bolts that are torqued to their yield points have been stretched beyond their elastic limit and require replacement whenever they are removed or loosened.

CAUTION:

If a torque procedure includes rotation of the bolt a certain number of degrees after the final torque, it is a torque-to-yield bolt. Most manufacturers require replacement of torque-to-yield bolts after one use, although some manufacturers allow a specified number of reuses. Always check the service information specific to the vehicle you are working on.

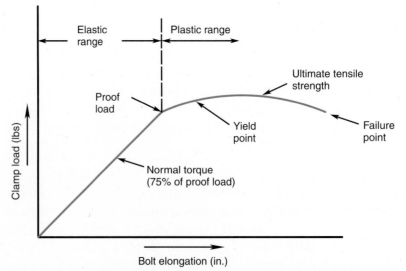

FIGURE 5-15 Relationship between proper clamp load and bolt failure.

GASKETS, SEALS, SEALANTS, AND ADHESIVES

Gaskets

There are three basic engine gasket classifications (Figure 5-16):

- Hard gaskets. These include gaskets made of metals or metals covering a layer of clay/fiber compound. Head gaskets, exhaust manifold, intake manifold, and exhaust gas recirculation (EGR) valve gaskets are examples of this type of gasket.
- Soft gaskets. These gaskets are made of rubber, cork, paper, and rubber-covered metal. Common usages include valve covers, oil pans, thermostat housings, water pumps, timing covers, and some intake manifolds.
- Sealants. These are usually a type of liquid material used to form gaskets. Silicones and anaerobics are examples of sealants.

Cylinder Head Gaskets. Perhaps the cylinder head gasket is subjected to the greatest demands. It must be capable of sealing combustion pressures up to 2,700 psi (18,616 kPa) and withstanding temperatures over 2,000°F (1,100°C). In addition, it must be able to seal coolant and oil under pressure. To complicate the matter further, the head gasket must be able to perform these functions while accomodating a shearing action resulting from expansion rates of the metals. Shearing results from the difference in thermal expansion between the cylinder head and block. As the two metals expand and contract at different levels, a scrubbing stress is created (Figure 5-17). The head gasket cannot move during these times.

The head gasket consists of a core, facing, and coating. The core is usually made from solid or clinched steel. The facing is usually constructed of graphite and rubber fiber. These materials allow sufficient compression to conform to minor surface irregularities. The coating may be Teflon or silicone based and work with the facing to seal minor surface irregularities and resist shearing. Some head gaskets use and elastomeric sealing bead to increase clamping forces around fluid passages (Figure 5-18).

One of the most recent developments in head gaskets is rubber-coated embossed (RCE) steel shim gaskets. A rubber coating is bonded to a steel shim gasket. The coating protects the shim from corrosive elements in the cooling and lubrication systems and provides good friction reduction. Because of their construction, RCE gaskets are also referred to as multi-layer steel (MLS) gaskets. There are many advantages to the multilayer design. First, it has a uniform thickness that prevents bore distortion during cylinder head installation. Also, the MLS gasket is very resilient in that, once compressed, it "bounces" back for a good seal.

Most cylinder head gaskets have a metal **fire ring** around the combustion chamber opening (Figure 5-19). The fire ring protects the gasket material from the high temperature to which it is exposed. Also, the fire ring increases the gasket thickness around the cylinder bore so that it uses up to 75 percent of the clamping force to form a tight seal against combustion pressure losses.

Intake Manifold Gaskets. A leak in the connection of the intake manifold and cylinder head creates a vacuum leak that allows unmeasured air into the engine. This can cause the mixture entering the combustion chamber to be too lean (not enough fuel), resulting in rough idle and/or detonation. The **intake manifold gasket** is designed to provide a good seal under the changing temperatures it will be subject to (Figure 5-20).

Many intake manifold gaskets are constructed of a solid or perforated steel core with a fiber facing. Some manufacturers use gaskets made from a rubber silicone bonded to a steel or high temperature plastic. On this style, the silicone provides the actual sealing. Some V-type engines use a **valley pan** style of intake manifold gasket. In addition, V-type engines use end strip seals to seal the connection between the ends of the manifold and the block. **Strip seals** are generally made from molded rubber or cork-rubber. Many

A **fire ring** is a metal ring surrounding the cylinder opening in a head gasket.

The **intake manifold gasket** fits between the manifold and cylinder head to seal the air-fuel mixture or intake air.

Valley pans prevent the formation of deposits on the underside of the intake manifold.

A **strip seal** provides a seal between the flat surfaces on the front and back of the engine block and the intake manifold.

Intake manifold

Gasket

Gasket

Gasket

Gasket

Delivery pipe

Gasket

Cold start
injector pipe

Adjusting shim

Valve filter

Valve keepers

Valve spring retainer

Valve spring

Valve stem oil seal

Spring seat

Valve guide bushing

Valve

Cylinder head cover

Gasket

Spark plug tube gasket

Camshaft
subgear

Snap ring

Exhaust camshaft

Camshaft
bearing cap

Intake camshaft

Water inlet housing

Distributor

Wave ring

Camshaft
gear spring

Oil seal

Cylinder head

Fan belt
adjusting bar

Water outer pipe

Gasket

Manifold heat
insulator (upper)

Gasket

Exhaust manifold

Manifold heat insulator
(lower)

FIGURE 5-16 Seals and gaskets are located at critical seal points within the engine.

fuel-injected engines use a plenum that requires a gasket between it and the intake manifold (Figure 5-21).

With the advancement of plastics and composites, some manufacturers are designing intake manifolds from these products. Some of these do not use the typical intake manifold gasket; instead, they use a series of O-rings to seal each port (Figure 5-22).

FIGURE 5-17 The coating on the gasket material allows for different expansion rates between metals and still provides a good seal without distorting the gasket.

FIGURE 5-18 Sealing beads work to prevent fluid leakage.

FIGURE 5-19 This head gasket has a sealing bead and fire rings around the cylinders.

FIGURE 5-20 Intake manifold gasket location.

Sealer

Intake manifold gaskets

Intake plenum

Gasket

Engine manifold

FIGURE 5-21 Intake plenum gasket location.

Intake manifold

Captive O-ring gaskets

FIGURE 5-22 Some intake manifolds use captive O-rings to seal against the cylinder head.

Exhaust Manifold Gaskets. The **exhaust manifold gasket** must prevent leakage under extreme temperatures (Figure 5-23). Leaks in this connection disrupt the flow of exhaust gases and can result in poor engine performance. In addition, exhaust leaks can lead to burned valves and objectionable noises. The use of exhaust manifold gaskets by engine manufacturers has declined in recent years with improved machining processes and better materials; however, when the exhaust manifold is removed, it may become warped and require the use of a gasket. Most technicians opt to install an exhaust manifold gasket even if the engine was not originally equipped with one.

Valve Cover Gaskets. **Valve cover gaskets** are common locations for external oil leakage. This is largely due to the wide spacing between the attaching bolts. The wide spacing allows the stamped steel cover to distort easily and provides less clamping forces (Figure 5-24).

The **exhaust manifold gasket** seals the connection between the cylinder head and the exhaust manifold.

FIGURE 5-23 Exhaust manifold gasket location.

FIGURE 5-24 The valve cover gasket can easily be deformed.

To seal properly, the valve cover gasket must be highly compressible yet have good torque retention. To perform these tasks, there are a variety of materials used to construct valve cover gaskets. Some of the most common are synthetic rubber, cork-rubber, and molded rubber. Synthetic rubber gaskets seal by deforming instead of compressing. The synthetic rubber has a tendency to "remember" its original shape and will attempt to return to it. When the valve cover is tightened against the gasket, the gasket deforms. The gasket attempts to return to its original shape and pushes back against the valve cover, creating a tight seal. One drawback to this gasket is it can be difficult to install. Cork-rubber gaskets compress very well and provide good sealing. Molded rubber gaskets are the easiest to install and provide the best sealing. Sealers or adhesives should not be used on molded rubber gaskets. In addition to the valve cover gasket, many overhead camshaft engines use molded rubber semicircular plugs (Figure 5-25).

Oil Pan Gaskets. **Oil pan gaskets** can be multipiece or a single-unit molded gasket, depending on the crankcase design. The most common type of multipiece gasket uses two side pieces made of cork and rubber and two end pieces made of synthetic rubber (Figure 5-26). Many modern engines use a single-unit molded gasket (Figure 5-27).

FIGURE 5-25 Camshaft plugs work with the valve cover gasket to seal oil.

FIGURE 5-26 Multipiece oil pan gasket.

FIGURE 5-27 A one-piece oil pan gasket.

Miscellaneous Gaskets. Many additional gaskets are used on the engine. Each of these must be replaced during the assembling process. Additional gasket applications include:

- Exhaust gas recirculation (EGR) valve
- Water pump
- Air cleaner
- Fuel injector mounting
- Timing cover
- Exhaust pipe

Each of these gaskets is designed for a particular purpose. The materials used must perform many of the tasks previously discussed, but under unique circumstances. Always refer to the service manual for the proper torque and sequence when installing these gaskets.

Seals

The most common seals in the engine are the **front main seal** (also called the timing cover seal) and the **rear main seal**. The front main seal seals around the harmonic balancer to prevent oil leakage from the front of the crankshaft, while the rear main seal prevents leakage from the rear of the crankshaft. Most modern engines use a type of **lip seal** to perform these functions. The lip seal is generally constructed of butyl rubber or neoprene. The rubber seal is attached to a metal case that is driven into the bore (Figure 5-28). The seal lip will use the pressure of the fluid between the seal and the case to force the lip tight against the shaft. To assist in providing a tighter seal, some lip seals use a garter spring behind the lip. If the seal is installed in a location where the front of the seal many be exposed to dirt, a dust lip may be formed into the seal to deflect dirt away from the outside of the seal. When installing a new lip seal, the lip always faces the hydraulic pressure side.

The front main and oil pump housing oil seals are generally one pipe lip seals with a steel outer ring (Figure 5-29). To aid in sealing during installation, apply silicone sealer to the outer diameter of the metal shell. Neoprene seals should always be lubricated prior to installation. This will prevent seal damage from overheating during initial startup. If the

FIGURE 5-28 The common lip seal has an outer case, rubber sealing lips, and a gator spring. Hydraulic pressure works to cause the lip seal to ride tight against the shaft.

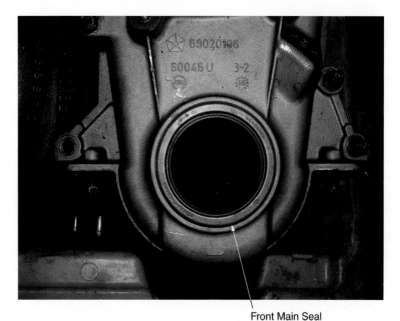

Front Main Seal

FIGURE 5-29 The front main seal fits into the timing cover.

surface of the harmonic balancer is damaged, a service sleeve can be installed to provide a new sealing surface.

There are three common seal designs for rear main seals; rope-type, two-piece molded synthetic rubber, and one-piece rubber with steel ring. The rope-type rear main seal is found on many older engines. These are difficult to replace without removing the crankshaft. They were replaced with molded two-piece seals. Rope-type seals are also two pieces. The lower piece is installed into the bearing cap, while the upper piece is installed into the block.

Adhesives, Sealants, and Form-in-Place Gaskets

There are a number for chemical sealing materials available designed to reduce labor and increase the probability of a good seal. Proper use of these chemical materials is required to assure good results.

To assist in gasket material removal, spray or brush-on gasket remover solvent is available. Adhesives are designed to bond different types of gasket materials in place prior to installation. These would be used to hold the gasket in place during installation of parts that may require a lot of wiggling into place. Some gaskets must be coated with a sealer prior to installation. This is true of a few MLS cylinder head gaskets. Some sealants can damage gasket coatings: be sure to check the instructions for the gasket you are installing.

In the 1980s, many manufacturers began switching to **sealants** and form-in-place gaskets instead of molded gaskets at many joint connections throughout the engine. When applying these materials, care must be taken to assure proper application. Bead size, continuity, and location are factors affecting proper sealing. If the bead is too thin, a lead can result. A bead that is too wide can result in spillover and clog the oil pump or oil galleries. Also, the manufacturer may require different types of sealers at specified locations in the engine. For this reason, always refer to the service manual for the correct type and usage of sealers.

There are two basic types of sealants:

- **Aerobic**
- **Anaerobic**

A BIT OF HISTORY

In the year 2000, sales of light trucks surpassed car sales. The light truck market consists of these vehicles.

1. Minivans—17.9 percent
2. Full-sized vans—5.5 percent
3. Full-sized pickups—26.6 percent
4. Compact pickups—14.7 percent
5. Full-sized sport-utility vehicles (SUVs)—8 percent
6. Compact SUVs—23.3 percent
7. Small SUVs—3.9 percent

Most aerobic sealants are silicone compounds and work well with metals and plastics. The most common type of silicone sealant is **room-temperature vulcanizing (RTV)** compound. This type of sealant forms a rubber seal by absorbing moisture from the air. Before attempting to use RTV, the sealing surfaces must be thoroughly cleaned. RTV begins to set in about 10 minutes, requiring the components to be assembled quickly. Some RTV suppliers use different colors to denote the temperature range the RTV is capable of withstanding. Use the correct RTC for the expected temperature. RTV cannot be used on the exhaust system components.

Anaerobic sealers are used between two machined surfaces. These sealants cure after the components have been assembled. Anaerobic sealants are capable of filling gaps up to 0.030 in. (0.8 mm) and withstand temperature up to 350°F (177°C). Some engines using a one-piece main bearing require the use of special anaerobic sealants that do not cure until a specified temperature is reached. This assures proper maintenance of oil clearances as the main bearing bolts are torqued, without fear of the sealant setting too soon.

SUMMARY

TERMS TO KNOW

Aerobic sealant

Alloys

Anaerobic sealant

Annealing

Bolt diameter

Bolt length

Carbon steel

Case hardening

Ceramics

Coke

Composites

Exhaust manifold

gasket

Ferrous metals

Fire ring

Front main seal

Gaskets

Grade

Grade marks

Gray cast iron

- Metals are divided into two basic groups: ferrous and nonferrous. Ferrous metals are those containing iron. Nonferrous metals contain no iron.
- Alloys are mixtures of two or more metals.
- Iron containing very low carbon (between 0.05 and 1.7 percent) is called steel.
- Composites are man-made materials using two or more different components tightly bound together. The result is a material consisting of characteristics that neither component possesses on its own.
- Powder metallurgy is the manufacture of metal parts by compacting and sintering metal powders.
- Metric and standard bolts are used on today's automobiles.
- Gaskets are used to prevent gas, coolant, oil, or pressures from escaping between two mating parts.
- Seals are used to seal between a stationary part and a moving one.
- Sealants are a liquid material used to form gaskets. Examples of sealants are anaerobics and silicones.
- Cylinder head gaskets must seal cylinder pressures up to 2,700 psi (18,616 kPa) and withstand temperatures over 2,000°F (1,100°C).
- Aerobic sealants require the presence of oxygen to cure. Anaerobic sealants will only cure in the absence of oxygen.

REVIEW QUESTIONS

Short Answer Essays

1. Describe the advantage of aluminum alloy engine components compared to cast iron or steel components.

2. List the methods available to locate cracks in castings.

3. Describe the purpose of torque-to-yield bolts.

4. Explain the bolt grade marking system.

5. Define the following terms used with bolts:
   Head
   Diameter
   Pitch
   Length
   Grade

6. Describe what an anaerobic sealant is.

7. What is the function of a seal?

8. Describe the demands to which a head gasket is subjected.

9. Explain the advantages of rubber-coated embossed or multilayer steel head gaskets.

10. List three different types of valve cover gaskets.

Fill-in-the-Blanks

1. _____ metals are those containing iron. _____metals contain no iron.

2. _____ _____ is the metal's resistance to being pulled apart.

3. _____ is the process of heating the metal to remove its hardness.

4. Most aluminum blocks are fitted with cylinder _____.

5. Bolt diameter is the measure across the _____ area.

6. _____ sealants require the presence of oxygen to cure.

7. When installing a lip seal, the lip always faces the _____ side.

8. Neoprene seals should always be _____ before installation.

9. Head gaskets must be able to seal _____ and _____ under pressure.

10. The most common lip seals used on a typical engine are the _____ _____ and the _____ _____.

Multiple Choice

1. Ceramic engine components have all these characteristics EXCEPT:
   A. Light weight
   B. Heat resistance
   C. Non-brittleness
   D. Ability to isolate sound and vibration

2. Forged engine components have these advantages and applications:
   A. Intake manifolds may be forged.
   B. Crankshafts may be forged.
   C. Rocker arm covers may be forged.
   D. Provide a softer metal that is more easily machined.

3. Engine components may be cracked by:
   A. Extreme temperature changes in a very short time.
   B. Extremely low, uniform temperatures.
   C. An excessive amount of antifreeze in the coolant.
   D. Engine operating temperature below normal.

4. The manufacturing process that yields the strongest finished part is:
   A. Casting
   B. Forging
   C. Machining
   D. Stamping

TERMS TO KNOW
(continued)

Intake manifold gasket

Lip seal

Metallurgy

Nodular iron

Nonferrous metals

Oil pan gaskets

Pitch

Rear main seal

Room-temperature vulcanizing (RTV)

Sealants

Seals

Shot peening

Sintering

Stainless steel

Steel

Strip seal

Tempering

Tensile strength

Thread depth

Torque-to-yield

Valley pan

Valve cover gasket

5. Each of the following statements is true EXCEPT:
   A. Tempering increases the toughness of the metal.
   B. Annealing makes a metal easier to be machined.
   C. Sintering bonds the metal powder together.
   D. Case hardening makes the metal's grain coarser.

6. When installing intake manifold gaskets:
   A. End strips are used to seal the ends of an OHV intake manifold to the block.
   B. A valley pan type intake gasket is used on in-line engines.
   C. A leaking intake manifold gasket will cause a rich air-fuel mixture.
   D. Silicone sealer should be placed on both sides of the intake manifold gasket before installation.

7. All of these statements about head gaskets are true EXCEPT:
   A. A conventional head gasket contains a core, facing, and coating.
   B. The core is usually made from silicone.
   C. The facing is usually made from graphite or rubber fiber.
   D. The coating may be made from Teflon.

8. The fire ring in a head gasket:
   A. Compensates for movement between metals with different expansion rates.
   B. Increases heat transfer to the cooling system.
   C. Protects the gasket material from high combustion temperatures.
   D. Seals coolant passages through the head gasket.

9. Which of the following statements is true for replacing valve cover gaskets?
   A. A sealant should be used with molded rubber valve cover gaskets.
   B. An anaerobic sealant should be used with cork valve cover gaskets.
   C. The exhaust manifold and the valve cover gasket are usually made of the same material.
   D. Some valve cover gaskets are made from a synthetic rubber.

10. When discussing exhaust manifold connections, each of the following statements is true EXCEPT:
    A. Some engines do not use an exhaust manifold gasket.
    B. Some exhaust manifolds are sealed with RTV.
    C. A damaged gasket can result in poor engine performance.
    D. An exhaust manifold gasket is a hard gasket.

Chapter 6

INTAKE AND EXHAUST SYSTEMS

UPON COMPLETION AND REVIEW OF THIS CHAPTER, YOU SHOULD BE ABLE TO:

- Describe the purpose of the air filter.

- Describe air filter design.

- List three different materials used to manufacture intake manifolds.

- Describe the purpose of the intake manifold.

- Explain the advantages of aluminum and plastic intake manifolds compared to cast iron.

- Describe the operation and advantages of intake manifolds with dual runners.

- Describe two different methods for operating the valves that open and close the intake manifold runners.

- Explain how the engine creates vacuum.

- Describe how vacuum is used to operate and control many automotive devices.

- Explain the operation of exhaust system components, including exhaust manifold, gaskets, exhaust pipe and seal, muffler, resonator, tailpipe, and clamps, brackets, and hangers.

- Explain the benefits and operation of turbochargers and superchargers.

- Explain the purpose and operation of the catalytic converter.

INTRODUCTION

The better an engine can breathe, the more power it can make. The intake and exhaust system is the breathing apparatus for the engine. To maximize volumetric efficiency, the intake and exhaust manifolds must be carefully designed. Both must allow adequate airflow to provide strong engine operation. A plugged air filter can seriously restrict airflow, causing the engine to run with reduced power or even not to start. A restricted exhaust system can cause the same symptoms; the engine must be able to freely take air in and easily push air out to breathe well. The intake system must be well sealed so that a strong vacuum is formed. This ensures that air will flow into the cylinder when the throttle and intake valves are open.

Some engines used a forced induction system to increase airflow into the engine. A turbocharger or supercharger pressurizes the air above atmospheric pressure to force more air in when the intake valves open. This improvement in volumetric efficiency can increase engine power dramatically.

AIR INDUCTION SYSTEM

The intake system consists of the air inlet tube, an air filter, ducting, a throttle bore, and the intake manifold (Figure 6-1). The opening and closing of the throttle plate in the throttle bore controls airflow into the engine. Some systems use a throttle cable to open and close the throttle plate, while some newer systems employ "drive by wire" technology. In these systems, the PCM responds to an input on the throttle pedal and controls a motor at the throttle plate to open it to the correct angle. The intake system should provide airflow

FIGURE 6-1 A late-model intake air distribution system.

at a high velocity and low volume when the engine is operating at low rpms. This ensures good fuel atomization and mixing when the injector sprays into the airstream in the manifold near the intake valve. The intake system should deliver a high volume of air when the engine is running at high rpms, to fill the combustion chamber as much as possible in a short time. This keeps power high even at higher engine speeds. Volumetric efficiency will naturally fall at higher rpms, but the design of the intake and exhaust systems can reduce the losses.

The intake system pulls in fresh air from outside the engine compartment. Ducts direct this air through the air filter and into the throttle body assembly to power the engine. The air filter is placed below the top of the engine to allow for aerodynamic body designs. Electronic sensors measure airflow, temperature, and density. The air induction system performs the following functions:

- Filters the air to protect the engine from wear
- Silences air intake noise
- Heats or cools the air as required
- Provides the air the engine needs to operate
- Monitors airflow temperature and density for more efficient combustion and a reduction of hydrocarbon (HC) and carbon monoxide (CO) emissions
- Operates with the positive crankcase ventilation (PCV) system to burn the crankcase fumes in the engine

AIR INTAKE DUCTWORK

The **mass airflow sensor (MAF) sensor** measures the mass of airflow entering the engine and sends an input to the PCM reflecting this value. This is a primary input for the PCM's control of fuel delivery.

Ductwork is used to direct the air into the throttle body. Cool outside air is drawn into the air cleaner assembly and, on some engines, warm air from around the exhaust is also brought in for cold engine operation. Many air ducts include a resonator chamber that looks like an irregularly shaped protrusion on the side of an intake duct. This is used to reduce the rumble of intake air as it rushes through the ductwork. Air then flows through the air filter, where particulate matter that could damage the engine is removed. Another leg of ducting connects the air filter housing to the throttle bore. Many vehicles have a **mass airflow sensor (MAF) sensor** in this ducting to measure the mass of air flowing into the engine. Most engines also use an intake air temperature (IAT) sensor in the air cleaner housing of intake ductwork (Figure 6-2). As the air temperature increases, the air density decreases. The PCM will reduce the amount of fuel delivered, as needed to ensure excellent fuel economy and low emissions.

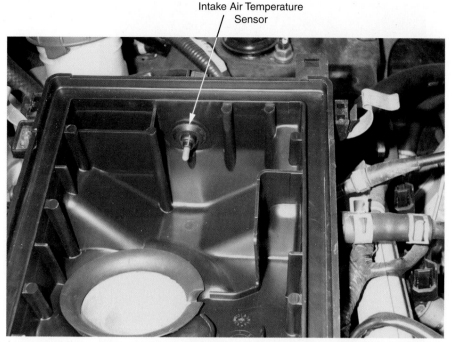

Intake Air Temperature Sensor

FIGURE 6-2 **The intake air temperature sensor helps the PCM determine the correct fuel delivery and spark timing.**

Be sure that the intake ductwork is properly installed and all connections are airtight, especially those between the MAF sensor or remote air cleaner and the throttle plate assembly. If outside air can sneak in through a leak in the ductwork, it is not filtered. Vehicles that use a MAF sensor measure the air coming into the engine from the air filter. The PCM uses this input as the primary indicator of how much fuel should be injected. If air is drawn into the engine beyond the MAF sensor through a leak in the duct, it will not be measured. The PCM will deliver less fuel, and the engine will run too lean. Generally, metal or plastic air ducts are used when engine heat is not a problem. Special paper/metal ducts are used when they will be exposed to high engine temperatures.

AIR CLEANER/FILTER

The primary function of the **air filter** is to prevent airborne contaminants and abrasives from entering the engine. Without proper filtration, these contaminants can cause serious damage and appreciably shorten engine life. All incoming air should pass through the filter element before entering the engine (Figure 6-3).

Air Filter Design

Air filters are basically assemblies of pleated paper supported by a layer of fine mesh wire screen. The screen gives the paper some strength and also filters out large particles of dirt. A thick plastic-like gasket material normally surrounds the ends of the filter. This gasket adds strength to the filter and serves to seal the filter in its housing. If the filter does not seal well in the housing, dirt and dust can be pulled into the airstream to the cylinders. In most air filters, the air flows from the outside of the element to the inside as it enters the intake system.

The shape and size of the air filter element depends on its housing; the filter must be the correct size for the housing or dirt will be drawn into the engine (Figure 6-4). On today's

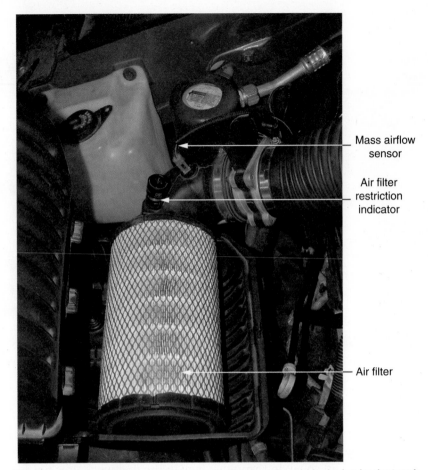

FIGURE 6-3 The air filter is located in the air filter housing within the intake ductwork.

FIGURE 6-4 The air cleaner assembly is shaped to fit in the crowded engine compartment.

engines, air filters are either flat (Figure 6-5) or round (Figure 6-6). Air filters must be properly aligned and closed around the filter to ensure good airflow of clean air.

Many air cleaners in recent years have an **airflow restriction indicator** mounted in the air cleaner housing (refer to Figure 6-7). If the air filter element is not restricted, a window in the side of the restriction indicator shows a green color. When the air filter element is restricted, the window in the airflow restriction indicator is orange and "Change Air Filter"

FIGURE 6-5 Typical flat air cleaner element.

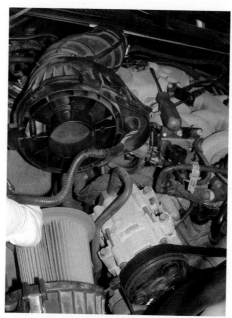

FIGURE 6-6 Typical round air filter for a late-model vehicle.

FIGURE 6-7 Airflow restriction indicator.

appears. Then the air filter must be replaced. After the air filter is replaced, a reset button on top of the airflow restriction indicator must be pressed to reset the indicator so it displays green in the window.

Some air cleaners have a combined MAF sensor and **intake air temperature (IAT) sensor** attached to the air outlet on the air cleaner housing. A duct is connected from the MAF sensor to the throttle body. The MAF sensor must be attached to the air cleaner so air flows through this sensor in the direction of the arrow on the sensor housing (Figure 6-8). If the airflow is reversed through the MAF sensor, the powertrain control module (PCM) supplies a rich air-fuel ratio and increased fuel consumption. Other air cleaners contain a separated IAT sensor (Figure 6-9).

INTAKE MANIFOLD

The **intake manifold** distributes the clean air as evenly as possible to each cylinder of the engine. Fuel is injected at the very end of the intake manifold runners near the combustion chamber.

The **intake air temperature (IAT) sensor** sends an analog voltage signal to the PCM in relation to air intake temperature.

An **intake manifold** is a cast iron, aluminum, or plastic casting with internal passages that conducts air or an air-fuel mixture from the throttle body or carburetor to the cylinder head intake ports.

MAF Sensor
airflow
direction arrow

FIGURE 6-9 The intake air
temperature sensor may be
mounted in the air cleaner
housing.

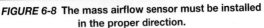

FIGURE 6-8 The mass airflow sensor must be installed
in the proper direction.

Modern intake manifolds for engines with port fuel injection are typically made of die-cast aluminum or plastic (Figure 6-10). These materials are used to reduce engine weight. A plastic manifold transfers less heat to the intake air and this results in a denser air-fuel mixture. Because intake manifolds for port-injected engines only deliver air to the cylinders, fuel vaporization and condensation are not design considerations. These intake manifolds deliver air to the intake ports, where it is mixed with the fuel delivered by the injectors (Figure 6-11). A primary consideration of these manifolds is the delivery of equal amounts of air to each cylinder.

Modern intake manifolds also serve as the mounting point for many intake-related accessories and sensors. Some include a provision for mounting the thermostat and

FIGURE 6-10 This die-cast aluminum manifold feeds the
five-cylinder engine through short runners off of the plenum.

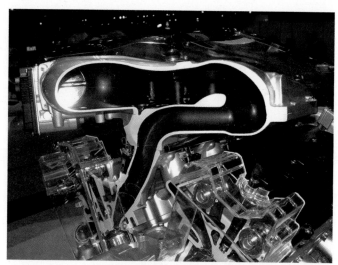

FIGURE 6-11 In a port fuel injection engine, the intake manifold delivers air to the intake ports.

thermostat housing. In addition, connections to the intake manifold provide a vacuum source for the exhaust gas recirculation (EGR) system, automatic transmission vacuum modulators, power brakes, and/or heater and air-conditioning airflow control doors. Other devices located on or connected to the intake manifold include the manifold absolute pressure (MAP) sensor, knock sensor, various temperature sensors, and EGR passages.

Most engines cannot produce the amount of power they should at high speeds because they do not receive enough air. This is the reason why many race cars have hood scoops. With today's body styles, hood scoops are not desirable because they increase air drag. However, to get high performance out of high performance engines, more air must be delivered to the cylinders at high engine speeds. There are a number of ways to do this; increasing the air delivered by the intake manifold is one of them.

Intake manifolds can be designed to improve the volumetric efficiency of the engine. This process is called intake manifold tuning. As the air rushes through the intake manifold into the cylinders, the opening and closing of the intake valves cause the airflow to pulse. If the intake manifold is divided into individual runners for each cylinder, the pulsing effect of the airflow pushes or rams more air into the cylinder. By adjusting or tuning the length of the individual runners, the manufacturer can design an intake manifold to supply the amount of air required by the particular engine. Smaller diameter, longer intake manifold runners ram more air into the cylinders at lower engine rpm. To push more air into the cylinders at higher rpm when air pulsing is faster, the intake manifold runners are designed larger in diameter and shorter. Runners must be curved to avoid sharp bends that create more airflow restriction.

Manufacturers have also developed manifolds that deliver more air only at high engine speeds. The design of manifolds is intended to keep air moving quickly. At low rpm, narrower diameter, longer runners keep the smaller volume of air moving quicker. At high rpm, shorter, wider runners offer less restriction and allow the air to keep moving quickly.

One such system (Figure 6-12) uses a **variable-length intake manifold**. At low speeds, the air travels through only part of the manifold on its way to the cylinders. When the engine speed reaches about 3,700 rpm, two butterfly valves open, forcing the air to take a longer route to the intake port. This increases the speed of the airflow, as well as increasing the amount of air available for the cylinders. As a result, more power is available at high speeds without decreasing low speed torque and fuel economy and without increased exhaust emissions.

The **variable-length intake manifold** has two runners of different lengths connected to each cylinder head intake port.

Runner valve open

Runner valve closed

FIGURE 6-12 A variable-length intake manifold.

Another approach is the use of two large intake manifold runners for each cylinder (Figure 6-13). Separating the intake runners from the intake ports is an assembly that has two bores for each cylinder. There is one bore for each runner. Both bores are open when the engine is running at high speeds. However, a butterfly valve in one set of the runners is closed by the PCM when the engine is operating at lower speeds. This action decreases airflow speed and volume at lower engine speeds and allows for greater airflow at high engine speeds. The butterfly valves in the intake manifold runners may be operated by a vacuum actuator, and the PCM operates an electric/vacuum solenoid that turns the vacuum

FIGURE 6-13 This manifold provides a short and a long runner for each cylinder.
Both are opened at higher engine rpms to improve engine breathing.

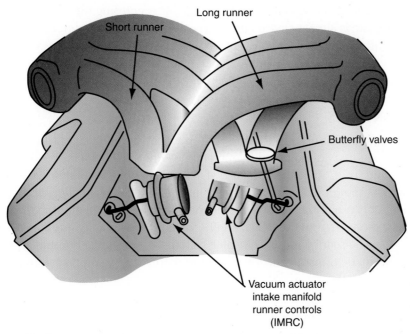

Short runner

Long runner

Butterfly valves

Vacuum actuator
intake manifold
runner controls
(IMRC)

FIGURE 6-14 The butterfly valves are controlled by the PCM through a vacuum
actuator to open the second set of runners at high rpms.

on and off to the actuator. In other intake manifolds, the valves that open and close some of
the runners are operated electrically by the PCM, much like a solenoid (Figure 6-14).

VACUUM BASICS

Vacuum refers to any pressure that is lower than the earth's atmospheric pressure at any
given altitude. Atmospheric pressure appears as zero on most pressure gauges. This does
not mean there is no pressure; rather, it means the gauge is designed to read pressures
greater than atmospheric pressure. All measurements taken on this type of gauge are given
in pounds per square inch and should be referred to as psig (pounds per square inch gauge).
Gauges and other measuring devices that include atmospheric pressure in their readings
also display their measurements in psi; however, these should be referred to as psia (pounds
per square inch absolute). There is a big difference between 12 psia and 12 psig. A reading of 12 psia is less than atmospheric pressure and therefore would represent a vacuum,
whereas 12 psig would be approximately 26.7 psia. When referring to the pressure in the
intake manifold, we use manifold absolute pressure (MAP) as the measure. MAP is a
reading of absolute pressure in the manifold. The MAP reading at idle is typically 7–9 psi.
Because vacuum is defined as any pressure less than atmospheric, vacuum is any pressure
less than 0 psig or 14.7 psia. Remember that vacuum is simply a difference in the pressures
of two areas. Vacuum is not normally measured using negative psi numbers. Vacuum had
to originally be measured in a tube filled with mercury, a metal which is in its liquid state at
room temperatures. The mercury would then be drawn by the vacuum up a tube, and the
amount drawn up was measured in inches. Today, we do not use mercury tubes but rather a
calibrated gauge that reads in inches (or millimeters) of mercury (in. Hg or mm. Hg). Other
units of measurement for vacuum are kilopascals and bars. Normal atmospheric pressure at
sea level is about 1 bar or 100 kilopascals.

An engine in good condition should develop 16–18 in. Hg vacuum at idle. The amount
of low pressure produced by the piston during its intake stroke depends on a number of
things. Basically it depends on the cylinder's ability to form a vacuum and the intake system's
ability to fill the cylinder. When there is high vacuum (16–18 inches [381 to 559 mm. Hg]),

we know the cylinder is well sealed and not enough air is entering the cylinder to fill it. At idle, the throttle plate is almost closed and nearly all airflow to the cylinders is stopped. This is why vacuum is high during idle. Because there is a correlation between throttle position and engine load, it can be said that load directly affects engine manifold vacuum. Therefore, vacuum will be high whenever there is no, or low, load on the engine.

VACUUM CONTROLS

Engine manifold vacuum is used to operate and/or control several devices on an engine. Prior to the mid-70s, vacuum was only used to operate the windshield wipers and/or a distributor vacuum advance unit. Since then, the use of vacuum has become extensive. While some emission control output devices are solenoid or linkage controlled, many operate on a vacuum. This vacuum is usually controlled by solenoids that are opened or closed, depending on electrical signals received from the PCM. Engine vacuum is used to control operation of certain accessories, such as air conditioner/heater systems, power brake boosters, speed-control components, automatic transmission vacuum modulators, and so on. Vacuum is applied to these systems through a series of hoses and tubes. Vacuum solenoids are widely used on today's engines to operate emission control devices. A vacuum solenoid either blocks or flows vacuum to actuate components or control airflow. The evaporative emissions control system and the EGR system often use electrically controlled vacuum solenoids (Figure 6-15).

TURBOCHARGERS

Another method of improving volumetric efficiency and engine performance is to compress the intake air before it enters the combustion chamber. This can be accomplished through the use of a turbocharger or a supercharger.

A turbocharger or supercharger changes the effective compression ratio of an engine, simply by packing in air at a pressure that is greater than atmospheric; for example, an engine

FIGURE 6-15 This vacuum solenoid is controlled by the PCM and uses the pull of vacuum to draw fuel vapors into the intake manifold.

that has a compression ratio of 8:1 and receives 10 pounds (69 kPa) of boost will have an effective compression ratio of 10.5:1.

A **turbocharger** is a blower or special fan assembly that utilizes the expansion of hot exhaust gases to turn a turbine and compress incoming air. In a typical car engine, a turbocharger may increase the horsepower approximately 20 percent. A typical turbocharger consists of the following components:

- Turbine wheel
- Shaft
- Compressor wheel
- Wastegate valve
- Actuator
- Center housing and rotating assembly (CHRA)

Basic Operation

The **turbine wheel** and the compressor wheel are mounted on a common shaft. Both wheels have fins or blades, and each wheel is encased in its own spiral-shaped housing. The shape of the housing works to control and direct the flow of the gases. The shaft is supported on bearings in the turbocharger housing. The expelled exhaust gases from the cylinders are directed through a nozzle against the blades of the turbine wheel. When engine load is high enough, there is enough exhaust gas flow to cause the turbine wheel and shaft to rotate at a high speed. This action creates a vortex flow. Since the compressor wheel is positioned on the opposite end of the shaft, the compressor wheel must rotate with the turbine wheel (Figure 6-16).

The **compressor wheel** is mounted in the air intake system. As the compressor wheel rotates, air is forced into the center of the wheel, where it is caught by the spinning blades and thrown outward by centrifugal force. The air leaves the turbocharger housing and enters into the intake manifold. Since most turbocharged engines are port injected, the fuel is injected into the intake ports. The rotation of the compressor wheel compresses the air and fuel in the intake manifold, creating a denser air-fuel mixture. This increased intake manifold pressure forces more air-fuel mixture into the cylinders to provide increased engine power. The turbocharger must reach a certain rpm before it begins to pressurize the intake manifold. Some turbochargers begin to pressurize the intake manifold at 1,250 engine rpm, and reach full boost pressure in the intake manifold at 2,250 rpm.

In a normally aspirated engine, air is drawn into the cylinders by the difference in pressure between the atmosphere and engine vacuum. A turbocharger pressurizes the intake charge to a point above normal atmospheric pressure. **Turbo boost** is the positive pressure increase created by the turbocharger.

FIGURE 6-16 Exhaust gases spin the turbine wheel. The compressor wheel is connected to the turbine through a shaft and spins at the same rate to pressurize the intake charge.

A **turbocharger** uses the expansion of exhaust gases to rotate a fan-type wheel which increases the pressure inside of the intake manifold.

A **turbine wheel** is a vaned wheel in a turbocharger to which exhaust pressure is supplied to provide shaft rotation.

A **compressor wheel** is a vaned wheel mounted on the turbocharger shaft that forces air into the intake manifold.

Turbo boost is a term used to describe the amount of positive pressure increase of the turbocharger; for example, 10 psi (69 kPa) of boost means the air is being induced into the engine at 24.7 psi (170 kPa). Normal atmospheric pressure is 14.7 psi (101 kPa); add the 10 psi (69 kPa) to get 24.7 psi (170 kPa).

A **wastegate** limits the maximum amount of turbocharger boost by directing the exhaust gases away from the turbine wheel. The wastegate valve is also referred to as a bypass valve.

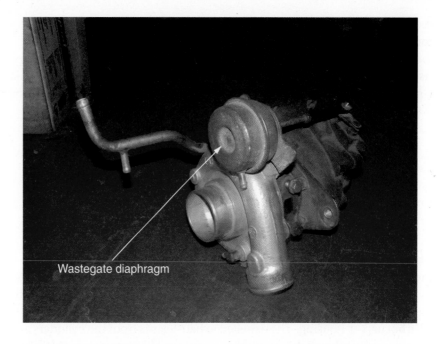

Wastegate diaphragm

FIGURE 6-17 The wastegate diaphragm is operated by a PCM-controlled solenoid, to protect the engine from dangerous overboosting.

Turbocharger wheels rotate at very high speeds, in excess of 100,000 rpm. Engineers design and balance the turbocharger to run in excess of 150,000 rpm, about twenty-five times the maximum rpm of most engines. Due to this high speed operation, turbocharger wheel balance and bearing lubrication are very important.

Boost Pressure Control. If the turbocharger boost pressure is not limited, excessive intake manifold and combustion pressure can destroy engine components. Also, if the amount of boost is too high, detonation knock can occur and decrease engine output. To control the amount of boost developed in the turbocharger, most turbochargers have a **wastegate** diaphragm mounted on the turbocharger. A linkage is connected from this diaphragm to a wastegate valve in the turbine wheel housing (Figure 6-17).

Under low to partial load conditions, the diaphragm spring holds the wastegate valve closed. This routes all of the exhaust gases through the turbine housing. Boost pressure from the intake manifold is also supplied to the wastegate diaphragm.

Under full load, when the boost pressure in the intake manifold reaches the maximum safe limit, the boost pressure pushes the wastegate diaphragm and opens the wastegate valve. This action allows some exhaust to bypass the turbine wheel, which limits turbo-charger shaft rpm and boost pressure (Figure 6-18).

On some engines, the boost pressure supplied to the wastegate diaphragm is controlled by a computer-operated solenoid. In many systems, the PCM pulses the wastegate solenoid on and off to control boost pressure. Some computers are programed to momentarily allow a higher boost pressure on sudden acceleration to improve engine performance.

Turbo Lag

Turbo lag occurs when the turbocharger compressor and turbine wheels are not spinning fast enough to create boost. It takes time to get the exhaust gases to bring the turbocharger wheels up to operating speed. The size and weight of the turbine and compression wheels, along with housing design, are factors that affect the amount of turbo lag. Smaller, lighter turbochargers result in less lag time but lower boost pressures. Some engines use two

FIGURE 6-18 As boost pressure increases toward the safe limit, the wastegate opens to divert exhaust gases away from the turbine wheel.

sequential turbochargers: a very small, lightweight one to help acceleration at low rpms and a bigger volume one to create plenty of boost to maximize power at higher rpms. Low-pressure turbochargers may provide boost of only 4–8 psi, but they reduce the lag time until the 8–15 psi from the full-size turbo kicks in. Also, variable nozzle turbine turbocharger systems have been developed to reduce the lag period (Figure 6-19).

Turbo lag is a short delay period before the turbocharger develops sufficient boost pressures.

Turbocharger Cooling

Exhaust flow past the turbine wheel creates very high turbocharger temperature, especially under high engine load conditions. Many turbochargers have coolant lines connected from the turbocharger housing to the cooling system (Figure 6-20). Coolant circulation through the turbocharger housing helps to cool the bearings and shaft. Full oil pressure is supplied from the main oil gallery to the turbocharger bearings and shaft to lubricate and cool the bearings. This oil is drained from the turbocharger housing back into the crankcase. Seals on the turbocharger shaft prevent oil leaks into the compressor or turbine wheel housings. Worn turbocharger seals

FIGURE 6-19 Variable nozzle turbine type turbocharger.

FIGURE 6-20 A water-cooled turbocharger uses coolant to control the temperature of the bearings and the shaft.

allow oil into the compressor or turbine wheel housings, resulting in blue smoke in the exhaust and oil consumption. Some heat is also dissipated from the turbocharger to the surrounding air.

Some turbochargers do not have coolant lines connected to the turbocharger housing. These units depend on oil and air cooling. On these units, if the engine is shut off immediately after heavy load or high speed operation, the oil may burn to some extent in the turbocharger bearings. When this action occurs, hard carbon particles, which destroy the turbocharger bearings, are created. The coolant circulation through the turbocharger housing lowers the bearing temperature to help prevent this problem. When turbochargers that depend on oil and air cooling have been operating at heavy load or high speed, idle the engine for at least one minute before shutting it off. This action will help to prevent turbocharger bearing failure.

Intercoolers

A disadvantage of the turbocharger (and supercharger) is that it heats the incoming air. The hotter the air, the less dense it is. As the air gets hotter, fewer air molecules can enter the cylinder on each intake stroke. Also, hotter intake air leads to detonation problems. To combat these effects, many turbocharger systems use an **intercooler** (Figure 6-21). The intercooler is like a radiator in that it removes heat from the turbocharger system by dissipating it to the atmosphere. The intercooler can be either air cooled or water cooled. Cooling the air makes it denser, increasing the amount of oxygen content with each intake stroke. The intercooler cools the air that leaves the turbocharger at about 100°F (38°C) before it enters the cylinders (Figure 6-22). For every 10°F (5.5°C) that the air is cooled, a power gain of about 1 percent is obtained. If the intercooler is capable of cooling the air by 100°F (38°C), then a 10 percent power increase is obtained.

Scheduled Maintenance

There are four main things that will reduce the life of a turbocharger: lack of oil, **oil coking**, contaminants in the oil, and ingestion of foreign material through the air intake. To prevent premature turbocharger failure, engine oil and filters should be changed at the vehicle manufacturer's recommended intervals. The engine oil level must be maintained at the specified level on the dipstick. The air cleaner element and the air intake system must be maintained in satisfactory condition. Dirt entering the engine through an air cleaner will damage the compressor wheel blades. When coolant lines are connected to the turbocharger housing,

The **intercooler** cools the intake air temperature to increase the density of the air entering the cylinders. It is also referred to as a charge air cooler.

Oil coking occurs after the engine is shut down and the turbocharger is still spinning. The oil is no longer circulating, and what remains inside the turbo can literally bake into a white, hard, caked restriction in the lines. Allowing the engine to idle for a minute before shutting the engine off allows the turbo compressor to wind down and prevents this.

FIGURE 6-21 The intercooler cools the pressurized air to create a denser intake charge.

FIGURE 6-22 Intercoolers can be air cooled (A), or water cooled (B).

the cooling system must be maintained according to the vehicle manufacturer's maintenance schedule to provide normal turbocharger life.

Turbocharged engines have a lower compression ratio than a normally aspirated engine, and many parts are strengthened in a turbocharged engine because of the higher cylinder pressure. Therefore, many components in a turbocharged engine are not interchangeable with the parts in a normally aspirated engine.

AUTHOR'S NOTE: During my experience in the automotive service industry, I encountered a significant number of turbocharged engines with low boost pressure. This low boost pressure was often caused by low engine compression. Because low engine compression reduces the amount of air taken into the engine, this problem also reduces turbocharger speed and boost pressure. Therefore, when diagnosing low boost pressure, always verify the engine condition before servicing or replacing the turbocharger.

SUPERCHARGERS

After falling into disuse for a number of years (except for racing applications), the **supercharger** started to reappear on automotive engines as original equipment manufacturer (OEM) in 1989 (Figure 6-23). The supercharger is belt driven from the crankshaft by a

FIGURE 6-23 A factory installed supercharger.

ribbed belt (Figure 6-24). A shaft is connected from the crankshaft pulley to one of the drive gears in the front supercharger housing, and the driven gear is meshed with the drive gear. The rotors inside the supercharger are attached to the two drive gears.

The drive gear design prevents the rotors from touching; however, there is a very small clearance between the drive and driven gears. In some superchargers, the rotor shafts are supported by roller bearings on the front and needle bearings on the back. During the manufacturing process, the needle bearings are permanently lubricated. The ball bearings are lubricated by a synthetic-based, high speed gear oil. A plug is provided for periodic checks of the front bearing lubricant. Front bearing seals prevent lubricant loss into the supercharger housing.

In a typical car engine, a supercharger increases horsepower approximately 20 percent compared to a normally aspirated engine with the same CID; for example, the General Motors normally aspirated 1997 3800 V-6 engine is rated at 200 HP. The supercharged

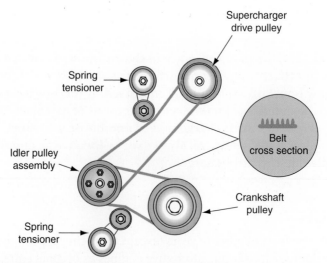

FIGURE 6-24 The supercharger is driven by a belt off of the crankshaft.

version of this engine is rated at 240 HP. The supercharger will operate around 10,000 to 15,000 rpm. Unlike the turbocharger, the amount of boost the supercharger produces is a function of engine rpm (not load). The advantage of the supercharger is that it will produce more torque at lower engine speeds than a turbocharger. Also, there is no lag time associated with a supercharger. The disadvantage of the supercharger is that it consumes horsepower as it is driven.

Supercharger Operation

While there have been a number of supercharger designs on the market over the years, the most popular is the **Roots-type supercharger**. The pair of three-lobed rotor vanes (Figure 6-25) in the Roots supercharger is driven by the crankshaft. The lobes force air into the intake manifold. The helical design evens out the pressure pulses in the blower and reduces noise. It was found that a 60-degree helical twist works best for equalizing the inlet and outlet volumes. Another benefit of the helical rotor design is it reduces carryback volumes—air that is carried back to the inlet side of the supercharger because of the unavoidable spaces between the meshing rotors, which represents a loss of efficiency. Many aftermarket superchargers are a centrifugal type.

To handle the higher operating temperatures imposed by supercharging, an engine oil cooler is usually built into the engine lubrication system. This water-to-oil cooler is generally mounted between the engine front cover and oil filter.

Intake air enters the inlet plenum at the back of the supercharger, and the rotating blades pick up the air and force it out the top of the supercharger. The blades rotate in opposite directions, acting as a pump as they rotate. This pumping action pulls air through the supercharger inlet and forces the air from the outlet. There is a very small clearance between the meshed rotor lobes and between the rotor lobes and the housing (Figure 6-26).

Air flows through the supercharger system components in the following order:

1. Air flows through the air cleaner and MAF sensor into the throttle body.
2. Airflow enters the supercharger intake plenum.
3. From the intake plenum the air flows into the rear of the supercharger housing (Figure 6-27).
4. The compressed air flows from the supercharger to the intercooler inlet.
5. Air leaves the intercooler and flows into the intercooler outlet tube.
6. Air flows from the intercooler outlet tube into the intake manifold adapter.
7. Compressed, cooled air flows through the intake manifold into the engine cylinders.
8. If the engine is operating at idle or very low speeds, the supercharger is not required. Under this condition, airflow is bypassed from the intake manifold adapter through a butterfly valve to the supercharger inlet plenum (Figure 6-28).

The **Roots-type supercharger** is a positive displacement supercharger that uses a pair of lobed vanes to pump and compress the air.

Three-lobed rotor vane (2)

FIGURE 6-25 Common design of the rotor.

Boost pressure pushed out into intake manifold

PSI developing

FIGURE 6-26 As the rotors mesh, they pump air out under pressure.

FIGURE 6-27 Airflow through a supercharger.

FIGURE 6-28 The bypass actuator can divert airflow from the intake manifold back into the supercharger plenum to reduce boost.

In one example, the bypass butterfly valve (Figure 6-29) is operated by an air bypass actuator diaphragm as follows:

1. When manifold vacuum is 7 in. Hg (23.6 kPa) or higher, the bypass butterfly valve is completely open and a high percentage of the supercharger air is bypassed to the supercharger inlet.
2. If the manifold vacuum is 3 to 7 in. Hg (10 to 23.6 kPa), the bypass butterfly valve is partially open and some supercharger air is bypassed to the supercharger inlet, while the remaining airflow is forced into the engine cylinders.
3. When the vacuum is less than 3 in. Hg (10 kPa), the bypass butterfly valve is closed and all the supercharger airflow is forced into the engine cylinders.

Elbow assembly

Bypass hose

Front of
Engine

Bypass valve

Supercharger and
plenum assembly

FIGURE 6-29 Supercharger bypass hose and intake elbow.

On some superchargers, the pulley size causes the rotors to turn at 2.6 times the engine speed. Since supercharger speed is limited by engine speed, a supercharger wastegate is not required. Belt-driven superchargers provide instant low speed action compared to exhaust-driven turbochargers, which may have a low speed lag because of the brief time interval required to accelerate the turbocharger shaft. Compared to a turbocharger, a supercharger turns at much lower speeds.

Friction between the air and the rotors heats the air as it flows through the supercharger. The intercooler dissipates heat from the air in the supercharger system to the atmosphere, creating a denser air charge. When the supercharger and the intercooler supply cooled, compressed air to the cylinders, engine power and performance are improved.

The compression ratio in the supercharged 3.8-liter engine is 8.2:1, compared to a normally aspirated 3.8-liter engine, which has a 9.0:1 compression ratio. The following components are reinforced in the supercharged engine because of the higher cylinder pressure:

■ Engine block
■ Main bearings
■ Crankshaft bearing caps
■ Crankshaft
■ Steel crankshaft sprocket
■ Timing chain
■ Cylinder head
■ Head bolts
■ Rocker arms

Superchargers can be enhanced with electrically operated clutches and bypass valves. These allow the same computer that controls fuel and ignition to kick the boost on and off precisely as needed. This results in far greater efficiency than a full-time supercharger.

EXHAUST SYSTEM COMPONENTS

The various components of the typical exhaust system include the following:

■ Exhaust manifold
■ Exhaust pipe and seal
■ Catalytic converter
■ Muffler
■ Resonator
■ Tailpipe
■ Heat shields

- Clamps, brackets, and hangers
- Exhaust gas oxygen sensors

All the parts of the system are designed to conform to the available space of the vehicle's undercarriage and yet be a safe distance above the road.

Exhaust Manifold

The **exhaust manifold** (Figure 6-30) collects the burnt gases as they are expelled from the cylinders and directs them to the exhaust pipe. Exhaust manifolds for most vehicles are made of cast or nodular iron. Many newer vehicles have stamped, heavy gauge sheet metal or stainless steel units.

In-line engines have one exhaust manifold. V-type engines have an exhaust manifold on each side of the engine. An exhaust manifold will have either three, four, or six passages, depending on the type of engine. These passages blend into a single passage at the other end, which connects to an exhaust pipe. From that point, the flow of exhaust gases continues to the catalytic converter, muffler, and tailpipe, then exits at the rear of the car.

V-type engines may be equipped with a dual exhaust system that consists of two almost identical, but individual, systems in the same vehicle.

Exhaust systems are designed for particular engine–chassis combinations. Exhaust system length, pipe size, and silencer size are used to tune the flow of gases within the exhaust system. Proper tuning of the exhaust manifold tubes can actually create a partial vacuum that helps draw exhaust gases out of the cylinder, improving volumetric efficiency. Separate, tuned exhaust headers (Figure 6-31) can also improve efficiency by preventing the exhaust flow of one cylinder from interfering with the exhaust flow of another cylinder. Cylinders next to one another may release exhaust gas at about the same time. When this happens, the pressure of the exhaust gas from one cylinder can interfere with the flow from the other cylinder. With separate headers, the cylinders are isolated from one another, interference is eliminated, and the engine breathes better. The problem of interference is especially common with V-8 engines. However, exhaust headers tend to improve the performance of all engines.

Exhaust manifolds may also be the attaching point for the air injection reaction (AIR) pipe (Figure 6-32). This pipe introduces cool air from the AIR system into the exhaust stream. The added air in the exhaust stream during engine warmup helps burn excess fuel

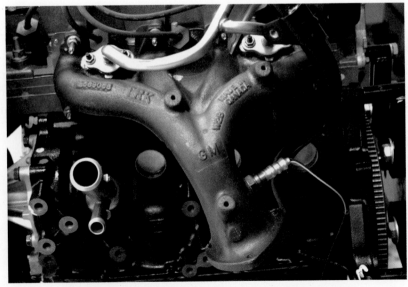

FIGURE 6-30 An exhaust manifold for a four-cylinder engine.

FIGURE 6-31 Engine efficiency can be improved with tuned exhaust headers.

Air injection
reaction pipe

FIGURE 6-32 AIR pipe mounting on an exhaust manifold.

in the exhaust and heat up the catalytic converter faster. Some exhaust manifolds have pro-
visions for the exhaust gas recirculation (EGR) pipe. This pipe takes a sample of the exhaust
gases and delivers it to the EGR valve. Also, some exhaust manifolds have a tapped bore that
retains the oxygen sensor (Figure 6-33).

Exhaust Pipe and Seal

The exhaust pipe is metal pipe, either aluminized steel, stainless steel, or zinc-plated heavy
gauge steel, that runs under the vehicle between the exhaust manifold and the catalytic
converter (Figure 6-34).

Catalytic Converters

A **catalytic converter** (Figure 6-35) is part of the exhaust system and a very important part
of the emission control system. Because it is part of both systems, it has a role in both. As
an emission control device, it is responsible for converting undesirable exhaust gases into
harmless gases. As part of the exhaust system, it helps reduce the noise level of the exhaust.
A catalytic converter contains a ceramic element coated with a catalyst. A catalyst is a

> A **catalytic converter** reduces tailpipe emissions of carbon monoxide, unburned hydrocarbons, and nitrogen oxides.

FIGURE 6-33 One oxygen sensor sits in this exhaust manifold. The catalytic converter mounts onto the exhaust manifold, and another oxygen sensor fits into the exhaust pipe.

FIGURE 6-34 The front exhaust pipe for a V-6 engine.

FIGURE 6-35 Monolythic-type catalytic converter.

substance that causes a chemical reaction in other elements without actually becoming part of the chemical change and without being used up or consumed in the process.

Catalytic converters may be pellet type or monolithic type. A pellet-type converter contains a bed made from hundreds of small beads. Exhaust gases pass over this bed. In a monolithic-type converter, the exhaust gases pass through a honeycomb ceramic block. The converter beads or ceramic block are coated with a thin coating of cerium, platinum, palladium, and/or rhodium, and are held in a stainless steel container. Modern vehicles are equipped with three-way catalytic converters, which means the converter reduces the three major exhaust emissions, hydrocarbons (HC), carbon monoxide (CO), and oxides of nitrogen (NO_x). The converter oxidizes HC and CO into water vapor and carbon dioxide (CO_2) and reduces NO_x to oxygen and nitrogen.

Many vehicles are equipped with a mini-catalytic converter that is either built into the exhaust manifold or is located next to it (Figure 6-36). These converters are used to clean the exhaust during engine warmup and are commonly called warmup converters. Some older catalytic converters have an air hose connected from the AIR system to the oxidizing catalyst. This air helps the converter work by making extra oxygen available. The air from the AIR system is not always forced into the converter; rather, it is controlled by the vehicle's PCM. Fresh air added to the exhaust at the wrong time could overheat the converter and produce NO_x, something the converter is trying to destroy.

OBD II regulations call for a way to inform the driver that the vehicle's converter has a problem and may be ineffective. The PCM monitors the activity of the converter by comparing the signals of an HO_2S located at the front of the converter with the signals from an

FIGURE 6-36 A warmup converter may be used before the primary converter.

FIGURE 6-37 Exhaust system for an OBD II vehicle.

HO_2S located at the rear (Figure 6-37). If the sensor outputs are the same, the converter is not working properly and the malfunction indicator lamp (MIL) on the dash will light.

Converter Problems

The converter is normally a trouble-free emission control device, but two things can damage it. One is leaded gasoline. Lead coats the catalyst and renders it useless. The difficulty of obtaining leaded gasoline has reduced this problem. The other is overheating. If raw fuel enters the exhaust because of a fouled spark plug or other problem, the temperature of the converter quickly increases. The heat can melt the ceramic honeycomb or pellets inside, causing a major restriction to the flow of exhaust. A plugged converter or any exhaust restriction can cause damage to the exhaust valves due to excess heat, loss of power at high speeds, or stalling after starting (if totally blocked).

MUFFLERS

The **muffler** is a cylindrical or oval-shaped component, generally about 2 feet (0.6 meters) long, mounted in the exhaust system about midway or toward the rear of the car. Inside the muffler is a series of baffles, chambers, tubes, and holes to break up, cancel out, or silence the pressure pulsations that occur each time an exhaust valve opens.

Two types of mufflers are commonly used on passenger vehicles (Figure 6-38). Reverse-flow mufflers change the direction of the exhaust gas flow through the inside of the unit. This is the most common type of automotive muffler. Straight-through mufflers permit exhaust gases to pass through a single tube. The tube has perforations that tend to break up pressure pulsations. They are not as quiet as the reverse-flow type.

There have been several important changes in recent years in the design of mufflers. Most of these changes have been centered at reducing weight and emissions, improving fuel economy, and simplifying assembly. These changes include the following:

1. *New materials.* More and more mufflers are being made of aluminized and stainless steel. Using these materials reduces the weight of the units as well as extending their lives.

> The **muffler** is a device mounted in the exhaust system behind the catalytic converter that reduces engine noise.

FIGURE 6-38 (A) Reverse-flow muffler, (B) Straight-through muffler.

Back pressure
reduces an
engine's volumetric
efficiency.

A **resonator** is a
device mounted
on the tailpipe to
reduce exhaust
noise further.

The **tailpipe**
conducts exhaust
gases from the
muffler to the rear
of the vehicle.

A BIT OF HISTORY

In the 1930s Charles
Nelson Pogue
developed the Pogue
carburetor. This car-
buretor used exhaust
heat to vaporize the
fuel before it was
mixed with the air
entering the engine.
Charles Pogue was
issued several patents
on this carburetor,
and claimed greatly
increased fuel
mileage. Mr. Pogue
never did sell his
invention to any of the
car manufacturers,
but rumors were
repeated for years
about the fantastic
fuel economy supplied
by this carburetor.

2. *Double-wall design.* Retarded engine ignition timing that is used on many small cars tends to make the exhaust pulses sharper. Many cars use a double-wall exhaust pipe to better contain the sound and reduce pipe ring.

3. *Rear-mounted muffler.* More and more often, the only space left under the car for the muffler is at the very rear. This means the muffler runs cooler than before and is more easily damaged by condensation in the exhaust system. This moisture, combined with nitrogen and sulfur oxides in the exhaust gas, forms acids that rot the muffler from the inside out. Many mufflers are being produced with drain holes drilled into them.

4. *Back pressure.* Even a well-designed muffler will produce some **back pressure** in the system. Back pressure reduces an engine's volumetric efficiency, or ability to breathe. Excessive back pressure caused by defects in a muffler or other exhaust system part can slow or stop the engine. However, a small amount of back pressure can be used intentionally to allow a slower passage of exhaust gases through the catalytic converter. This slower passage results in more complete conversion to less harmful gases. Also, no back pressure may allow intake gases to enter the exhaust.

Resonator

On some older vehicles, there is an additional muffler, known as a **resonator** or silencer. This unit is designed to further reduce or change the sound level of the exhaust. It is located toward the end of the system and generally looks like a smaller, rounder version of a muffler.

Tailpipe

The **tailpipe** is the last pipe in the exhaust system. It releases the exhaust fumes into the atmosphere beyond the back end of the car.

Heat Shields

Heat shields are used to protect other parts from the heat of the exhaust system and the catalytic converter (Figure 6-39). They are usually made of pressed or perforated sheet metal. Heat shields trap the heat in the exhaust system, which has a direct effect on maintaining exhaust gas velocity.

Clamps, Brackets, and Hangers

Clamps, brackets, and hangers are used to properly join and support the various parts of the exhaust system. These parts also help to isolate exhaust noise by preventing its transfer through the frame (Figure 6-40) or body to the passenger compartment. Clamps help to secure exhaust system parts to one another. The pipes are formed in such a way that one slips inside the other. This design makes a close fit. A U-type clamp usually holds this connection tight (Figure 6-41). Another important job of clamps and brackets is to hold pipes to the bottom of the vehicle. Clamps and brackets must be designed to allow the exhaust system to vibrate without transferring the vibrations through the car.

There are many different types of flexible hangers available. Each is designed for a particular application. Some exhaust systems are supported by doughnut-shaped rubber rings between hooks on the exhaust component and on the frame or car body. Others are supported at the exhaust pipe and tailpipe connections by a combination of metal and reinforced fabric hanger. Both the doughnuts and the reinforced fabric allow the exhaust system to vibrate without breakage that could be caused by direct physical connection to the vehicle's frame.

Some exhaust systems are a single unit in which the pieces are welded together by the factory. By welding instead of clamping the assembly together, car makers save the weight of overlapping joints as well as that of clamps.

FIGURE 6-39 Typical locations of heat shields in an exhaust system.

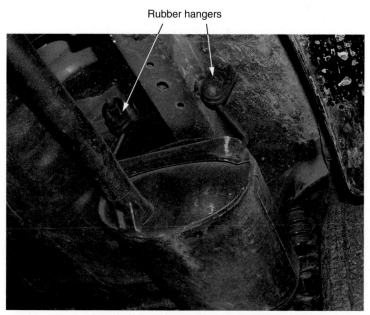

FIGURE 6-40 Rubber hangers are used to keep the exhaust system in place without allowing it to contact the frame.

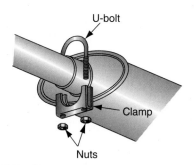

FIGURE 6-41 A U-clamp is often used to secure two pipes that slip together.

AUTHOR'S NOTE: During my experience in the automotive service industry, I encountered several cases where restricted exhaust or intake systems were misdiagnosed and confused with ignition system or fuel system defects. Defective fuel system or ignition system components may cause a loss of engine power and reduced maximum speed, but these symptoms are accompanied by cylinder misfiring and/or engine surging. When the exhaust or intake system is restricted, the maximum speed is reduced but the engine does not misfire or surge.

SUMMARY

TERMS TO KNOW

Air filter

Airflow restriction indicator

Back pressure

Catalytic converter

Compressor wheel

Exhaust manifold

Intake air temperature (IAT) sensor

Intake manifold

Intercooler

Mass airflow (MAF) sensor

Muffler

Oil coking

Resonator

Roots-type supercharger

Supercharger

Tailpipe

Turbine wheel

Turbocharger

Turbo boost

Turbo lag

Variable-length intake manifold

Wastegate

- The air induction system allows a controlled amount of clean, filtered air to enter the engine. Cool air is drawn in through a fresh air tube. It passes through an air cleaner before entering the throttle body.
- The air intake ductwork conducts airflow from the remote air cleaner to the throttle body mounted on the intake manifold.
- The air cleaner/filter removes dirt particles from the air flowing into the intake manifold, to prevent these particles from causing engine damage.
- The intake manifold distributes the air or air-fuel mixture as evenly as possible to each cylinder. Intake manifolds are made of cast iron, plastic, or die-cast aluminum.
- The vacuum in the intake manifold operates many systems such as emission controls, brake boosters, heater/air conditioners, cruise controls, and more. A diagram of emission system vacuum hose routing is located on the underhood decal. Loss of vacuum can create many driveability problems.
- Turbochargers and superchargers create forced induction systems that improve the volumetric efficiency and power of the engine.
- Turbochargers use the heat energy in exhaust gases to spin the turbine and pressurize intake air on the compressor side.
- A turbocharger wastegate regulates boost pressure based on engine load to prevent engine-damaging overboost conditions.
- A belt-driven supercharger supplies air to the intake manifold under pressure to increase engine power.
- A vehicle's exhaust system carries away gases from the passenger compartment, cleans the exhaust emissions, and muffles the sound of the engine. Its components include the exhaust manifold, exhaust pipe, catalytic converter, muffler, resonator, tailpipe, heat shields, clamps, brackets, and hangers.
- The exhaust manifold is a bank of pipes that collects the burned gases as they are expelled from the cylinders and directs them to the exhaust pipe.
- The catalytic converter reduces HC, CO, and NO_x emissions.
- The muffler consists of a series of baffles, chambers, tubes, and holes to break up, cancel out, and silence pressure pulsations.

REVIEW QUESTIONS

Short Answer Essays

1. Explain the operation of an airflow restriction indicator.

2. Explain the purposes of the intake manifold.

3. Explain the advantages of plastic intake manifolds.

4. Explain the operation of an intake manifold with dual runners at low and high engine speeds.

5. Explain how a turbocharger or supercharger creates more engine power.

6. Describe basic turbocharger operation.

7. What is the purpose of the intercooler?

8. Describe the advantages of tuned exhaust manifolds compared to conventional exhaust manifolds.

9. Explain two catalytic converter operating problems.

10. When referring to a turbocharger system, describe what oil coking is, how it is caused, and how to prevent it.

Fill-in-the-Blanks

1. Without proper intake air filtration, contaminants and abrasives in the air will cause severe _____ damage.

2. If the air filter is restricted, the airflow restriction indicator window appears _____ in color.

3. When an intake manifold has dual runners, both runners are open at _____ engine speeds.

4. The exhaust flows past the _____ wheel in a turbocharger.

5. Turbocharged or supercharged engines have _____ compression ratios compared to naturally aspirated engines.

6. The air flows through the intercooler _____ it flows through the supercharger.

7. In place of cast- or nodular-iron exhaust manifolds, newer vehicles have manifolds manufactured from stamped, heavy gauge sheet metal or _____ _____.

8. A tuned exhaust manifold prevents exhaust flow from one cylinder from interfering with the _____ _____ from another cylinder.

9. A catalytic converter may be overheated by a _____ air-fuel ratio.

10. The exhaust pipe between the engine and the catalytic converter may be a _____ _____ design.

Multiple Choice

1. The intake manifold is typically made of:
   A. Plastic
   B. Stainless steel
   C. Cast iron
   D. Fiber composites

2. An engine should develop _____ in. Hg vacuum at idle.
   A. 5
   B. 8
   C. 18
   D. 25

3. A variable-length intake manifold is designed to increase:
   A. Compression
   B. Airflow at idle
   C. Intake turbulence
   D. Airflow at high rpm

4. The catalytic converter reduces _____ emissions.
   A. HC, CO_2, O_2
   B. HC, CO, NO_x
   C. CO, CO_2, NO_x
   D. N, HC, O_2

5. *Technician A* says that a turbocharger may spin at speeds of 100,000 rpm.
   Technician B says that oil coking is a common cause of turbocharger failure.
   Who is correct?
   A. A only
   B. B only
   C. Both A and B
   D. Neither A nor B

6. On a turbocharged engine, when the _____ opens, boost pressure is reduced.
   A. Compressor bypass
   B. Turbine dump valve
   C. Wastegate
   D. Intercooler router

7. A supercharger typically uses _____ to pressurize the intake charge.
   A. Two rotors
   B. Four rotors
   C. A compressor pump
   D. Three helix gears

8. A mini-converter is used
   A. On small engines where a normal converter will not fit properly.
   B. On engines that used leaded fuels.
   C. In conjunction with EGR systems to supply clean exhaust for the cylinders.
   D. To reduce emissions during engine warmup.

9. A restricted exhaust system can cause
   A. Stalling.
   B. Backfiring.
   C. Loss of power.
   D. Acceleration stumbles.

10. All of these statements about an intake manifold dual runner system are true EXCEPT:
    A. Both runners to each cylinder are open at 1,400 engine rpm.
    B. The butterfly valves in one set of runners may be operated by a vacuum actuator.
    C. The butterfly valves may be operated electrically by the PCM.
    D. One runner to each cylinder is open at 900 engine rpm.

Chapter 7

FACTORS AFFECTING ENGINE PERFORMANCE

UPON COMPLETION AND REVIEW OF THIS CHAPTER, YOU SHOULD BE ABLE TO:

■ Describe common engine mechanical causes of improper engine performance.

■ Describe the components that seal the combustion chamber.

■ Explain the effects of combustion chamber sealing problems on engine performance.

■ Define fuel volatility and explain its effect on engine performance.

■ Define octane rating and explain its effects on combustion.

■ Explain the different forms of abnormal combustion.

■ Describe the potential causes and effects of abnormal combustion.

INTRODUCTION

Proper engine performance requires that the engine is mechanically sound and that its support systems are functioning as designed. One of the most important tasks in engine repair is determining the precise cause of the concern. Before you make engine mechanical repairs, you must gather as much information as possible. There are many different reasons why an engine can perform inadequately. You will need to understand the requirements for proper engine performance before you can successfully diagnose the reasons for performance failures. In this chapter, we will look at the areas of the engine that can affect performance and create a need for mechanical repairs.

Components as common as spark plugs can cause an engine to perform as though an engine has a "dead" cylinder(s). Problems that occur related to the combustion chamber may require special test procedures.

Smoke coming from the tailpipe is never a good sign (Figure 7-1). Some causes may be relatively simple and inexpensive to rectify, while others may require a complete engine overhaul or replacement. You will need to understand the possible causes of engine smoking to correctly diagnose the cause of the unwanted emission of smoke.

As engine components wear, clearances become greater, or parts break, and noises develop in the engine. These can be warning signs that a component is wearing, or they can predict imminent engine failure. Training, experience, and careful listening can help you narrow down the possible causes of unusual noises and recommend effective repairs (Figure 7-2).

SPARK PLUGS

Spark plugs are the workhorses of the ignition system. All the buildup of electrical energy of the coil(s) comes to a head at the spark plug gap (Figure 7-3). The electrodes of the spark plug must provide an adequate electrical path. If the spark is well timed and strong enough, it jumps the gap with a powerful burst. Then, it is the conditions within the combustion chamber that determine whether the spark will ignite the air-fuel mixture well enough to foster good combustion.

FIGURE 7-1 A lot of thick blue smoke can require serious engine mechanical repairs or relatively inexpensive and simple cures. Your task will be to properly diagnose the cause to make an effective repair.

FIGURE 7-2 You will listen in different engine areas to identify unusual noises.

FIGURE 7-3 The spark plugs must be functioning well in order for the engine to provide good performance.

Inspecting the spark plugs is a method you will use to gather information about how the engine is operating. You may use this during a routine tune up, when there is a drivability concern, and as part of gathering information about an engine mechanical failure. Spark plugs can give you tremendous insight into what has been occurring within the combustion chambers. When you have a drivability concern, one of the first areas to begin investigating is the spark plugs. This can help identify if one or more cylinders are acting up or if all cylinders are affected. Spark plugs can also help guide your diagnosis toward a fuel, ignition, or mechanical issue, for example.

> **AUTHOR'S NOTE:** Many technicians will use nothing but OEM (original equipment from the manufacturer) spark plugs on late-model vehicles. Today's PCMs and their monitoring systems are very sensitive; using aftermarket spark plugs is a very common cause of drivability problems. Manufacturers spend a lot of time and money designing the spark plug that will deliver the perfect spark for the individual combustion chamber. Many aftermarket companies produce excellent spark plugs. Unfortunately, they are usually designed to fit several applications. This does not guarantee that they will operate exactly like original equipment in each vehicle. If you do find quality aftermarket spark plugs that perform properly in an engine, it is appropriate to use them. Many technicians have a spark plug brand they have come to trust and use it on many, if not all, applications successfully.

COMBUSTION CHAMBER SEALING

Proper combustion is the key to the engine's power production. Fuel must be delivered in the correct quantity; a strong spark must be ignited at precisely the right instant, and the combustion chamber must be full of a fresh charge of air. Then, the combustion chamber must be properly sealed in order for the engine to be able to harness the power of combustion. The spark plug, the engine valves, the headgasket, as well as the piston, rings, and cylinder walls seal the combustion chamber (Figure 7-4). Each of those components must be functioning properly or the piston strokes cannot achieve their optimum result and engine performance will deteriorate.

FIGURE 7-4 A leak through any of these combustion chamber sealing points will lead to engine performance problems.

Throttle body

Intake manifold runner

FIGURE 7-5 **The engine needs a strong vacuum from the cylinder to pull air through the intake runners and into the cylinder.**

On the intake stroke, the engine relies on engine vacuum to pull a fresh charge of air into the cylinder. If any of the components that form a seal for the combustion chamber are leaking, adequate vacuum will not develop. Without a strong pull of vacuum, the cylinder will not be filled with a fresh charge of air. The ratio of air and fuel will be incorrect, and the cylinder will not be able to contribute its share of power if it can't pull in enough air (Figure 7-5). Remember, the more air an engine can take in, the more power it can put out. If an exhaust valve is leaking, residual pressure will be present in the cylinder when the piston moves down on the intake stroke. Exhaust from the exhaust manifold will join with the fresh air from the intake manifold and dilute the charge. The cylinder will not be full either; there is not enough vacuum formed to pull a full charge in. These effects will significantly create rough engine operation and a loss of engine power. Any other combustion chamber leaks would have a similar effect on the intake stroke.

The compression stroke should develop somewhere between 125 and 200 psi of pressure while the engine is cranking. This is a widely generalized estimate; always refer to the manufacturer's specifications when evaluating a specific engine. Compression of the air and fuel mixture adds heat and pressure to the mix to make it more combustible. If compression is low, the spark may not be able to develop a strong flame front. Or, the flame may start but sputter out as it hits loosely held areas of cooler air and fuel. This would cause incomplete combustion or even a total **misfire**. A misfire is when combustion does not take place. Any leaks past the valves, headgasket, piston, rings, and cylinder walls would lower the amount of compression a cylinder can develop.

Leaks in the combustion chamber reduce the power of combustion. All the force of the expanding gases should be exerted on the top of the piston. This pushes the piston down and turns the crankshaft through the connecting rod. A cylinder has the potential to develop roughly 2,000 lbs. of force on top of the piston. If one quarter of that force leaks past worn piston rings and cylinder walls, the driver will definitely notice that lack of power. This is a typical scenario of a worn, high mileage engine. The engine could also leak those forces past a burned intake or exhaust valve (Figure 7-6). The power loss would be evident, but the consumer would also likely notice a popping noise in the intake or exhaust, depending on which valve(s) had failed.

A **misfire** means that combustion is either incomplete or does not occur at all.

FIGURE 7-6 A burned valve will significantly reduce the power output of an engine and cause rough running.

Ring and Cylinder Wall Wear

Piston rings will normally develop wear as the engine accumulates miles. Out of the box they have very sharp edges; these wear down over time and compromise their ability to seal against the cylinder walls (Figure 7-7). When piston rings wear, they may allow combustion gases to leak down into the crankcase. They may also let oil track up the cylinder walls into the combustion chamber. This is one of the primary causes of bluish smoke exiting the tailpipe. It is called oil consumption when oil is burning in the combustion chamber. It is important to remember that some oil consumption is considered normal. The piston's rings will not completely seal all of the oil from the combustion chamber. The amount of oil consumption (oil burning) varies with engine load, driving style, type of oil, and amount

FIGURE 7-7 The cylinder walls must be in excellent condition to allow the rings to seal the combustion chamber properly.

FIGURE 7-8 A pitted valve face allows leakage that will eventually cause the valve to burn.

of wear of the engine. One of the primary functions of an engine overhaul is replacing the piston rings and refurbishing the cylinder walls.

Valve Wear

Intake and exhaust valves can wear in a few different ways. The most dramatic problem they encounter is when they burn. Valves generally burn when they are open during combustion and exposed to the extreme combustion temperatures. This can happen when the valve springs become weak and cannot close the valves at the proper time, particularly when the engine is spinning at high speeds. When the engine is revving high, the momentum of the valve opening is greater and a slightly weak spring may not be able to close the valve at the proper time. With the valve face and margin exposed to combustion temperatures, the valve will rapidly burn.

The valve faces and seats can also become pitted from the corrosive byproducts of combustion or can develop carbon buildup on the seats (Figure 7-8). In either case, leakage past the valve face and the seat can occur. Again, once combustion temperatures can leak past parts of a valve, the parts tend to burn rapidly.

Misadjusted valves can also cause valve leakage and burning. If a valve is adjusted too tightly, it will be held open longer than it was designed to. This can cause a reduction in performance through poor sealing of the combustion chamber during the appropriate strokes. In extreme cases, it can also cause the valves to burn if they are exposed to excessive temperatures.

Headgasket Damage

When a headgasket leaks, it can present a whole host of different symptoms. In the context of combustion chamber sealing, a failure usually results in a leak between two adjoining cylinders. This will lower the compression and combustion of both cylinders significantly. This will cause rough running and a lack of power. Combustion gases can also leak out into the

A BIT OF HISTORY

As recently as the mid 1980s it was not uncommon for engines to require valve reconditioning as early as 60,000 or 75,000 miles. Now, due to advances in materials and machining, most of today's engines can run at least 150,000 miles before requiring valve service.

FIGURE 7-9 A blown head gasket often results in a cloud of white smoke.

cooling system. This can cause the cooling system pressure cap to release pressure and coolant. The most common symptom of a blown headgasket is coolant leaking into the combustion chamber. Another common symptom of head gasket failure is the presence of coolant in the oil. The oil dipstick will show signs of coolant mixing with the oil and it will look foamy and brownish, like a coffee milkshake. In these situations, a complete engine rebuild may be required. This burning coolant causes clouds of white, sweet-smelling exhaust to exit the tailpipe (Figure 7-9).

There is a significant difference in cost, labor, and technique, depending on what problems exist with combustion chamber sealing. It will be your job to recognize the possible causes of low performance, in order to offer the customer responsible repair options with a realistic estimate.

FUEL AND COMBUSTION

When combustion does not occur normally, severe engine damage can result. It is important that you are aware of the causes of abnormal combustion. When you repair or replace an engine with a catastrophic failure, you need to find the source of the problem so it doesn't happen again (Figure 7-10). The fuel that customers use can have a significant impact on their engine performance and durability. Your customers may ask your advice about what type of fuel to use and why. You will also see drivability problems caused by fuel issues affecting combustion.

Octane Rating

The **octane rating** of a fuel describes its ability to resist spontaneous ignition or engine knock. Engine knocking (detonation) or pinging (preignition) results when combustion occurs at the wrong time or at the wrong speed. Low octane fuel can be a cause of preignition and detonation. If combustion begins before the spark, for example, combustion pressures may try to push the piston backwards at the end of the compression stroke. This results in a rattling noise from the piston. This is called preignition, and the sound is often described as pinging. Fuel is generally available for automobiles as regular, 87 octane,

FIGURE 7-10 This bent connecting rod could have been the result of detonation. You will need to uncover the source of the problem to ensure a complete repair.

FIGURE 7-11 Make sure your customer is using the correct octane fuel to prevent abnormal combustion.

mid-range, or 89 octane, or as premium, 92 octane or 93 octane (Figure 7-11). The higher the octane number, the greater its resistance to knock. Octane is tested in two ways, by the research method and by the motor method. The advertised octane rating is the average of the two ratings. You will often see this described on the pumps as:

$$\text{Octane} = \frac{\text{RON} + \text{MON}}{2}$$

Volatility

Fuel **volatility** is the ability of the fuel to vaporize (evaporate). The **Reid Vapor Pressure (RVP)** defines the volatility of the fuel. The RVP is the pressure of the vapor above the fuel in a sealed container heated to 100°F. The higher the pressure of the vapor, the greater the volatility of the fuel. This means that the fuel will more readily vaporize.

Fuel **volatility** is the ability of the fuel to vaporize. The higher the volatility, the easier it is for the fuel to evaporate; this allows easier cold starts.

Reid Vapor Pressure (RVP) is a method of describing the volatility of the fuel. You can test the RVP to rate the fuel's volatility.

159

Fuel volatility is adjusted seasonally in many parts of North America. A higher volatility fuel is allowed in the winter to help the engine start when it is cold. The fuel vaporizes more easily during compression rather than puddling along the cool walls of the combustion chamber. Too low a volatility fuel will cause hard starting and rough running at startup. A lower volatility fuel is used in the summer to reduce the amount of evaporative emissions and to prevent vapor lock in the fuel lines. The maximum volatility level is regulated by the Environmental Protection Agency (EPA) between June 1 and September 15. Depending on the state and month, gasoline RVP may not exceed either 9.0 psi or 7.8 psi. This is controlled because if the volatility is too high, excess vapors will form on top of the fuel in the tank and the excess pressure will have to be relieved through a fuel tank vent. These hydrocarbon (HC), raw fuel, emissions are a significant portion of overall vehicle emissions. Too high a volatility fuel used in the summer months can cause extended cranking times. This occurs because the fuel actually boils in the fuel lines, allowing air in the fuel stream after the vehicle is shut off. When the vehicle is restarted, it takes a significant time for the pump to develop fuel pressure.

Combustion Issues

Combustion is the chemical reaction between fuel and oxygen that creates heat. It is a closely controlled burning of the air and fuel. Spark ignition occurs before top dead center on the compression stroke. The hot, compressed, air-fuel mixture is ignited, and a flame front develops. For normal combustion to occur, the air-fuel mixture must be delivered in the proper proportions and mixed well, the spark must be timed precisely, and the temperatures inside the combustion chamber must be controlled. The flame can then move quickly and evenly (propagate) across the combustion chamber, harnessing the power of the fuel as heat. Pressure builds steadily as the gases expand from heat. The peak of this pressure develops around 10° ATDC to push the piston down on the power stroke (Figure 7-12).

Abnormal Combustion

Normal combustion takes about 3 milliseconds (3/1,000 of a second). One form of abnormal combustion, **detonation**, is more like an explosion, occurring as fast as 2 millionths of a second (2/100,000 of a second). The explosive nature of detonation can potentially cause engine damage. Preignition is a form of abnormal combustion when part of the compressed

Detonation occurs when two powerful flame fronts collide inside the combustion chamber and create a potentially catastrophic knocking.

FIGURE 7-12 Normal combustion allows a flame front to travel rapidly across the cylinder and develop the peak force of combustion at about 10° after top dead center.

Hot carbon deposit

FIGURE 7-13 The preignition process is started when an uncontrolled flame front is started too soon. (A) An uncontrolled flame front is started before the spark by a hot carbon deposit. (B) The spark plug is fired. (C) The flame fronts collide.

air-fuel mixture ignites before the spark. Engine misfire is another form of abnormal combustion; it causes the engine to run poorly and lack power.

Preignition

Preignition means that a flame starts before the spark. This can happen when a hot spot in the combustion chamber auto ignites the fuel. A flame front develops and starts moving across the chamber (Figure 7-13). Then the spark occurs, and the normal flame front develops. When these two flame fronts collide, a pinging or knocking is heard. Preignition causing pinging can lead to the more damaging detonation by overheating the cylinder. Hot spots in the combustion chamber, the wrong spark plug, or metal edges sticking out from the head gasket are some of the many reasons preignition can occur. Minor and occasional pinging is normal in some vehicles, particularly those that are driven by a low power engine, and does not cause engine damage.

Detonation

Detonation occurs when combustion pressures develop so fast that the heat and pressure will "explode" the unburned fuel in the rest of the combustion chamber. Before the primary flame front can sweep across the cylinder, the end gases ignite in an uncontrolled burst (Figure 7-14). The dangerous knocking results from the violent explosion. This can be caused by hot spots in the combustion chamber, spark plugs with the wrong heat range, low octane fuel, and other factors affecting the combustion chamber. Detonation can cause serious engine damage, such as bent connecting rods and badly damaged pistons (Figure 7-15).

Knock Sensor

Most modern engines are equipped with a **knock sensor (KS)**. The knock sensor creates an electrical signal when it senses a particular frequency of knocking or detonation. This signal serves as an input to the engine computer, the powertrain control module (PCM). When knocking is detected, the PCM modifies the spark timing to reduce the potentially dangerous knocking (Figure 7-16).

A **knock sensor (KS)** screws into the engine and creates an electrical input for the PCM when abnormal knocking is detected.

Misfire

Misfire is another type of abnormal combustion. When an engine misfires, it means that one or more of the cylinders are not producing their normal amount of power. The cylinder(s) is unable to burn the air-fuel mixture properly and extract adequate energy from

A   B   C

FIGURE 7-14 Detonation results from an uncontrolled flame front that is started after the spark plug is fired. **(A)** Combustion begins. **(B)** Detonation or postspark begins a second flame front. **(C)** The two flame fronts collide to create a knocking sound.

Detonation
damage

Preignition
damage

FIGURE 7-15 Results of the extra stress placed on the piston due to detonation or preignition.

FIGURE 7-16 The knock sensor screws into the engine to detect abnormal engine knocking.

the fuel. The misfire may be total, meaning that a flame never develops and the air and fuel are exhausted out of the cylinder unburned. Hydrocarbon emissions increase dramatically. Misfire may also be partial when a flame starts but sputters out before producing adequate power (due to a lack of fuel, compression, or good spark). When an engine is misfiring, the engine bucks and hesitates; it is often more pronounced under acceleration. Technicians normally call this a miss or a skip. On modern vehicles (1996 and newer) equipped with the OBD II system, a misfire DTC will be set when the PCM detects a single- or multiple-cylinder misfire. This can help guide your diagnosis.

A P0300 DTC means that the PCM has detected a multiple-cylinder misfire. A DTC P0301 means the PCM has detected a misfire on cylinder number 1, and right up to DTC P0312, a misfire on cylinder number 12.

Support Systems' Contribution to Abnormal Combustion

When the cooling, lubrication, intake, exhaust, or fuel system is not functioning properly, it can cause or contribute to improper combustion. If the engine is running too hot due to a malfunctioning cooling system, detonation and preignition are much more likely to occur. A buildup of deposits in a corner of a cooling passage near the combustion chamber can have devastating effects on the engine due to detonation and preignition. The lubrication

FIGURE 7-17 The bearings on the left are worn down to the copper underlayer; these created engine knocking from the bottom end.

system must also be reducing engine friction and heat enough to keep the cylinders cool enough to allow normal combustion.

A restriction in the intake or exhaust system can cause the engine to misfire from a lack of air in the cylinders. If an engine cannot breathe, it cannot produce normal combustion. When the fuel system is failing to deliver adequate fuel, the engine can suffer misfire and preignition from running too hot. A lean air-fuel mixture causes hot engine temperatures; the fuel actually helps to cool the mixture. If the compressed air-fuel mixture is too hot, a portion of it can auto ignite either before or after the spark begins.

ENGINE NOISES

The engine can create a wide variety of noises due to worn or damaged parts. It will be your job to diagnose the causes of these noises and to correct them. Most noises occur because of worn components causing excessive clearances. A typical wear item is crankshaft bearings, which cause expensive damage heard as a deep lower-end knock (Figure 7-17). Valve lifters also commonly wear with higher mileage and cause a higher pitch clatter. These are just a couple of examples of possible noises; in the *Shop Manual*, you will learn to isolate noises and their causes.

SUMMARY

- Proper engine performance requires that the engine is mechanically sound and that its support systems are functioning as designed.
- The combustion chamber must be properly sealed to provide good engine performance.
- The valves, spark plug, rings, and headgasket seal the combustion chamber.
- The octane rating of gasoline describes its ability to resist knock; the higher the number, the greater the resistance to knocking.
- Higher volatility fuel should be used in the winter to assist cold starts; lower volatility fuel should be used in the summer to prevent excessive HC emissions and vapor lock.
- Preignition and detonation are two types of abnormal combustion that can cause serious engine damage.
- Failures in the cooling or lubrication systems can cause abnormal combustion or serious engine defects.
- Normal engine wear will eventually lead to abnormal noises and reduced performance.

TERMS TO KNOW

Detonation

Knock sensor (KS)

Misfire

Octane rating

Reid vapor pressure (RVP)

Volatility

REVIEW QUESTIONS

Short Answer Essays

1. What components seal the combustion chamber?

2. What problems can occur from improper combustion chamber sealing?

3. What effect will a burned valve have on engine performance?

4. What does a gasoline's octane rating describe?

5. What problems can occur when fuel with an inappropriate volatility is used?

6. Define preignition.

7. Define detonation.

8. Describe some causes of abnormal combustion.

9. What engine problems can detonation lead to?

10. What causes abnormal engine noises?

Fill-in-the-Blanks

1. A leak past the _____ gasket can cause compromised engine performance.

2. The _____ _____ ,
   _____ _____ ,
   _____ _____ , and
   _____ seal the combustion chamber.

3. Malfunctions in the _____ system and the _____ system can cause engine mechanical problems.

4. Two likely causes of preignition are _____ and _____ .

5. Preignition occurs when a flame front starts _____ the spark.

6. _____ is likely to burn a hole in the top of the piston.

7. _____ describes a fuel's ability to vaporize.

8. _____ and _____ can lead to preignition.

9. A common cause of clatter from the top end of a high mileage engine is worn _____ _____ .

10. Two likely causes of engine misfire are _____ and _____ .

Multiple Choice

1. Fuel is generally available in octane ratings of all of the following Except:
   A. 87 octane   C. 92 octane
   B. 89 octane   D. 99 octane

2. Normal combustion takes approximately:
   A. Three milliseconds   C. Two nanoseconds
   B. Thirty milliseconds   D. Twenty nanoseconds

3. Misfire means:
   A. The cylinder fires at the wrong time.
   B. The cylinder does not fire.
   C. Two sparks occur.
   D. The wrong spark plug is installed.

4. Each of the following is a likely cause of detonation Except:
   A. Carbon buildup in the combustion chamber
   B. Engine overheating
   C. A lean air-fuel mixture
   D. Fuel with too high of an octane rating

5. A _____ informs the PCM about unusual frequency vibrations in the engine:
   A. Coolant temperature sensor
   B. Vibrator actuator
   C. Knock sensor
   D. PCM mounting sensor

6. *Technician A* says that worn piston rings can cause a lack of power.

   Technician B says that a burned valve can cause a popping noise out of the intake or exhaust.

   Who is correct?
   A. A only   C. Both A and B
   B. B only   D. Neither A nor B

7. Spark ignition occurs:
   A. ATDC on the power stroke
   B. ATDC on the compression stroke
   C. ATDC on the intake stroke
   D. BTDC on the compression stroke

8. *Technician A* says that hydrocarbon emissions typically increase when an engine misfires.
   Technician B says that low compression can cause misfire.
   Who is correct?
   A. A only
   B. B only
   C. Both A and B
   D. Neither A nor B

9. Too low a fuel volatility used in the winter will cause:
   A. Excessive evaporative emissions
   B. Hard starting and rough running when cold
   C. Extended cranking times when the engine is hot
   D. Detonation

10. A burned intake valve will cause:
    A. Weak engine vacuum
    B. Lower compression pressures
    C. Reduced engine power
    D. All of the above

Chapter 8

ENGINE CONFIGURATIONS, MOUNTS, AND REMANUFACTURED ENGINES

UPON COMPLETION OF THIS CHAPTER, YOU SHOULD BE ABLE TO:

- Describe different engine configurations.
- Describe the methods of mounting the engine.
- Explain the function of different engine mounts.
- List the common attaching points for the engine.
- List the common accessories found on the engine assembly.
- Describe the benefits and limitations of a remanufactured engine.

A **remanufactured engine** is one that has been professionally rebuilt by a manufacturer or engine supplier.

A **crate engine** is a new or remanufactured engine that is built with fresh components, such as bearings, rings, and lifters. You can purchase crate engines with or without cylinder heads.

The **differential** allows the wheels to turn at different speeds as the vehicle turns corners. It is internal to a transaxle and external to a transmission.

INTRODUCTION

When performing major engine mechanical repairs, you will be confronted with different engine configurations. The engine may be mounted transversely or longitudinally, or it may be a mid-engine, which is mounted in the rear. You will need to understand the operation of motor mounts as part of your assurance that the engine will perform smoothly. When overhauling or replacing an engine, you will have to remove many engine accessories and brackets; this requires some familiarity with these components. In many cases, you, your shop management, or your customer may choose to install a **remanufactured engine**. You will need to understand the benefits and limitations of choosing a "reman" or **crate engine** versus you overhauling the engine in-house.

ENGINE CONFIGURATIONS

The most common configuration for a front-wheel-drive vehicle is to have a transverse-mounted engine (Figure 8-1). The engine is mounted sideways in the engine compartment, so that the front of the engine is mounted toward one side of the vehicle. This is a popular design, in part because it can reduce the size of the engine compartment. The transaxle is attached to the side of the engine, and faces the other side of the vehicle. The transaxle contains the **differential** assembly, located in the rear of the vehicle on rear-wheel-drive vehicles. The compact size of the transaxle lowers overall vehicle weight. This is always a goal of vehicle manufacturers as they try to improve fuel economy.

A transverse engine is often removed from the bottom of the engine compartment with the transaxle attached. This often requires parts of the **engine cradle** or suspension cradle to be removed (Figure 8-2). The engine and the suspension components are attached to the engine cradle, so disassembly of parts of the suspension may be required to remove the engine.

Many rear-wheel-drive vehicles install their engines **longitudinally** (Figure 8-3). A few front-wheel-drive vehicles also mount their engines in this fashion, though it is not

FIGURE 8-1 This transverse-mounted engine has many accessories and attachments that would have to be removed during an engine overhaul or an engine swap.

Cradle toward the rear of the vehicle   Cradle mount

Engine and suspension cradle
front of the vehicle

FIGURE 8-2 This engine cradle comes out as one piece to remove the engine through the bottom. The suspension will have to be disconnected.

The **engine cradle**, which may also be called the suspension cradle, is the front frame member(s) in a unibody body design. The engine is mounted to this frame, as is the suspension.

A **longitudinally mounted** engine is placed in the vehicle lengthwise, usually front to rear.

common. The front of the engine faces the front of the vehicle, and the transmission mounts on the back of the engine. The engine and transmission are bolted together and mounted to the frame through a rear engine mount (Figure 8-4). A longitudinal engine may be removed from the top through the hood opening or through the bottom of the engine compartment.

FIGURE 8-3 A longitudinal engine mounted in an all-wheel drive SUV.

Transmission

Rear engine mount

Cross member

Engine mount bracket

FIGURE 8-4 The engine and transmission use a rear engine mount
to help stabilize the components.

A **mid-mounted engine** is located between the rear of the passenger compartment and the rear suspension. It may also be called a mid-engine.

An **engine mount** attaches the engine to the chassis and absorbs the engine vibrations.

A few vehicles mount their engines in the rear of the vehicles. Some are located between the passenger compartment and the rear suspension. These are called **mid-mounted engines** (or simply mid-engines) and are typically mounted transversely. They are used in compact, high performance, rear-wheel-drive sports cars. The front of the vehicle can be low and sloped to maximize aerodynamics. The weight of the powertrain near the center of the vehicle places the center of gravity near the vehicle midpoint. This improves steering and handling for fine sports car performance. The most rare configuration is an engine mounted at the true end of the vehicle.

ENGINE MOUNTS

Engines usually have two or three **engine mounts** to attach the engine to the frame and to isolate the vibration of the engine from the chassis and passenger compartment (Figure 8-5). There may be additional mounts that hold the transmission or transaxle and also help support the engine. The engine will have a top-engine mount that may look quite different from the one shown in the previous figure. Some mount at the very front of the

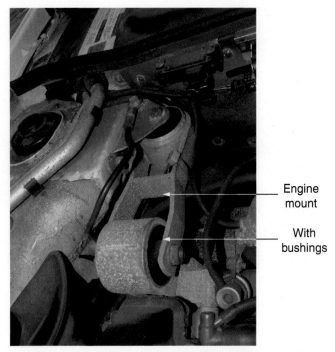

Engine mount

With bushings

FIGURE 8-5 This top-engine mount has two bushings to support the engine vertically and horizontally.

FIGURE 8-6 This top-engine mount attaches to the front cross member and the front of the engine.

engine and attach to the front radiator cross member (Figure 8-6). A lower engine mount on a transverse-mount engine may attach to either the back of the engine or the front of the transaxle (Figure 8-7).

Types of Engine Mounts

Many engine mounts are simply rubber bushings in a metal casing. The rubber bushing absorbs the engine vibrations. The rubber is attached to the housing with some slack to allow movement within the case (Figure 8-8). The rubber may have voids to allow more movement within the mount.

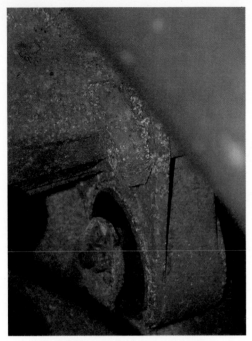

FIGURE 8-7 This lower engine mount actually attaches at the front of the transaxle, very close to the engine.

The rubber bushing has voids

FIGURE 8-8 The voids in the bushings allow a looser attachment to the case. This allows the engine to vibrate and the mount to isolate those vibrations from the chassis.

Hydraulic engine mounts use two chambers filled with fluid and an orifice in between to dampen the vibration of the engine. It takes some time and energy to displace the fluid from one chamber to the other as the engine moves. This helps absorb the engine vibration. They look and function very similarly to suspension shock absorbers (Figure 8-9).

Hydraulic engine mount

FIGURE 8-9 This hydraulic engine mount looks very much like a shock absorber. Its function is quite similar as well.

Some hydraulic engine mounts are electronically controlled. When the engine is idling, the engine mounting should be very soft to allow the engine to run as smoothly as possible during its low load condition. Once the engine is driving under a load, stiffer dampening must occur to isolate the harsher movements from the chassis. To achieve that exceptionally smooth idle, more and more vehicles are using electronically controlled hydraulic engine mounts.

In one example, the two hydraulic chambers are connected by a large and a small orifice. When the PCM determines that the engine is idling, it actuates a vacuum motor that opens the normally closed larger orifice. This allows more fluid flow and softer dampening to absorb the shaking of the engine at idle. Only the smaller valve is open during engine-loaded conditions, to provide the stiffer dampening needed to isolate engine vibration.

REMANUFACTURED ENGINES

It is a trend in today's automotive industry to replace seriously damaged engines with remanufactured or new units rather than overhaul the existing engine in the shop. Remanufactured engines are available from vehicle manufacturers and other suppliers (Figure 8-10). They will come with the internal components of the engine fully assembled and adjusted. Beyond that, the engines you receive may vary significantly. You can purchase a crate engine with or without cylinder heads. Some will come with new oil and water pumps and covers. In other cases, it is your responsibility to replace these components as warranted by the vehicle. Many crate engines will come as relatively bare engines with no covers or new accessories. All sensors, brackets, hoses, and tubing will be transferred from the old engine to the new.

There are several advantages to installing a remanufactured engine. A crate engine comes fully assembled and carries a written warranty. That means that if the "new" engine suffers damage within the warranty period, the supplier will provide a new unit. If a technician rebuilt an engine in the shop and it came back within a few thousand miles with another engine mechanical problem, the shop would be responsible for making it right for the customer. This would cost the shop the technician's labor and the cost of all the new parts to repair the damage.

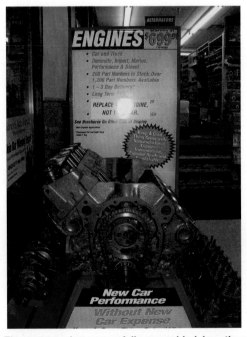

FIGURE 8-10 The crate engine comes fully assembled, less the accessories.
Some engines come more complete than others.

A Bit of History

The idea of supplying remanufactured engines rather than overhauling existing ones is relatively new. In the 1980s, for example, most shops were rebuilding all their engines. Technicians were well practiced at the fine measuring required for professional overhauls. Many shops were still performing cylinder head service or "valve jobs" in-house as well. Now several

It is also often more cost-effective to install a reman engine rather than overhaul one in-house. A supplier that provides crate engines has a production shop of sorts. The technicians who work there overhaul engines every day and all day. They are very efficient at this. The supplier is also able to source parts in high volume for significant discounts. A technician in a shop that overhauls an engine a few, or even a dozen, times a year will take longer to do the same work. The shop may not be able to source the discount parts that are available, especially to a high volume consumer. The machining work that may need to be done at a local machine shop also adds cost and time to the overhaul.

> **AUTHOR'S NOTE:** Many students are disappointed to learn that engine overhaul is not a common task in most general automotive repair shops. If engine overhaul is something you are interested in, you can make that a priority when looking for work. There are still many shops out there that specialize in engine repair. You can also consider working for an engine manufacturer.

Sometimes when you are beginning the disassembly of an engine, it becomes apparent that the damage is so extensive that a remanufactured engine would definitely be cheaper (Figure 8-11). A hole in the engine block from a piston that let loose would make that engine cheaper to replace than to repair. You may not make a decision until you have taken the engine apart and evaluated it. Other shops simply follow the policy of installing crate engines whenever possible.

There are reasons not to install remanufactured engines. For some vehicles, reman engines are not readily available or may be cost prohibitive (Figure 8-12). It is difficult to locate remanufactured engines for many low volume import engines, for example. The condition of the vehicle and the desire of the customer may well warrant the cost of a

FIGURE 8-11 After finding piston and ring pieces in the oil pan, this shop decided that a crate engine would be more cost-effective.

FIGURE 8-12 This high mileage engine needs some major mechanical work. A reman engine is not readily available. The customer likes the vehicle so much he has chosen to give the engine a thorough in-house overhaul.

thorough in-house overhaul. In some cases, the customer may want a new engine from the manufacturer despite the cost. This may be either a truly new or a remanufactured engine, as made available by the particular manufacturer. In other cases, the vehicle may be in such poor condition that it is not worth it to install a new crate engine. The customer could also be in a financial situation that does not allow him to purchase a new engine. As we will discuss during our short block service discussions in a later chapter, you may be able to perform satisfactory, though minimal, repairs on a high mileage engine. This sort of work may buy the customer the time he needs to be able to purchase a new engine or a new vehicle. By the time you complete your study of this text, you should feel confident in your ability to make these sorts of judgment calls.

A BIT OF HISTORY
(continued)

shops have let much of the engine overhaul machining tools deteriorate or have sold them to machine shops where most of the engine machining is now done. Many "old-timers" and young enthusiasts alike miss the more frequent opportunity to build an engine.

SUMMARY

- Most front-wheel-drive engines are mounted transversely.
- Many rear-wheel-drive engines are mounted longitudinally.
- Some sports cars have mid-mounted engines.
- Engine mounts absorb the engine vibrations.
- Many engine mounts are either rubber or hydraulic.
- Electronically controlled hydraulic engine mounts can produce a very smooth idle.
- The engine has mechanical attachments to many other systems.
- Remanufactured engines are available for many popular engines.
- Crate engines may be a cost-effective alternative to an in-house overhaul.

TERMS TO KNOW

Crate engine

Differential

Engine cradle

Engine mount

Longitudinally mounted

Mid-mounted engine

Remanufactured engine

Transverse-mounted engine

REVIEW QUESTIONS

Short Answer Essays

1. List the types of engine configurations.

2. Explain the advantages of a transverse-mounted engine.

3. Explain the advantages of a mid-mounted engine.

4. Explain the purposes of engine mounts.

5. Describe the operation of the hydraulic engine mount.

6. Explain the operation of the electronically controlled engine mount.

7. Explain the benefit of the electronically controlled engine mount.

8. Explain two examples when you would be likely to install a remanufactured engine.

9. Describe two benefits of a remanufactured engine.

10. Describe why you might choose to overhaul an engine in-house.

Fill-in-the-Blanks

1. A _____ engine is mounted between the rear of the passenger compartment and ahead of the rear suspension.

2. A transverse engine assembly allows better _____ because it is lighter.

3. A longitudinally mounted engine is usually used on _____ vehicles.

4. The two most common types of engine mountings are the _____ and the _____.

5. The _____ engine mounts work like a shock absorber.

6. Rubber engine mounts have _____ to make them softer.

7. Electronically controlled hydraulic engine mounts improve _____ quality.

8. The engine is usually detached from the exhaust system at the _____.

9. A remanufactured engine may also be called a _____ engine.

10. The decision to install a remanufactured engine or overhaul an engine in-house may be made by _____, the _____, or _____.

Multiple Choice

1. The most common type of engine installation in a front-wheel-drive vehicle is:
   A. Longitudinal
   B. Transverse
   C. Mid-mount
   D. Rear

2. The most common type of engine installation in a rear-wheel-drive vehicle is:
   A. Rear
   B. Mid-mount
   C. Transverse
   D. Longitudinal

3. *Technician A* says that a transverse engine is usually coupled to a transaxle.
   Technician B says that most front-wheel-drive vehicles use longitudinally mounted engines.
   Who is correct?
   A. A only
   B. B only
   C. Both A and B
   D. Neither A nor B

4. *Technician A* says that a mid-engine is naturally faster than a longitudinally mounted engine.
   Technician B says that a mid-engine installation can improve vehicle handling.
   Who is correct?
   A. A only
   B. B only
   C. Both A and B
   D. Neither A nor B

5. A rubber engine mount has voids to make the mount:
   A. Softer
   B. Stiffer
   C. Last longer
   D. Resist twisting

6. A hydraulic engine mount moves fluid through a:
   A. Pintle
   B. Solenoid
   C. Canister
   D. Orifice

7. *Technician A* says that an electronically controlled hydraulic engine mount can improve idle quality.
   Technician B says these mounts can help isolate engine vibration during hard acceleration.
   Who is correct?
   A. A only
   B. B only
   C. Both A and B
   D. Neither A nor B

8. *Technician A* says that it may be a shop policy to install a remanufactured engine rather than overhaul an engine in-house.
   Technician B says that some engines are damaged badly enough that it is not cost-effective to repair them.
   Who is correct?
   A. A only
   B. B only
   C. Both A and B
   D. Neither A nor B

9. A crate engine may come with a new _____ installed.
   A. A/C compressor
   B. Power-steering pump
   C. Oil pump
   D. PCM

10. A reason to install a remanufactured engine may be:
    A. Cost
    B. Warranty
    C. Time savings
    D. All of the above

Chapter 9

CYLINDER HEADS

UPON COMPLETION AND REVIEW OF THIS CHAPTER, YOU SHOULD BE ABLE TO:

- Explain valve design, including an explanation of the stem, head, face, seat, margin, and fillet.

- Explain the reasons for burned valves.

- Describe the causes of valve channeling.

- Describe two different types of valve seats.

- Explain the reasons for adding Stellite to valve seats.

- Explain valve seat recession, and explain the causes of this problem.

- Explain the purposes of valve stem seals, and list three different types of these seals.

- Explain the results of worn valve stems and guides.

- Describe different combustion chamber designs.

- Explain the combustion process.

- Explain how combustion chamber design can affect exhaust emissions and engine performance.

INTRODUCTION

Proper combustion requires the engine to "breathe" properly. The air intake system directs air to the throttle body. From the throttle body, the intake air is directed through the intake manifold, cylinder head, and past the intake valves into the combustion chambers. Once in the combustion chambers, the air-fuel mixture is ignited, and the resulting power output is affected by chamber design. Finally, the spent gases are expelled from the combustion chambers, past the exhaust valves, and out through the exhaust manifold and exhaust system.

The intake and exhaust manifolds, cylinder head, valves, and rocker arms must all work together to move the air-fuel mixture in and out of the combustion chambers and expel the exhaust gases. In this chapter, we discuss cylinder heads and related components such as valves, valve seats, and valve guides. Discussion of the typical system failures and their causes is also included. Today's technician must be able to identify a failed component, and determine the cause of the failure. Although lack of lubrication is one of the most common causes of component failure, all engines will exhibit some form of normal wear between moving parts. A technician must be able to classify component wear as normal or abnormal. If abnormal wear is determined, the cause must be identified and corrected before the engine is reassembled.

CYLINDER HEADS

On most engines, the cylinder head contains the valves, valve seats, valve guides, valve springs, and the upper portion of the combustion chamber (Figure 9-1). In addition, the cylinder head has passages to allow coolant and oil flow through the head. Many modern engines have overhead camshafts (OHCs). In these cases, the cylinder head also houses the camshaft(s) and all of the valvetrain components.

Exhaust camshaft

Hydraulic lash adjuster

Rocker arm

Intake camshaft

Valve spring retainer

Valve spring

Valve stem seal

Exhaust port

Intake port

Exhaust valve

Intake valve

FIGURE 9-1 Typical OHC cylinder head components.

The cylinder head is attached to the block above the cylinder bores. This mating surface must be flat and have the correct surface finish. The cylinder head gasket is installed between the cylinder head and the block. It has the formidable task of sealing in the powerful forces of combustion and preventing internal or external leaks of oil or coolant.

The cylinder head is made of cast iron or aluminum. It is common for engine manufacturers to use aluminum alloys to cast the head and mate it to a cast-iron block. The difference in thermal expansion between the cylinder head and block creates a scrubbing stress that must be withstood by the head gasket. Head-gasket failures can cause serious driveability concerns and engine damage. When the head gasket fails, it frequently allows coolant to enter the combustion chamber and also admits combustion gases into the cooling system. The burning coolant causes white, sweet-smelling smoke to be emitted from the tailpipe and will often cause misfire in the affected cylinder. In extreme cases, it can hydraulically lock the engine and cause extensive piston and cylinder damage. In most cases, the coolant passing across the oxygen sensors will destroy them. Many technicians replace the oxygen sensors after a head-gasket failure or at least warn the customer that the sensors may fail in the near future. The combustion gases in the cooling system cause pressure and air pockets in the system that frequently result in engine overheating and boiling over. Other times the symptoms may include: erratic heater operation and engine temperature, consistent or intermittent overheating, rough running, misfiring (especially at start up), excessive pressure in the cooling system, bubbles in the coolant overflow and coolant overflow from the tank, coolant loss without an external leak, or low compression on two adjacent cylinders. A blown head gasket can also allow coolant to mix in the oil, creating a fluid in the lubrication system that closely resembles a coffee milkshake. The oil's lubricating and corrosion protection properties are diminished. Driving the vehicle with this condition can cause extensive engine damage.

A cracked cylinder head can cause the same symptoms as a blown head gasket. Occasionally, the crack in the head gasket or cylinder head is very small and will cause less dramatic symptoms. When the engine is cool, the metals in either of the suspect parts contract and enlarge the crack. Coolant may leak into the affected cylinder and foul the spark plug or cause small amounts of white smoke to exit the tailpipe. A classic symptom of a cracked cylinder head (or a small crack in the head gasket) is misfire on a cylinder accompanied by a little white smoke for about one minute after startup. Then the engine heat can cause enough expansion in the metal to seal the gap, and the engine may run perfectly for the rest of the day.

Prompt repair of a vehicle exhibiting any of these symptoms is essential to prevent more severe engine damage. Overheating typically causes failure of the head gasket or a crack in the cylinder head. Often, the white smoke will billow from the tailpipe, making diagnosis straightforward. In other cases, very little smoke will be visible, and further testing will be needed to determine the cause of the overheating or misfire.

The combination of an aluminum cylinder head and a cast-iron block with coolant circulating between them presents the problem of electrolysis. This is the chemical action producing a small electrical charge in the coolant. As the coolant ages, it becomes acidic and electrolysis accelerates. Electrolysis will eat away at the aluminum components in the cooling system, including the cylinder head. Proper cooling system maintenance reduces the impact of electrolysis on aluminum components.

Intake and Exhaust Ports

The intake and exhaust ports cast into the cylinder head are carefully machined to match the flow of air from the intake manifold and out to the exhaust manifold. Some cylinders have Siamese ports, meaning that two cylinders share the same opening. This is not an ideal design but may be used to save space or cost. Intake ports may use a hump in them to increase air velocity, thus increasing volumetric efficiency. When the intake ports are on one side of the head and the exhausts on the other, the head is called a crossflow cylinder head design. This adds to the volumetric efficiency of the head, especially during valve overlap. As the intake valve opens, the incoming air is drawn into the moving air across the chamber from the exhaust flow (Figure 9-2). Non-crossflow designs have a much harder time filling the combustion chamber fully with air. Exhaust airflow out of the chamber may actually provide resistance to incoming air as it is pushing out in the opposite direction of the incoming air. The proximity of the hot exhaust gases to the intake charge also has a tendency to heat (and therefore reduce the density of) the intake charge, reducing the volumetric efficiency.

Exhaust valve

Intake valve

Air out

Air in

FIGURE 9-2 The momentum of airflow helps draw in a fresh charge; in a cross-flow chamber this is particularly effective.

Camshaft Bore

The camshaft(s) on an overhead-cam engine has a bore machined into the head to house the camshaft. The top halves of the bores are caps or a cap housing that bolts to the head (Figure 9-3). On some engines, the bores may be fixed, and you can slide the camshaft into place from either the front or rear of the head. Overhead camshafts sometimes use bearings. More often the integral bearing surface is the soft aluminum of the head. Oil passages lead into these bores to provide essential lubrication. If the cam bores are damaged, the head usually requires replacement, though in some cases the bores are machined oversize and inserts can be installed.

Valves

Modern automotive engines use **poppet valves** (Figure 9-4). The poppet valve is opened by applying a force that pushes against its tip. The camshaft lobes work to open the valves through the components of the valvetrain. Closing of the valve is accomplished through spring pressure.

The valve as a whole has many parts (Figure 9-5). The large-diameter end of the valve is called the **head**. The angled outer edge of the head is called the **face** and provides the contact point to seal the port. The seal is made by the valve face contacting the **valve seat** in the cylinder head. The valve seat is press fit into the cylinder head or an integral part of the head. To provide a positive seal, the valve face is cut at an angle. The angle is usually

Poppet valves control the opening or passage by linear movement.

The **head** is the enlarged part of the valve.

The **face** is the tapered part of the section of the valve head that makes contact with the valve seat.

The **valve seat** is the point at which the seal is made by the valve face.

Camshaft bearing cap

Cylinder head

Camshaft

FIGURE 9-3 These soft aluminum camshaft caps serve as the cam bearing surface. In this case, there are no actual cam bearings or bearing inserts.

FIGURE 9-4 Typical poppet valve configuration using an overhead camshaft.

FIGURE 9-5 The parts of the valves.

45°, though some manufacturers may use a 30-degree angle on some valves and seats. The angle causes the seal to wedge tighter when the valve is closed and allows free flow of the air and fuel mixture when the valve is open. The valve face and seat must both be precisely machined to the correct angle and width to provide good sealing. Pits, corrosion, or carbon on either surface can result in leakage of compression and combustion forces.

The area between the valve face and the valve head is referred to as the **margin**. The margin allows for machining of the valve face and for dissipation of heat away from the face and stem. The margin adds structure to the valve. If it is cut too thin during refinishing, the valve will burn or become misshapen and leak cylinder pressure. The **valve stem** guides the valve through its linear movement. The stem also has valve keeper grooves machined into it close to the top. These grooves are used to hold keepers that retain the valve springs. The stem rides in a valve guide located in the cylinder head.

The head of the intake valve is typically larger than the exhaust valve's. This is because maximizing intake valve size is considered more important than the exhaust valve. There is only so much area in the combustion chamber to locate the valves, and the larger the intake valve, the smaller the exhaust valve has to be. Smaller exhaust valves are able to work efficiently because the exhaust is being forced out of the chamber by the upward movement of the piston and by the heat generated by combustion; thus they are less sensitive to restrictions, whereas the air-fuel charge is being pulled into the cylinder by vacuum alone, and restrictions will reduce the engine's efficiencies. Larger intake valves reduce the restrictions and improve volumetric efficiency.

Most valves are constructed of high strength steel; however, some manufacturers are experimenting with ceramics. Most valve heads are constructed of 21-2N and 21-4N stainless steel alloys; 21-2N alloy is commonly used in original equipment exhaust valves, and 21-4N is a higher grade of stainless steel containing more nickel. One of the primary concerns in valve construction and selection of alloys is the operating temperatures to which they are subjected. Combustion chamber temperatures can range from 1,500°F to 4,000°F (82°C to 220°C). The types of material used to construct the valve must be capable of withstanding these temperatures and dissipating the heat rapidly. For every 25°F (14°C) reduction in valve temperature, the valve's burning durability is doubled. There are several alloys available for valve construction that will increase the valve's burning durability. An alloy that is being used by many

The **margin** is the material between the valve face and the valve head.

The **valve stem** guides the valve through its linear movement.

Shop Manual
Chapter 9

manufacturers is Inconel. Inconel is constructed from a nickel base with 15 to 16 percent chromium and 2.4 to 3.0 percent titanium.

Valve construction design is also an important aspect for controlling heat. A valve can be constructed as one piece or two piece. One-piece valves run cooler since the weld of a two-piece valve inhibits heat flow up the stem. Two-piece valves allow the manufacturer to use different metals for the valve head and stem.

In addition to alloy selection, manufacturers design the valve and cylinder head to dissipate heat away from the valve (Figure 9-6). Most of the heat is dissipated through the contact surface of the face and seat. The heat is then transferred to the cylinder head. About 76 percent of the heat is transferred in this manner. Most of the remaining heat is dissipated through the stem to the valve spring and on to the cylinder head. To increase the transfer of heat through the stem, some stems are filled with sodium. Turbocharged, supercharged, or other high performance engines may use sodium-filled valves for their excellent heat dissipation quality and decreased weight. A sodium-filled valve has a hollow stem that is partially filled with metallic sodium (Figure 9-7). When the valve opens, the sodium splashes down toward the head and picks up heat. When the valve closes, the sodium moves up the stem taking heat away from the hottest area of the valve, the head. The heat is transferred from the stem to the guide and then dissipated in the coolant passages around the guide. Sodium-filled valves should not be machined; if you were to cut through the valve, the sodium could explode. These valves should be replaced when reconditioning the head.

Many manufacturers use chrome-plated stems to provide additional protection against wear resulting from initial engine starts when oil is not present on the valve stem. In addition, chrome works to protect against galling when cast-iron guides are used. If the valve stem is reground, the chrome plating will be removed. In this case, either the valve will have to be replated or a bronze guide must be used.

The **fillet** is the curved area between the stem and the head. It provides structural strength to the valve. The fillet shape also affects the flow of the air-fuel mixture around the valve. A valve with a steeply raked fillet (called a tulip valve) has better flow than a valve with a flatbacked fillet. Fillet shape is only a factor while the valve is opening; after the valve is completely open, the fillet shape has very little effect on airflow.

Although not normally considered a wear area, the fillet must be thoroughly cleaned. Carbon will build up in this area, causing the flow to be disrupted. Excessive carbon in this location can be an indication of worn valve stem seals or guides. A wire brush is usually required to remove carbon buildup on the fillet.

FIGURE 9-6 Heat must be removed from the valve.

FIGURE 9-7 The hollow valve is partially filled with sodium. When the valve closes, the sodium splashes up to help take heat away from the head and into the stem.

Aluminized Valves. Aluminized valves provide increased valve life by reducing the effects of corrosion. Small particles of aluminum are fused to the valve head, causing a reaction that makes the surface corrosion resistant. Aluminized valves should not be reconditioned because this process removes the coating from the head.

Determining Valve Malfunctions

It is not enough to be able to recognize a failed valve. It is very important to determine the cause of the malfunction to prevent a repeat failure. Different types of failures are usually caused by identifiable conditions.

Deposits on Valves. Excessive carbon buildup on valve stems is a relatively common problem on modern vehicles (Figure 9-8). This can be caused by excessive valve stem temperatures caused by increased clearance between the valve guide and the valve stem. Deposits may increase as the valve seals wear and allow more oil into the guide, which can then bake onto the back of the valve. The use of improper or contaminated oil also leads to the formation of carbon deposits on the valve stems.

The carbon buildup shown on the valve in Figure 9-8 affects engine performance. First, the mass of carbon restricts airflow into the combustion chamber and the loss of air intake reduces power. Secondly, the weight of the carbon commonly causes **valve float** at higher rpms. Valve float occurs when the springs can no longer hold the valve closing consistent with the profile of the cam lobe. Instead of closing swiftly with the drop of the camshaft lobe, the valve floats over the ramp and hangs open at higher rpm, when the momentum is greater. This causes engine misfire. Another common symptom of carbon buildup affects starting in the morning or when the engine is cold. The engine will typically start and stall two or three times and then start and run roughly for a few seconds. What happens is that the fuel injected into the intake is absorbed by the sponge-like carbon during the first couple of starting attempts. Then, when the carbon becomes saturated with fuel, the vehicle starts and runs roughly, until the excess fuel is drawn out of the carbon.

Valve float is a condition that allows the valve to remain open longer than intended. It is the effect of inertia on the valve.

Burned Valves. A **burned valve** is identifiable by a notched edge beginning at the margin and running toward the center of the valve head (Figure 9-9). The most common cause of a burned valve is a leak in the sealing between the valve face and seat. At the location of the leak, the valve will not cool efficiently. This is because the heat from the valve head cannot be transferred to the valve seat and cylinder head. As the area increases in temperature, it begins to warp. The warpage results in less contact at the valve seat and compounds the problem of heat transfer. The resultant heat buildup melts the edge of the valve.

Burned valves are actually valves that have warped and melted, leaving a groove across the valve head.

FIGURE 9-8 Carbon buildup on the neck of the valve interferes with breathing.

FIGURE 9-9 Burned valves are characterized by a groove from the face toward the fillet.

FIGURE 9-10 Excessive heating of the valve can lead to cupping.

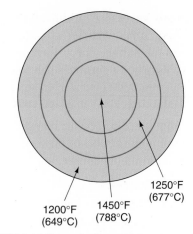

1250°F
(677°C)

1200°F   1450°F
(649°C)   (788°C)

FIGURE 9-11 Normal valve temperatures as heat is dissipated through the face and seat.

Since heat transfer is dependent upon full contact of the valve face to the seat, anything preventing this contact can result in a burned valve. This includes weak valve springs, carbon buildup, worn valve guides, and valve clearance adjusted too tight. Additional causes of burned valves include preignition, exhaust system leaks, and failure of the cooling system.

Not all overheating conditions result in burned valves. The metal of the valve head may become soft and form into a cup shape (Figure 9-10). The force applied by the springs to seat the valve, combined with combustion pressures, deforms the valve head.

Channeling. Under normal conditions, the fillet area of the valve should be the hottest area (Figure 9-11). The temperature around the valve edge should be the coolest. If the valve does not seat properly, the temperature is not transferred to the cylinder head. The area not contacting the seat becomes the hottest area of the valve head (Figure 9-12). The area with the hottest temperature will melt away, developing a channel. **Channeling** usually occurs just prior to valve burning. Channeling is generally due to four main causes:

1. Flaking off of deposits from the valve
2. Extensive valve face peening due to foreign material lodged between the valve face and seat
3. Corroded valve face
4. Thermal stress of the cylinder head, resulting in radial cracking

Metal Erosion. Metal erosion is identified by small pits in the valve face that usually cannot be removed by valve grinding (Figure 9-13). This erosion is generally caused by coolant

Area of leak

1550°F
(843°C)
1650°F
(899°C)   1450°F   1250°F   1200°F
(788°C)   (677°C)   (649°C)

FIGURE 9-12 Improper face and seat contact results in improper heat dissipation.

Excessive pitting

Heat damage

Pitting not removed

Margin ground off

FIGURE 9-13 Metal erosion of the valve.

entering the combustion chamber. The chemical reaction etches away the face material. This condition can also occur if some chemicals enter the combustion chamber. These chemicals can come in the form of coolant system conditioners, oil additives, and fuel additives.

Breakage. Breakage of the valve can occur due to valve and piston contact. In a **free-wheeling engine**, valve lift and angle prevent valve-to-piston contact if the timing belt or chain breaks. In an **interference engine**, this can be the result of timing chain or belt breakage. Interference designs are commonly used because they offer improved fuel efficiency, increased power output, and lower emissions. However, if the valve timing components fail, the valve can contact the top of the piston or another valve. This contact may cause the valve to break or be drawn into the combustion chamber. If this occurs, major engine damage results. If the engine you are working on has experienced this type of failure, inspect the head and block carefully for cracks and other damage. The head must usually be replaced with this type of failure.

Other than valve-to-piston contact, valves may also experience fatigue and impact breaks. A fatigue break is a gradual breakdown of the valve due to excessive heat and pressure. Impact breakage can result from the valve face being forced into the valve seat with excessive force. Impact breakage can be the result of excessive valve clearance, valve springs with excessive pressure, loss of valve keepers, or any condition that can result in **stem necking**.

Necking is identified by a thinning of the valve stem just above the fillet (Figure 9-14). The head pulls away from the stem, causing the stem to stretch and thin. This condition is caused by overheating or excessive valve spring pressures. In addition, necking can be caused by exhaust gases circling around the valve stem. The corrosive nature of these gases and the action resulting from the swirl eat away the metal of the valve stem. Necking resulting from this action is caused by anything resulting in increased exhaust temperatures, including:

- Improper ignition timing
- Improper air-fuel mixtures
- Overloading
- Improper gear ratio in the final drive
- Exhaust restriction

An additional cause of valve breakage is referred to as **off-square** seating or tipping. If the valve does not seat properly, it may flex as it attempts to set into the valve seat. This flexing weakens the stem and eventually may lead to its breakage (Figure 9-15). Off-square seating can be caused by worn valve guides, distortion of the valve seat, improper rocker arm alignment, and weak valve springs.

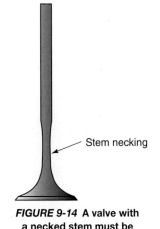

FIGURE 9-14 A valve with a necked stem must be replaced.

FIGURE 9-15 Tipping of the valve in the guide can keep the valve from seating properly, resulting in breakage near the fillet.

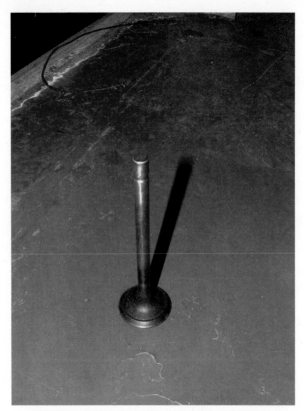

FIGURE 9-16 A valve with a scored stem must be replaced to prevent premature failure of the valve seal.

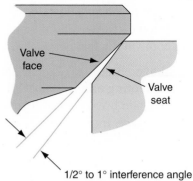

Valve face

Valve seat

1/2° to 1° interference angle

FIGURE 9-17 The valve face contacts the valve seat to seal the combustion chamber. The face and seat may be cut at different angles to create an interference angle. This allows the valve to seal with greater seating pressure.

Valve Stem Wear. When visually inspecting the valve stem, look for galling and scoring (Figure 9-16). Either of these can be caused by worn valve guides or carbon buildup or off-square seating. Also look for indications of stress. A stress crack near the fillet of the valve indicates excessive valve spring pressures or uneven seat pressure. Stress cracks near the top of the stem can be caused by excessive valve lash, worn valve guides, or weak valve springs.

Inspect the tip of the valve stem for wear and flattening. These conditions can be caused by improper rocker arm alignment, worn rocker arms, or excessive valve lash. Separation of the tip from the stem can be caused by these same conditions.

Valve Seats

When the valve closes, it must form a seal to prevent loss of compression and combustion pressures. The valve face rests against the valve seat to accomplish this task (Figure 9-17). In addition to sealing compression pressures, the seat and face provide a path to dissipate heat from the valve head to the cylinder head.

There are two basic types of **valve seats**: integral and removable inserts. **Integral valve seats** are machined into the cylinder head. The seat is induction hardened to provide a durable finish. Most aluminum cylinder heads (and many cast-iron cylinder heads) use **valve seat inserts** that are pressed into a recessed area (Figure 9-18). Common materials used in seat insert construction include cast iron, bronze, and steel alloys. Cast-iron inserts are used to repair cast-iron heads not originally equipped with seat inserts. Bronze inserts are typically used in aluminum cylinder heads. Bronze inserts are claimed to have better heat transfer, thus prolonging valve life. Steel alloys are available in several grades of stainless steel and with several grades of hardness. One of the hardest seats is made of **Stellite**. The Stellite is applied to the valve seat to protect against oxidation and corrosion, and for added wear resistance. The seat is also induction hardened to provide a durable finish. Seat hardening is done by an electromagnet using induction to heat the valve seat to

Shop Manual
Chapter 9

The **valve seat** provides the mating surface for the valve face. **Integral valve seats** are part of the cylinder head. **Valve seat inserts** are pressed into a machined recess in the cylinder head.

Stellite is a hard facing material made from a cobalt-based material with a high chromium content.

FIGURE 9-18 A valve seat insert.

a temperature of about 1,700 °F (930 °C). This hardens the seat to a depth of about 0.060 in. (1.5 mm). The greatest advantage of integral seats is their ability to transfer heat. Integral seats run about 150 °F (83 °C) cooler than seat inserts.

The valve seat must be finished properly and cut to the specified width. Typically, the intake seats are approximately 1/16 in. and the exhaust seats about 3/32 in. Always check the manufacturers' specifications when refinishing valve seats. The exhaust seat is a little wider because it has to dissipate more heat after being open during the end of the power stroke. The seat width is critical to proper cooling and sealing. The seat should be narrow enough to provide a high seating force from the valve spring. If the seat is too wide, the valve will not seat with enough force to chip away any carbon developing on the seat. It can also allow the valve to run cooler than designed, which can allow carbon to develop on the seat. The seat must be wide enough to allow extreme heat to dissipate during the short time that the valve is closed. If the seat is too narrow, the valve will burn or cup over time because it will run too hot.

Seat Recession. Over time, valve seats have a tendency to recede further down into the head. This places the valves further into the head so that when the valves open there is not as much open area for air to flow; power is reduced. You can inspect for this before disassembling a cylinder head; simply look at each of the valves and make sure they are at the same level in the head. This measurement is called the installed height. There are three common causes of **valve seat recession** (Figure 9-19). The first is due to the pressure and heat in the combustion chamber causing local welding of the valve face and valve seat. When the valve is opened again, it forces a breaking of this weld, resulting in the removal of metal. In the past, this action of local welding was controlled by adding lead to the fuel.

Valve seat recession is the loss of metal from the valve seat, causing the seat to recede into the cylinder head.

FIGURE 9-19 Valve seat recession.

The lead would be the subject of the welding and breaking instead of the valve face and seat. To counteract the damage to valves and valve seats as a result of the loss of lead, most manufacturers use very hard valve seats.

The second leading cause of valve recession is sustained high engine speeds. The third cause is weak valve springs. Both of these cause valve recession in the same manner. Under these two conditions, the valve rotates faster than normal. This results in excessive grinding between the valve face and seat. If the valve face is a harder material than the seat, the seat will recede.

The solution for seat recession is to install hard valve seats and use high quality valves. In addition, spring dampeners should be used to reduce chatter.

Valve Guides

Valve guides may be integral to the head or they may be pressed in. Cast-iron cylinder heads typically use integral valve guides. Holes are drilled down through the head where the valves will ride. On aluminum heads, the holes are bored oversized and guide inserts are pressed into the head. The guide inserts are usually cast iron or bronze (Figure 9-20). While bronze is a softer material, it has a natural lubricating quality that helps it resist wear.

The guide assures that the valve rides straight up and down and contacts the valve seat accurately. When valve guides wear, the valve can wobble in the head and the valve face-to-seat sealing is compromised. The wobbling of the stem also reduces valve seal contact and distorts the seals. Guide wear allows leakage past the valves and seals and reduces compression and power. Wear can also cause stress cracks and breakage near the fillet. The guides are machined to provide just the right amount of clearance between the guide and the stem of the valve. The clearance is typically 0.001 to 0.003 in., though some manufacturers may specify more clearance. If the fit is too loose, the valve will wobble and provide poor seating, and if it is too tight, the valve can stick. The small clearance allows a thin film of oil to lubricate the valve and guide. Too much clearance allows oil to be pulled past the intake valve seal as the valve wobbles, resulting in oil consumption. This is most noticeable at startup, during extended idling, or on deceleration when vacuum is high. The guides have coolant passages around them, helping the valves dissipate heat.

Valve Seals

To control the amount of oil allowed to travel down the valve stem to provide lubrication, valve stem seals are used. These seals are constructed of synthetic rubber or plastic and are designed to control the flow of oil, not prevent it. There are three basic types of valve stem seals: O-ring, umbrella, and positive seal.

The O-ring seal is basically a square-cut seal made of rubber (Figure 9-21). The valve stem has a groove machined near the top to accept the O-ring. The O-ring forms a seal between the valve stem and the spring retainer. A shield is also used to deflect oil away from the valve stem.

Valve guides
support and
guide the valve
stem through the
cylinder head.

FIGURE 9-20 Insert valve guides.

Keepers
(valve locks)

Spring
retainer

O-ring

Spring

FIGURE 9-21 O-ring valve
stem seal.

FIGURE 9-22 The umbrella seal is installed over the top of the valve stem.

FIGURE 9-23 Positive valve seal designs.

Umbrella seals fit over the valve stem (Figure 9-22). The seal provides a tight fit around the valve stem and a loose fit around the guide. This design allows the seal to move up and down with the valve.

Positive seals are generally constructed of rubber with a Teflon® insert (Figure 9-23). The design fits the valve guide snugly with either a light spring, spring clip, or press fit. Positive seals are considered by many to be the best type of valve stem seal. It is possible to replace umbrella or O-ring seals with positive seals. If the original guide was not cut for a positive seal, it can be cut to accept the seal.

CYLINDER HEAD COMPONENT RELATIONSHIPS

All components of the cylinder head associated with valve opening and seat (the valve, spring, guide, and seat) must work together to accomplish their designed tasks. Wear, damage, or problems in one component will have an effect on the other components. If the valve stem or guide is worn, there will be excessive movement of the valve inside the guide (Figure 9-24). This movement will translate to improper seating and possible burning of the valve. In Figure 9-24, "A" shows oil leaking past a worn intake valve guide. Atmospheric pressure pushes the oil into the air-fuel mixture. When the intake valve opens and the mixture is drawn into the cylinder, the oil is burned with the air-fuel charge. This results in oil consumption and blue smoke from the tailpipe. In Figure 9-24, "B" shows an exhaust valve with a worn guide allowing oil to be sucked by the guide due to venturi action. This too will result in oil consumption and blue smoke.

FIGURE 9-24 Worn valve guides can translate to improper valve sealing and oil consumption.

A valve that is not seating properly will cause poor engine performance. Both compression and combustion gases will be lost through the leakage. Lost compression reduces the potential power of combustion, and when combustion gases leak past a valve, they are not exerting force on top of the piston to produce power. A small leak by the valve seat will rapidly turn into a burned valve. You will feel poor valve sealing as misfire and a loss of power in the engine. You may also hear popping in the intake or exhaust system as the pressurized combustion gases are allowed in through a leaking valve.

Thorough reconditioning of the cylinder head components is often called a valve job. It consists of repairing or replacing all of the necessary components in the valvetrain to provide trouble-free operation for extended mileage, typically 100,000 miles. A complete valve job requires that the cylinder head(s) be removed from the engine. If the valve seals and/or valve springs are faulty, a repair may be made with the cylinder head still installed on the engine. Replacing these valve seals with the head installed was common with older engines where the seals did not last very long.

A valve job includes refinishing or replacing the valves and seats and any other worn components in the valvetrain.

> **AUTHOR'S NOTE:** During the cylinder head reconditioning process, all valvetrain components must be machined or replaced so they meet the engine manufacturer's specifications. An unsuccessful cylinder head reconditioning job may be defined as one that did not provide a reasonable mileage interval before some valvetrain component failed. It has been my experience that unsuccessful cylinder head reconditioning jobs are usually caused by failure to make sure that all valvetrain components are within manufacturer's specifications. For example, the technician may have done an excellent job of reconditioning the valves, seats, and guides, but the condition and measurements on the valve springs were overlooked.

COMBUSTION CHAMBER DESIGNS

The size, shape, and design of the combustion chamber affect the engine's performance, fuel efficiency, and emission levels. Manufacturers can use several different combustion chamber designs to achieve desired engine efficiency results.

Wedge Chamber

The most common method of creating a **wedge chamber** is to use a flat-topped piston and cast a wedge into the cylinder head (Figure 9-25). Wedge chamber design locates the spark plug between the valves in the widest portion of the wedge. The air-fuel mixture is compressed into the **quench** and **squish areas**, resulting in a turbulence that mixes the air-fuel

Wedge chamber design locates the spark plug between the valves in the widest portion of the wedge in the cylinder head.

Quenching is the cooling of gases as a result of compressing them into a thin area. The quench area has a few thousandths of an inch clearance between the piston and combustion chamber. Placing the crown of the piston this close to the cooler cylinder head prevents the gases in this area from igniting prematurely.

FIGURE 9-25 Wedge combustion chamber design.

mixture. The squish area squishes the air-fuel mixture into the center of the combustion chamber as the piston moves up toward TDC. This added **turbulence** helps mix the air-fuel mixture, cools it, and helps promote a faster burn once the ignition system begins combustion. The quench area cools the end gases in the chamber to prevent them from igniting before the intended flame front engulfs them. The squish and quench area of the combustion chamber gives it a lot of surface area compared to the volume. This high **surface-to-volume (S/V) ratio** of the combustion chamber has a negative impact on emissions. A higher S/V ratio contributes to greater hydrocarbon emissions as the fuel clings to the relatively cooler walls and exits the combustion chamber unburned in the form of hydrocarbons. The wedge combustion chamber design does not promote high volumetric efficiency. Air does not easily reach the hidden pocket of the wedge and the added turbulence reduces airflow, particularly at higher rpm. Another design method is to use flat chambers in the cylinder head and to offset the cylinder bores. This results in the piston approaching the cylinder head at an angle, and produces the same basic functions of a wedge chamber.

Hemispherical Chamber

The **hemispherical chamber** is designed in a half-circle with the spark plug located in the center of the dome (Figure 9-26). The valves are located on an incline of 60° to 90° from each other.

The hemi-head design provides an even flame front because the spark plug is centered. The design also allows for easy "breathing" of the engine because the valves are located across from each other. During the period of valve overlap, the stream of airflow out the exhaust valve helps pull air in through the intake valve. The area inside the hemi chamber is wide open and fills easily. The hemi design has a very low S/V ratio. A disadvantage to this design is that the combustion chamber creates little turbulence. While this helps volumetric efficiency at higher rpm, it can allow the hot gases to ignite without a spark, causing uncontrolled combustion. To prevent this, some manufacturers may use a domed piston.

Pentroof Chamber

Pentroof combustion chambers are a very common design for today's multivalve engines. They were first used in the 1980s and have grown in popularity since. The inverted-V shape of the chamber provides a low S/V ratio, resulting in lower emissions (Figure 9-27). These chambers are similar to the hemispherical head and share most of the advantages of that design. The pentroof chamber allows for larger valves and provides some turbulence.

FIGURE 9-26 Hemispherical combustion chamber design.

FIGURE 9-27 Pentroof combustion chamber design.

The **squish area** of the combustion chamber is the area where the piston is very close to the cylinder head. The air-fuel mixture is rapidly pushed out of this area as the piston approaches TDC, causing a turbulence and forcing the mixture toward the spark plug. The squish area can also double as the quench area.

Turbulence is the movement of air inside the combustion chamber.

The term "hemi" may be used for an engine with **hemispherical combustion chambers**.

A high **surface-to-volume (S/V) ratio** results in higher HC emissions. The S/V ratio is a mathematical comparison between the surface area of the combustion chamber and the volume of the combustion chamber. A typical S/V ratio is 7.5:1.

Pentroof combustion chambers are a common design for multivalve engines using two intake valves with one or two exhaust valves and provide good S/V ratio.

High swirl and turbulence of air-fuel mixture as piston moves downward

Combustion chamber

Exhaust valve closed

Intake valve open

3 mm (0.112 in.) mask to valve clearance

Masked area promotes swirling action of air-fuel mixture

FIGURE 9-28 Swirl combustion chamber design.

Swirl Chamber

Swirl chambers create an air flow that is in a horizontal direction.

To improve combustion efficiencies, **swirl chambers** are designed with a curved port that causes the air-fuel mixture to swirl in a corkscrew fashion (Figure 9-28). The compression of this swirling mixture results in a thorough mixing of the gases, resulting in increased fuel economy and reduced exhaust emissions. Location of the valve and design of the chamber causes the swirl action.

Chamber-in-Piston

A **chamber-in-piston** is a piston with a dish or depression in the top of the piston.

The **chamber-in-piston** design locates the combustion chamber in the piston head instead of the cylinder head (Figure 9-29). The advantage of this chamber design is that the piston remains hot enough to provide proper vaporization of the air-fuel mixture. This design is used on many diesel engines.

Fast Burn

A faster combustion can be achieved by directing the incoming air-fuel mixture tangentially through the intake valve (Figure 9-30). This results in a turbulence that mixes the gases.

Valve spring

Intake valve

Chamber

Piston

FIGURE 9-29 Chamber-in-piston combustion chamber design.

A

B

FIGURE 9-30 (A) Standard uniform-flow and (B) fast-burn tangential-flow combustion chambers.

FIGURE 9-31 Comparison of conventional and tumble port designs.

An additional benefit of this design is the compression ratio can be increased without an increase in octane requirements. This is because the fast-burn design has less potential for knock.

Tumble Port

Tumble port combustion chambers use a modified intake port design, resulting in a tumbling vortex mixture burning technique (Figure 9-31). This design uses the principle of **eddy currents** by using a volume of air that runs against the main current to provide a tumbling condition. Benefits include increased fuel economy and reduced hydrocarbon emissions.

Stratified-Charge Chambers

The **stratified-charge** combustion chamber is actually two chambers contained within the cylinder head (Figure 9-32). The smaller chamber (prechamber) is located above the main chamber and has its own intake valve. A very rich air-fuel mixture is delivered to the prechamber, where the spark plug is located. The rich mixture is very easy to ignite. At the same time, a very lean mixture is delivered to the main chamber. At the completion of the compression stroke, the spark plug fires to ignite the rich mixture in the precombustion chamber. The burning rich mixture will ignite the lean mixture in the main combustion chamber.

The first manufacturer to mass produce the stratified-charge engine was Honda. It used the engine from 1975 to 1987 with the compound vortex controlled combustion (CVCC) design. A three-barrel carburetor was used. Two barrels supplied the lean mixture for the main combustion chamber, while the third barrel supplied the prechamber with the rich mixture.

Tumble port combustion chambers use a modified intake port design.

An **eddy** is a current that runs against the main current.

A **stratified-charge** engine has two combustion chambers. A rich air-fuel mixture is supplied to a small auxiliary chamber, and the very lean air-fuel mixture is supplied to the main combustion chamber.

Intake valve
rich mixture

Intake valve
lean mixture

Intake
manifold

Precombustion
chamber

Combustion
chamber

FIGURE 9-32 Stratified charge combustion chamber.

THE COMBUSTION PROCESS

Combustion is the chemical reaction between fuel and oxygen that creates heat and pressure. It is a closely controlled burning of the air and fuel. Spark ignition occurs before top dead center on the compression stroke. The hot, compressed air-fuel mixture is ignited, and a flame front develops. For normal combustion to occur, the air-fuel mixture must be delivered in the proper proportions and mixed well, the spark must be timed precisely, and the temperatures inside the combustion chamber must be controlled. The flame can then move quickly and evenly (propagate) across the combustion chamber, harnessing the power of the fuel as heat. Pressure builds steadily as the gases expand from heat. The peak of this pressure develops around 10° ATDC to push the piston down on the power stroke (Figure 9-33).

The heat produced inside the combustion chamber must be controlled to reduce the possibility of preignition or detonation. The expansion of gases forces the unburned gases to be packed tighter and made hotter. Heat is dissipated through the top of the piston, the walls of the combustion chamber, and the engine coolant. The design of the combustion chamber has a huge impact on the process of combustion.

FIGURE 9-33 During proper combustion, the flame spreads
steadily across the chamber and heats all of the gases.

FIGURE 9-34 Quenching and turbulence are used to cool the temperatures of the end gases to prevent autoignition.

The heat absorption of the **end gases** can be increased by creating some turbulence within the combustion chamber and by proper design of the squish and quench areas (Figure 9-34). As the gases swirl around, they come into contact with the walls of the combustion chamber or the top of the piston. Turbulence prevents the gases from becoming stagnant. Stagnant gases will only dissipate heat from the outer edges, allowing the center to become very hot.

Turbulence also causes the flame to burn faster. The turbulence can be promoted by piston top design and the squish area of the chamber. The squish area causes the gases to be squirted out of an area of the chamber as the piston compresses the gases.

As the gases burn, the end gases are packed into the quench area of the chamber. This thins the gases and allows them to dissipate heat faster. The quench area design affects the emission output of the engine. As we discussed, to prevent the end gases from igniting on their own, heat is removed from them. This unburned mixture has a thickness of about 0.002 to 0.010 in. (0.1 to 0.3 mm). This thin layer of gases never burns because the temperature is too low, and it is expelled into the exhaust system as unburned hydrocarbons. The larger the combustion chamber, the more the surface area for the formation of the **skin effect**.

Manufacturers have reduced hydrocarbon emissions by redesigning the quench area. Reduction of hydrocarbons can be accomplished by increasing the height of the quench area. This increases the temperature of the end gases and reduces the thickness of the skin effect. Another means is by eliminating undesirable quench areas. This is done by removing sharp corners and pockets. Cylinder head gasket redesign and improved fit have eliminated many of these undesired quench areas.

Finally, skin effect has been reduced by lowering the compression ratio of the engine. This is not desirable from a performance point of view, but it was necessary to reduce emissions. Most engine manufacturers reduced the output of their engines during the late 1970s to the mid 1980s. Better understanding of the combustion process, chamber designs, fuels, and emission controls has made a return of higher compression engines possible.

Preignition

Preignition occurs when a flame is ignited in the combustion chamber prior to the spark plug firing. Preignition results in a light knocking or rattling sound during acceleration at operating temperature. It is usually the result of hot spots in the chamber. This can be caused by excessive combustion chamber pressures. An excessively high compression ratio causes the temperature of the fuel charge to be raised above its flash point. A change in compression ratio can occur any time there is a decrease in combustion chamber volume. This can be caused by excessive carbon buildup. Other things that change compression ratio include piston design and milling of the cylinder head and/or the block. Another cause of

The unburned gases are called **end gases**.

Skin effect is a small layer of unburned gases formed around the walls of the combustion chamber.

A BIT OF HISTORY

With the great amount of moving parts within the engine, it is truly surprising how well it works. Unfortunately, engines requiring carbon-based fuels pollute the air and use a nonreplenishable fuel source. Today, most manufacturers are attempting to design a reliable, cost-efficient electric vehicle. This idea is not new. During the early years of the "horseless carriage," over one-third were driven electrically.

preignition is the use of excessively low octane fuel. The octane required to properly operate the engine is based on the compression ratio of the engine. The higher the ratio, the higher the octane required.

The result of preignition is high pressures within the combustion chamber before the piston reaches TDC. This pressure forces the piston to reverse direction too soon, wasting useful combustion energy. It is possible to blow a hole in the top of the piston near the outer edge as a result of preignition (Figure 9-35).

Detonation

Normal combustion takes about 3 milliseconds (3/1,000 of a second). Abnormal combustion, detonation, is more like an explosion, occurring as fast as 2 millionths of a second (2/1,000,000 of a second). The explosive nature of detonation can cause serious engine damage. Detonation occurs when combustion pressures develop so fast that the heat and pressure will "explode" the unburned fuel in the rest of the combustion chamber. Before the primary flame front can sweep across the cylinder, the end gases ignite in an uncontrolled burst (Figure 9-36). The dangerous knocking results from the violent explosion. Detonation causes piston and ring damage, bent connecting rods and worn bearings, top ring groove wear, blown head gaskets, and possibly complete engine failure.

Common causes of preignition and detonation are:

- Deposits in the cooling system around the combustion chamber
- Engine overheating
- Too hot a spark plug
- An edge of metal or gasket hanging into the combustion chamber
- Fuel with too low an octane rating
- An inoperative or clogged exhaust gas recirculation (EGR) system
- Too much ignition timing advance
- Lean air-fuel mixtures (when there is less-than-desired fuel mixed with air)
- Carbon buildup in the combustion chamber
- A faulty knock sensor
- Excessive boost pressure from a turbocharger or supercharger

FIGURE 9-35 Piston damage caused by abnormal combustion.

FIGURE 9-36 End gases explode before the flame can expand across the chamber. The hot pressure spike knocks the piston so hard it can blow a hole in it.

Misfire

Misfire is another type of abnormal combustion. When an engine misfires, it means that one or more of the cylinders are not producing their normal amount of power. The cylinder(s) is unable to burn the air-fuel mixture properly and extract adequate energy from the fuel. The misfire may be total, meaning that a flame never develops and the air and fuel are exhausted out of the cylinder unburned. Hydrocarbon emissions increase dramatically. Misfire may also be partial when a flame starts but sputters out before producing adequate power due to a lack of fuel, compression, or good spark. When an engine is misfiring, the engine bucks and hesitates; the misfire is often more pronounced under acceleration. Technicians normally call this a miss or a skip. Extended misfire can destroy the catalytic converter by overloading it with fuel and causing it to overheat. Misfire can be caused by ignition or fuel-system faults as well as engine mechanical problems. Typical engine mechanical faults that cause misfire are:

- Burned or leaking valves
- Valve-stem or valve-back carbon buildup
- Weak or broken valve springs
- Worn valve guides or lifters
- Broken, worn, or bent rocker arms or pushrods
- Worn camshaft(s)
- Improper valve timing
- Worn pistons or rings

MULTIVALVE ENGINES

Many of today's engines are designed with more than two valves per cylinder. An engine may use multiple valves to increase volumetric efficiency. The easier an engine can breathe, the more efficient and powerful it can be. A multivalve design allows more air-fuel mixture to enter the combustion chamber faster, and helps the engine expel the spent exhaust gases. The most common multivalve engine is the four-valve-per-cylinder design (Figure 9-37).

Other designs use an extra intake valve, either a three-valve or a five-valve combustion chamber. One such variation is the use of a smaller second intake valve used as a jet valve. The jet valve directs air from above the throttle plates into the combustion chamber. At low engine speeds, this air causes a swirling action in the combustion chamber (Figure 9-38). The resultant turbulence increases the efficiency of the burn. Since the air inlet to the jet valve is above the throttle plates, the amount of air entering the valve decreases as engine speed increases.

FIGURE 9-37 Multivalve cylinder head.

FIGURE 9-38 Some multivalve engines use a jet valve to increase the burn effectiveness at low engine speeds.

SUMMARY

TERMS TO KNOW

Burned valve

Chamber-in-piston

Channeling

Eddy currents

End gases

Face

Fillet

Free-wheeling engine

Hemispherical

chamber

Integral valve seats

Interference engine

Margin

Off-square

Pentroof cumbustion

chambers

Poppet valves

Quenching

Skin effect

Squish area

Stellite

Stem necking

Stratified charge

Surface-to-volume

(S/V) ratio

Swirl chambers

Tumble port

Valve float

Valve guides

Valve job

Valve seat

Valve seat insert

Valve seat recession

Valve spring

Valve stem

Wedge chamber

- On most engines, the cylinder head contains the valves, valve seats, valve guides, valve springs, and the upper portion of the combustion chamber.
- The poppet valve is opened by applying a force that pushes against its stem. Closing of the valve is accomplished through spring pressure.
- The valve as a whole has many parts. The large-diameter end of the valve is called the head. The angled outer edge of the head is called the face and provides the contact point to seal the port. The area between the valve face and the valve head is referred to as the margin. The stem guides the valve through its linear movement.
- Manufacturers use several methods to dissipate the heat from the valve. First, the contact surface of the face and seat is used to transfer the heat to the cylinder head. The second method is through the stem, valve guide, valve spring, and on to the cylinder head.
- A burned valve is identifiable by a notched edge beginning at the margin and running toward the center of the valve head.
- Metal erosion is identified by small pits in the valve face and is generally caused by a chemical reaction in the combustion chamber.
- Breakage of the valve can occur due to valve and piston contact, valve springs that are too tight, loss of valve keepers, any condition that can result in stem necking, and off-square seating.
- The continual movement of the valve may result in wear on the portion of the stem traveling in the guide. As the stem wears, the oil clearance increases, resulting in excessive oil being digested into the combustion chamber and the formation of deposits on the valve.
- The valve seat is located in the cylinder head and provides a seal when the valve face contacts it. It also provides a path to dissipate heat from the valve head to the cylinder head.
- There are two basic types of valve seats: integral and removable inserts. Integral valve seats are machined into the cylinder head. Most aluminum cylinder heads (and many cast-iron cylinder heads) use valve seat inserts that are pressed into a recessed area.
- Valve guides support and guide the valve stem through the cylinder head. There are two types of valve guides used: insert and integral. Insert guides are removable tubes that are pressed into the cylinder head; integral guides are machined into the head.
- Valve seals control the amount of oil allowed to travel down the valve stem to provide lubrication.
- The size, shape, and design of the combustion chamber affect the engine's performance, fuel efficiency, and emission levels.
- The wedge and the pentroof-combustion-chamber designs are popular on modern gasoline engines.
- Preignition is a form of abnormal combustion that occurs when a hot spot in the combustion chamber causes autoignition of the air-fuel mixture before the spark.
- Detonation occurs when the end gases autoignite after the spark. Detonation's damaging force can blow holes in the top of pistons.

REVIEW QUESTIONS

Short Answer Essays

1. List the purpose of the following components:
   Cylinder head   Exhaust valve
   Intake valve   Valve seat

2. List and describe the common types of combustion chambers.

3. List the common causes for the following valve failures: burned valves, channeling, metal erosion, and breakage.

4. Explain the advantages of the pentroof combustion chamber.

5. List the causes of valve seat recession.

6. What are the differences between interference and free-wheeling engines?

7. List and describe the functions of the parts of a valve.

8. Explain why you should not refinish or machine sodium-filled valves.

9. List and describe the three common types of valve stem seals.

10. Describe the purpose of the valve guides and the difference between integral and insert types.

Fill-in-the-Blanks

1. The area between the valve head and the valve face is called the valve _____ .

2. _____ control the flow of gases into and out of the engine cylinder.

3. Valve stem _____ control the amount of oil allowed to travel down the valve stem and provide lubrication.

4. When the valve closes, it must form a seal to prevent loss of compression. The _____ _____ rests against the valve seat to accomplish this task.

5. _____ guides are removable tubes that are pressed into the cylinder head. _____ guides are machined into the head and are a part of the cylinder head.

6. Most of the heat is dissipated from the valve through the contact area between the valve _____ and _____ .

7. The most common cause of a burned valve is a leak in the sealing between the _____ and _____ .

8. Metal erosion is generally caused by _____ entering the combustion chamber.

9. Necking is identified by a _____ of the valve stem just above the _____ .

10. An engine may use multiple valves to _____ volumetric efficiency.

Multiple Choice

1. The most common refinishing valve face angle is:
   A. 15°   C. 45°
   B. 30°   D. 60°

2. Valve stem seals:
   A. Eliminate oil leakage into the valve guides
   B. Are not used on newer engines
   C. Keep combustion gases from corroding the valve stem
   D. Control oil flow to the valve stem

3. Each of the following is a likely cause of valve burning EXCEPT:
   A. A faulty thermostat
   B. A leak between the valve face and seat
   C. Worn valve guides
   D. A thick valve margin

4. *Technician A* says that the intake valve is often smaller than the exhaust valve.
   Technician B says that the exhaust valves can operate more efficiently because the air is under pressure when the exhaust valve opens.
   Who is correct?
   A. A only   C. Both A and B
   B. B only   D. Neither A nor B

5. Most of the valve's heat is dissipated through the:
   A. Face   C. Stem
   B. Margin   D. Fillet

6. The advantages of the pentroof combustion chamber include each of the following EXCEPT:
   A. Good turbulence
   B. A high S/V ratio
   C. High volumetric efficiency
   D. Centrally located spark plug

7. The skin effect causes:
   A. Premature valve burning
   B. Preignition
   C. Flaking of the Stellite seats
   D. High HC emissions

8. _____ occurs when another flame starts after the spark.
   A. Combustion   C. Detonation
   B. Preignition   D. Misfire

9. *Technician A* says that preignition is harmless.
   Technician B says that detonation can cause severe engine damage.
   Who is correct?
   A. A only   C. Both A and B
   B. B only   D. Neither A nor B

10. Each of the following is a likely cause of detonation EXCEPT:
   A. Carbon buildup in the combustion chamber
   B. Engine overheating
   C. A lean air-fuel mixture
   D. Fuel with too high an octane rating

Chapter 10

CAMSHAFTS AND VALVETRAINS

UPON COMPLETION AND REVIEW OF THIS CHAPTER, YOU SHOULD BE ABLE TO:

- Explain the function of the valvetrain.
- Explain the differences between overhead valve valvetrains and overhead camshaft valvetrains.
- List the components of the valvetrain.
- Explain the purpose and function of the camshaft.
- Describe the relationship among the camshaft lobe design and lift, duration, and overlap.

- Explain the benefits and limitations of valve overlap.
- Explain the purpose of the lifters.
- Describe the operation of hydraulic lifters.
- Explain the purpose of pushrods.
- List and explain the common methods of mounting the rocker arms.
- Explain rocker arm geometry.

INTRODUCTION

Every component that works to open and close the valves is part of the **valvetrain**. Air must flow efficiently into and out of the cylinder head to supply maximum power. Proper breathing of the engine depends on the valves opening and closing at the correct time. All of the components that make up the valvetrain work together to perform this task. The camshaft(s) open the valves for the correct amount of time and to the proper distance. The springs close the valves and must overcome tremendous momentum when the engine is spinning near the red line to hold the valves against the profile of the camshaft. The camshaft is mounted in the block in overhead valve engines. When the camshaft is mounted in the cylinder head, it is called an overhead camshaft engine. The other valvetrain components—pushrods, rocker arms, lifters, keepers, and valve rotators and retainers—will differ slightly depending on the engine design. Everything in the valvetrain is related and dependent upon each other; thus, a failure in the system can result in poor performance, engine noise, or mechanical failures. One stray movement in the valvetrain, a stretched component, or a snapped lock can throw the whole assembly into a wrecked heap of parts.

The **valvetrain** is made up of components that open and close valves to allow the engine to breathe correctly.

Shop Manual
Chapter 10

VALVETRAIN COMPONENTS

The valvetrain components work together to open and close the valves. For proper engine operation, this function must be performed at the precise time without loss of motion. The components of the valvetrain vary depending on engine design. The common components include:

- Camshaft
- Lifters or lash adjusters
- Pushrods
- Rocker arms or followers

The valvetrain components work together to open and close the valves.

- Valves
- Valve springs
- Valve keepers
- Valve spring retainers
- Valve rotators
- Timing chain, belt, or gears
- Camshaft and crankshaft sprockets

Camshafts

Shop Manual
Chapter 10

The **lobes** of the camshaft push the valves open as the camshaft is rotated.

Camshaft lift is the valve lifter movement created by the rotating action of the camshaft lobe. Typical lobe lift of original equipment in the block camshafts is between 0.240 and 0.280 in. (6 and 6.6 mm).

Valve lift is the total movement of the valve off of its seat, expressed in inches or millimeters.

Simply stated, the camshaft is used to control valve opening (Figure 10-1). However, its function is actually more complicated. The camshaft also controls the rate of opening and closing, and how long the valve is open. The camshaft is precisely timed to the crankshaft to be sure the valves open at the correct time during their respective strokes. The camshaft spins at half the speed of the crankshaft to allow each valve to open only once during a complete engine cycle. One complete engine cycle means that the engine has rotated 720° and that each of the cylinders has had one firing event. The camshaft may be driven by a gear, belt, or chain off the camshaft. Camshafts are designed differently depending on the type of lifter to be used with them. There are roller camshafts, hydraulic camshafts, and mechanical camshafts. The difference in the camshaft may be in the material used or in the shape of the lobe. Camshafts are usually constructed of nodular cast iron, although some high performance engines use cast steel. Since roller lifter camshafts have the highest amount of load per square inch, most of these style camshafts are manufactured from forged steel billets.

Some late-model engines have camshafts made from a steel tube (Figure 10-2). The lobes are bonded to the steel tube, and a nose piece is friction welded to the tube. Camshaft end play is controlled by a thrust wall at the front of the camshaft. This type of camshaft is more rigid than a camshaft made from cast or forged steel. The increased camshaft rigidity reduces engine vibration.

The **lobes** of the camshaft push the valves open as the camshaft is rotated. There is usually one lobe for each intake and exhaust valve. The lobe design is a factor in determining how far the valve will open and for how long (Figure 10-3). The height of the lobe defines the amount of valve lift or opening.

The amount of **camshaft lift** affects the amount of **valve lift**. Additional factors of valve lift include rocker arm ratio, pushrod deflection, and lifter condition. Camshaft lift is a measurement of the height of the lobe above the base circle. When an overhead camshaft engine does not use rocker arms, the camshaft lobe opens the valve directly; camshaft lift

FIGURE 10-1 Typical camshaft design.

FIGURE 10-2 A lightweight tubular camshaft used in a late-model V6 DOHC engine.

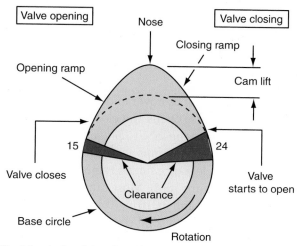

FIGURE 10-3 The lobe design determines how far the valve will open and for how long.

is then the same as the valve lift. Valvetrains that use rocker arms use mechanical advantage in the rocker arm design to multiply camshaft lift. Valve lift is an important factor in determining how well the engine can breathe (ingest the air-fuel mixture and expel it). The valve is a restriction to airflow into the combustion chamber. This is true until the valve is lifted about 0.300 in. (8 mm) off its seat. At this point, the valve offers very little restriction to airflow. However, there is still an advantage to increasing valve lift over 0.300 in. (8 mm). A typical amount of valve lift is 0.4–0.55 in. (10.16 mm–13.97 mm).

The longer the valve remains open after reaching the 0.300 in. (8 mm) lift, the better the airflow. Many manufacturers have increased the amount of lift in their camshafts to increase the amount of valve opening. Lift does not adversely affect fuel economy, power, or emissions (as does excessive **duration**). For this reason, some manufacturers have maintained moderate duration of the camshafts and increased lift in an attempt to increase airflow. One of the limitations on maximum valve lift is the amount of travel of the valve spring. If the valve spring is required to travel a greater distance, its coils can bind.

The camshaft duration is determined by the length of the slope of the lobe beyond the base circle. The wider and longer the lobe is, the longer the valve stays open. Camshaft duration is measured in degrees of crankshaft rotation. Intake valve duration may differ from exhaust valve duration.

If a valve opens at top dead center (TDC) and closes at bottom dead center (BDC), the duration of the valve is 180°. This is not practical for real-world engine operation. In actuality, the intake valve must open before TDC and close after BDC. The additional degrees of rotation must be calculated into the total duration degrees. If the intake valve

Camshaft **duration** is the amount of time the camshaft holds the valve open. It is measured in degrees of *crankshaft* rotation.

opens 6° before TDC and closes 12° after BDC, total duration is 198° of crankshaft rotation.

Since duration allows more time for the air-fuel mixture to fill the combustion chamber, duration has more effect on engine performance and efficiency than any other valve timing specification. There are instances when an engine may require longer durations to perform a specified function; for example, the high performance engines used in most racing applications require a camshaft with higher durations. This camshaft allows the engine to breathe at higher rpm. The drawback is that the engine performs poorly at lower engine rpms. Variable valve timing technology allows a manufacturer to have the ability to design the engine for good performance, low emissions, and good economy. All of this can be done with only one design for the camshaft. By changing the timing of the camshaft, the lift and duration will change when compared to their relationship with the piston's position.

If a replacement camshaft is to be purchased, there are two methods used to measure duration. Society of Automotive Engineers (SAE) specifications are measured after 0.006 in. (0.15 mm) of lift on hydraulic lifter camshafts. Performance camshafts are usually measured for duration at 0.050 in. (1.25 mm) lift. An SAE-measured camshaft will have approximately 40° to 50° more duration than a camshaft measured at 0.050 in. (1.25 mm) lift. This means that a camshaft with 220° of duration at 0.006 in. has less-effective duration than a camshaft with 195° of duration at 0.050 in.

The design of the lobe also determines how rapidly the valve opens. The steepness of the ramp dictates how quickly the valve opens and closes.

Clearance ramps are ground into the lobe to prevent the valvetrain components from hammering on each other. The clearance ramps are located between the base circle and the opening or closing ramps. The clearance ramps remove clearance in the valvetrain at a slower rate than the rate of lift (Figure 10-4). Most camshafts used with hydraulic lifters have a clearance ramp between 0.005 and 0.010 in. (0.13 and 0.25 mm). This ramp assures the lifter check valve is closed to provide a solid connection.

At the end of the exhaust stroke and the beginning of the intake stroke, both valves are open at the same time. This is called valve overlap. The amount of overlap is defined by the camshaft **lobe separation angle** (Figure 10-5). The larger or wider the separation angle, the less overlap. A tighter or lower separation angle means that the valves will be open together for a longer time, providing more overlap. A wider separation angle, used on most passenger cars, is 112° to 116°. A high performance engine or performance camshaft may have a tighter angle of 106° to 110°. The lobe separation angle or spread is determined by finding the lobe centers of the intake lobe and the exhaust lobe and averaging the two values. The lobe center can be found by using a degree wheel and a dial indicator (Figure 10-6). The degree wheel indicates the number of degrees after TDC where the maximum amount of valve lift is achieved for the intake valve. The exhaust lobe centerline is determined in the same manner. The degree wheel is used to measure the number of degrees

The area of the mechanical lifter camshaft from base circle to the edge comprises the **clearance ramp**.

The **lobe separation angle** is the average of the intake and exhaust lobe centers. It defines the amount of valve overlap. The tighter the angle (lower number of degrees), the more overlap the camshaft provides.

FIGURE 10-4 The clearance ramp removes valvetrain clearance at a slower rate than valve lift.

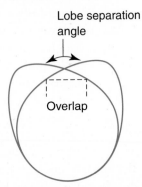

FIGURE 10-5 The time, in crankshaft degrees, that both valves are open is overlap. This is defined by the lobe separation angle.

FIGURE 10-6 A degree wheel can be used to find the lobe centers.

before or after TDC where the valve is fully opened. The lobe separation angle is the average of the two readings. For example, if the intake lobe center is 110° after TDC and the exhaust lobe center is 120° before TDC, the lobe separation angle is 115°. If the intake lobe center is less than the lobe separation angle, the camshaft is advanced. An **advanced camshaft** has the intake valve(s) open more than the exhaust valve(s) at TDC. This type of camshaft is used to increase lower rpm performance for greater torque. **Retarded camshafts** are designed to have the exhaust valve open more than the intake at TDC.

Valve overlap helps the engine breathe better, particularly at higher rpm. The movement of airflow out of the exhaust valve helps draw the fresh air in through the intake valve. The momentum of airflow helps to fully clean the spent exhaust gases out of the chamber even after TDC. This stream of airflow also helps pull the intake charge into the chamber. This is called scavenging. Too much overlap at a lower rpm, however, can allow some of the fresh air and fuel mixture to flow out of the exhaust valve and some exhaust to flow back into the intake. This is called reversion and leads to a rough idle, increased emissions, and reduced fuel economy. When you have exhaust gases in the intake, it reduces engine vacuum and counteracts the intended effects of overlap. Production vehicles use a wider separation angle to prevent the negative effects of reversion. Some engines using variable valve timing systems use two sets of camshaft lobes. At higher rpm, the camshaft lobes in use will provide greater overlap to improve volumetric efficiency and power.

AUTHOR'S NOTE: Aftermarket performance camshafts often produce an uneven or lopey idle accompanied by low vacuum. This is the result of reversion. If the engine is being designed for higher rpm performance, long overlap is essential, but be aware that you will sacrifice some low rpm torque to achieve this goal. The fuel economy will suffer greatly, and emissions will typically be higher. Most camshaft upgrades are not legal for street use. They have a significant impact on emissions because of the long period of overlap and its resulting flow of intake charge out the exhaust. If you live in a state that performs tailpipe emissions testing, it is unlikely that a vehicle with a significantly modified camshaft will pass the emissions test. Any engine modification that increases the emissions of the vehicle is illegal for street use.

An **advanced camshaft** has the intake valve open more than the exhaust valve at TDC. It is used to improve low speed torque.

A **retarded camshaft** has the exhaust valve open more than the intake at TDC.

CAUTION: The Automotive Engine Rebuilders Association (AERA) recommends only a stock camshaft replacement on computer-controlled engines due to poor performance at some rpms.

FIGURE 10-7 The taper of the camshaft lobes will provide lifter rotation to reduce wear.

Full round

FIGURE 10-8 Full round camshaft bearings are used in engines with the camshaft in the block and in some overhead camshaft engines.

The camshaft lobes generally have a very slight taper of 0.0007 to 0.002 in. (0.0178 to 0.0508 mm) across their face (Figure 10-7). Along with offsetting the centerlines of the **lifter** and the lobe, this taper provides for lifter rotation. The bottom of the lifter assembly is machined with a spherical shape. This prevents early wear, since the lifter's edge is not riding on the edge of the camshaft lobe. The rotation of the lifter and designs of the camshaft and lifter assemblies spread the load of the valvetrain against more of the lobe face. Rotating the lifter is desirable because it helps to dissipate the nearly 100,000 psi (690 mpa) load the lifter is subjected to.

Camshaft Bearings. The camshaft(s) rides in bearings or bearing surfaces on their journals. Oil pressure is fed through ports in the camshaft bearing bores. Like crankshaft main and rod bearings, cam bearing clearances are essential to good lubrication and adequate oil pressure. Camshafts that are installed within the engine block are usually fitted with full round bearings (Figure 10-8). These are one-piece bearings that are pressed into the block, and the journal of the camshaft is slid into place. Many overhead camshaft engines use split bearings with camshaft caps. Modern machining and oil improvements have made it possible for many manufacturers to eliminate the use of camshaft bearings in their overhead camshaft engines. In this case, the soft aluminum bore and caps serve as the bearing surface.

Lifters

Lifters are mechanical (solid) or hydraulic connections between the camshaft and the valves. Lifters follow the contour of the cam lobes to lift the valve off its seat. Solid lifters provide a rigid connection, while hydraulic lifters use oil to absorb the resultant shock of valvetrain operation. Each lifter design has its advantages and disadvantages.

Solid lifters (Figure 10-9) require a specified clearance between the components of the valvetrain. Periodic adjustment is required to maintain this clearance. If the clearance is too tight, there is no room for expansion due to heat. This can result in loss of power and value burning. If the clearance is too great, excessive noise, or valvetrain clatter, may be generated. In addition to noise, a change of 0.001 in. (0.03 mm) in valve lash may result in about a 3-degree change in duration. The lifter must be adjusted periodically to compensate for wear in the valve and valvetrain components. This is often recommended every 30,000 or 60,000 miles. Check the specific maintenance schedule. You will adjust the valve to have some lash between the rocker arm and the valve tip. This allows for a margin of error in valvetrain operation and also allows for the diminishing clearance as the valvetrain components heat up and expand. Many consumers see periodic valve adjustment as a disadvantage; it represents more costly maintenance.

Hydraulic lifters (Figure 10-10) use oil to automatically compensate for the effects of engine temperature and valvetrain wear. Some engine manufacturers use a roller-type hydraulic lifter in an effort to reduce friction (Figure 10-11). This type of lifter is fitted with

FIGURE 10-9 A solid lifter provides a rigid connection between the camshaft and the pushrod.

FIGURE 10-10 A hydraulic lifter compensates for temperature and wear by automatically adjusting valve lash.

FIGURE 10-11 A roller lifter from a late-model engine.

FIGURE 10-12 Roller lifters reduce friction and wear by using a rolling ball on the camshaft.

a roller that rides on the camshaft lobe. Friction is reduced since the lifter body is not rubbing against the camshaft lobe (Figure 10-12). Roller lifters can also be designed as solid lifters for high performance and heavy duty diesel applications. Roller lifters are accompanied by a roller camshaft, which typically has steeper opening and closing ramps. Roller lifters do not need to rotate in the lifter bore because they operate with lower friction.

The hydraulic lifter automatically maintains **zero lash**. When the hydraulic lifter is resting against the base of the camshaft lobe, the valve is in its closed position (Figure 10-13). In this position, the plunger spring maintains a zero clearance in the valvetrain (Figure 10-14). Oil supplied by the oil pump is pushed through the oil feed holes of the lifter bore and into the lifter body. Oil is able to enter the lifter body only when the lifter is in this position, since the feed through the engine block is not aligned with the lifter body when it is on top of the lobe. The oil passes through the body and enters the inside of the plunger through an oil passage on its side. The check valve allows oil to flow from the plunger into the body, but not from the body to the plunger. Oil continues to flow into the lifter until it is full. Pressurized oil is trapped in the lifter by the closed check valve. The spring under the plunger pushes the lifter body toward the camshaft lobe and the plunger toward the pushrod, removing any play in the valvetrain. The position of the check valve is determined by the pressure applied on the pushrod cap. This pressure is transferred from the closed valve through the pushrod to the cap. When the camshaft lobe attempts to raise the lifter body, the pressure on the cap and plunger is increased and the check valve closes, preventing the oil in the area below the plunger from escaping. Since the oil cannot be compressed, a rigid connection is formed, causing the plunger to raise with the lifter body and open the valve.

Zero lash is the point where there is no clearance or interference between components of the valvetrains.

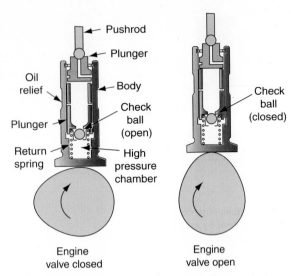

FIGURE 10-13 Hydraulic valve lifter operation comparison between valve-open position and valve-closed position.

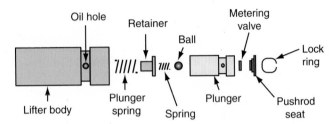

FIGURE 10-14 The internal components of a hydraulic lifter.

As the temperature of the engine increases, parts begin to expand, reducing the amount of clearances. This causes the plunger of the lifter to be held lower in the lifter body, causing some of the oil that was trapped below the plunger to leak out from the lifter body. In addition, when the lifter is moved up the block bore by the camshaft lobe, the pressures of attempting to open the valves push the plunger down in the body. This results in leakdown. When the lifter follows the cam lobe back to the base, the oil feed passages align and the lifter is filled with oil again. This process of filling and leakdown automatically adjusts the hydraulic lifter to compensate for temperature changes and component wear. You can hear faulty lifters; they cause significant valvetrain clatter as the valvetrain components clash against each other after taking up the unintended clearance. You can hear the noise that they make when they first start up; this is normal. The clattering should disappear within a few seconds as oil pressure fills the body. When lifters wear, it takes longer for them to fill after start up. Eventually they will become unable to hold oil as the camshaft tries to open the valve. This will cause noisy valvetrain clatter and loss of power because the valves will not open fully.

Solid lifters are often used in high performance engines because they cannot **pump up**. Since hydraulic lifters adjust automatically, they can overcompensate at higher engine speeds. At high engine rpm, valvetrain inertia causes the valves to open farther than what is accomplished by the cam lobe and rocker arm ratio. This valve float causes additional clearance to occur in the valvetrain. The lifter compensates for the additional clearance as it would under normal clearance corrections because the spring pushes the plunger up to take up the additional clearance. At this time, the body cavity fills with oil from the plunger cavity. The filling of the plunger cavity causes the lifter to be slightly longer than normal as it approaches the heel of the camshaft lobe. Again, the lifter compensates for this. On the next valve opening, the cavity is filled again and the cycle starts over. At high rpm, the lifter is unable to return to its normal size when it is approaching the heel of the lobe. The result is that the valve can be held open. When the engine speed is reduced, lifter operation returns to normal.

Lifter **pump-up** occurs when excessive clearance in the valvetrain allows a valve to float. The lifter attempts to compensate for the clearance by filling with oil. The valve is unable to close since the lifter does not leak down fast enough.

OHC Valvetrain Designs

Some overhead camshaft (OHC) engines use hydraulic lash adjusters (Figure 10-15). The lash adjuster works in the same manner as the hydraulic lifter. In some V6 DOHC valvetrains, the camshaft journals ride in cylinder head bores, and holders and holder plates retain the camshafts to the cylinder heads. The rocker arm shafts and rocker arms are mounted below the camshafts in the cylinder heads (Figure 10-16).

FIGURE 10-15 Some overhead camshaft engines use hydraulic valve lash adjusters.

FIGURE 10-16 This modern valvetrain uses hydraulic valve lash adjusters, roller rocker arms, and a tubular camshaft.

FIGURE 10-17 This follower sits directly under the camshaft and has a shim that can be changed to make valve clearance adjustments.

FIGURE 10-18 The valve lifter transfers force to the pushrod.

Cam followers are often used on overhead camshafts where the cam sits directly over the valves. The follower, often called tappet, is just a round cup that sits on top of the valve and provides a wider base for the camshaft to push on (Figure 10-17). Often there are shims either under or on top of the followers that can be fitted to size to make the proper adjustment. These lifters also need periodic adjustment. There are a few manufacturers that use hydraulic followers. The hydraulic body sits under a cup (like a follower) and works just like a hydraulic lifter to constantly adjust valve clearance to zero.

Pushrods

Overhead valve (OHV) engines, with the camshaft located in the block, use **pushrods** to transfer motion from the lifter to the rocker arms (Figure 10-18). Besides being a link to the rocker arms, some pushrods are also used to feed oil from the lifter to the rocker arms (Figure 10-19). The oil is sent up the hollow portion of the pushrod.

Pushrods connect the valve lifters to the rocker arms.

FIGURE 10-19 The pushrod can be used to deliver oil to the rocker arm.

FIGURE 10-20 This OHC head uses rocker arms with integral hydraulic lifters to open the valves.

Most pushrods are constructed from seamless carbon steel tubing. To decrease wear, some pushrods may have a hardened tip insert at each end. High performance engines may use pushrods that are made of chromium-molybdenum tubular steel for increased strength. Chrome-moly pushrods may also have a thinner wall thickness to reduce weight and inertia.

The bottom of the pushrod fits into a socket in the lifter. The recesses of the socket are deep enough to prevent the pushrod from falling out during operation. The top of the pushrod fits into a small cup in the rocker arm.

Tolerances within the valvetrain generally allow some machining of the cylinder head without the need to correct pushrod length; however, if the deck surface of the cylinder head and/or the engine block is machined or if valve stem length is increased, it may be necessary to change the pushrods to maintain the correct valvetrain geometry. This is especially true on engines with nonadjustable valvetrains.

Rocker Arms

The **rocker arm** is a pivoting lever used to transfer the motion of the pushrod to the valve stem.

The **rocker arm** is a pivoting lever used to transfer the motion of the pushrod to the valve stem. In overhead camshaft engines, the camshaft may operate the rocker arms directly (Figure 10-20). Rocker arms change the direction of the cam lift and spring closing forces, and they provide a leverage during valve opening. Most rocker arms are constructed of cast iron, cast aluminum, or stamped steel.

The difference in the rocker arm dimensions from center to valve stem and center to pushrod is called the **rocker arm ratio**. It is a mathematical expression of the leverage being applied to the valve stem.

The design of the rocker arm is such that the side to the valve stem is usually longer than the side to the pushrod (Figure 10-21). This design gives the rocker arms a mechanical advantage. The **rocker arm ratio** works with the camshaft lobe to provide the desired lift. The ratio of the distance is calculated by comparing the distance from the rocker pivot to the point where the rocker hits the valve tip to the distance from the rocker pivot to the point where the rocker contacts the pushrod. Divide the longer distance to the valve tip by the shorter distance to the pushrod to find the ratio. For example, if the distance to the pushrod is 1.00 in. (25.4 mm) and the distance to the valve tip is 1.50 in. (38.1 mm), divide 1.50 in. by 1.00 in. (38.1 mm by 25.4 mm). The answer is 1.50 in. (1.5 mm); the rocker ratio is stated as 1.50:1 or 1.5 mm:1 mm. If the camshaft lift is 0.3 in. (7.62 mm), multiply that by the rocker ratio to find the valve lift: 0.3 in. ∞ 1.50 in. = 0.45 in. (7.62 mm × 1.5 mm = 11.43 mm) of valve lift.

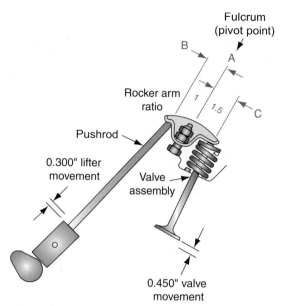

Fulcrum
(pivot point)

B

A

Rocker arm
ratio

1

1.5

C

Pushrod

0.300" lifter
movement

Valve
assembly

0.450" valve
movement

FIGURE 10-21 Rocker arms are designed to provide a mechanical advantage. The ratio is determined by comparing the distance from A to C to the distance from A to B. In this example, the ratio is 1.5:1.

There are three common methods of mounting the rocker arm (Figure 10-22):

- Shaft-mounted
- Stud-mounted
- Pedestal-mounted

Shaft-mounted rocker arms are located on a heavy shaft running the length of the cylinder head. The shaft is supported by stands. **Stud-mounted rocker arms** are used on overhead valve engines. Each rocker arm is mounted on a stud. A split ball is used as a pivot point for the rocker. **Pedestal-mounted rocker arms** are similar to the stud-mounted assemblies, except the rocker pivots on a split shaft.

Rocker Arm Geometry. When the valve is opened 50 percent of its travel, the rocker arm-to-stem contact should be centered on the valve stem (Figure 10-23). If the rocker arm is not centered, it will cause excessive side thrust on the stem. Rocker arm geometry does not usually need to be checked or corrected; however, if a different camshaft with a higher lift than the original camshaft is installed, rocker arm geometry will be affected. The higher lift causes the valve to move farther, causing the rocker arm tip to swing down farther in its arc. As the tip travels a greater distance, the tip moves away from the stem center. Other causes of incorrect rocker arm geometry include:

- Cylinder head resurfacing
- Cylinder block deck resurfacing
- Sinking the valve seat
- Worn or incorrect pushrods
- Worn rocker arms

In engines that use pushrods, the angle of the rocker arm in relation to the pushrod is also important. Incorrect rocker arm geometry may result in the pushrod hitting the side of its passage in the cylinder head or the pushrod guide plate.

In addition, on engines that use hollow pushrods to supply oil to the upper engine, incorrect rocker arm geometry may prevent proper lubrication. These engines have a small hole in the recessed cup that accepts the pushrod tip. When the opening in the pushrod aligns with the hole in the rocker arm, oil is sprayed over the rocker arms and valve springs. The holes should align when the rocker arm is in the lash position. As the rocker arm is moved to open

Shaft-mounted rocker arms are used in a mounting system that positions all rocker arms on a common shaft mounted above the cylinder head.

A **stud-mounted rocker arm** is mounted on a stud that is pressed or threaded into the cylinder head.

A **pedestal-mounted rocker arm** is mounted on a threaded pedestal that is an integral part of the cylinder head.

Shaft-mounted nonadjustable rocker arm

Rocker arm

Spacer

Oil hole

Rocker shaft

Stand

Stud-mounted rocker arm

Rocker arm stud nut

Fulcrum seat (ball pivot)

Rocker arm

Hollow stud

Oil hole

Adjusting screw

Locknut

Rocker arm

Stand

Compression spring

Rocker shaft

Shaft-mounted adjustable rocker arm

Attaching bolt

Fulcrum

Oil hole

Rocker arm

Fulcrum guide

Threaded pedestal

Pedestal-mounted rocker arm

FIGURE 10-22 Common methods of mounting the rocker arm.

Contact is centered on stem tip

50 percent open

FIGURE 10-23 Proper rocker arm geometry places the arm center on the stem and reduces side thrust.

the valve, the holes are no longer aligned and oil flow stops. Incorrect rocker arm geometry may cause the holes not to align properly and result in too much or too little oil flow.

Any machining operations that alter the distance between the camshaft and the rocker arm may affect rocker arm geometry. Even changes to camshaft lift and duration can change the geometry. Usually the tolerances in the valvetrain will allow for some changes and machining without problems, but the geometry should be checked while rebuilding the engine.

There are two basic functions of the **valve spring**: first, it closes the valve against its seat; second, it maintains tension in the valvetrain when the valve is open, to prevent float. The spring must be able to perform these functions while withstanding temperature changes and vibrations. A valve spring is another vital component in the operation of the valvetrain. In today's high speed, high performance engines, the valve spring becomes a very significant factor in efficiency and power output. The valve spring is strapped with the task of opening and closing the valves fifty times per second at 6,000 rpm. After miles of use, the spring may lose its tension or break. The reciprocating motion of the valve produces a considerable inertia effect. Worn or weak valve springs may not have sufficient tension for the lifter to follow the camshaft lobe, resulting in valve float.

Valve float causes the lifters to leave the camshaft lobe surface. In turn, this results in an abrupt seating of the valve and the lifter impacting the camshaft lobe. Valve float can result in excessive wear of the valve seat, lifter, camshaft, and valve. In addition, weak valve springs can prevent good heat transfer, resulting in burned valves and valve seat damage by bouncing the valve face against the seat. Weak valve springs usually have visible wear on their ends.

To accomplish its functions, the valve spring can be designed with different characteristics and features. The most common designs include: dual springs, dampers, and variable rate springs.

Dual springs are basically a set of two springs, with the smaller spring set inside the larger coil. The two springs have different diameters and lengths. The difference in sizes creates two different harmonic vibrations that cancel each other.

Some manufacturers use a damper inside the valve spring to reduce harmonic vibrations (Figure 10-24). The damper is a flat, wound coil designed to rub against the valve spring. The rubbing effect reduces the vibrations.

Variable-rate springs are designed with unequally spaced coils (Figure 10-25). This results in a spring that increases its rate of pressure as it is compressed. The closely wound coils are placed down toward the head. The more the spring is compressed, the more pressure it exerts for the distance compressed. For example, a variable-rate spring may exert an additional 40 pounds of pressure the first tenth of an inch it is compressed; the next tenth of an inch of compression will result in an additional 50 pounds of pressure; the next tenth of an inch of compression will result in an additional 60 pounds of pressure, and so forth.

Along with different shapes and designs, different metals are used to construct valve springs. The steels used to construct a valve spring can vary from the very expensive H-11 tool steel used in Pro Stock drag racing to various grades of stainless steel. The type of steel determines the strength of the valve spring and its ability to transfer heat.

The **valve spring** is a coil of specifically constructed metal used to force the valve to follow the profile of the camshaft and close fully. This provides a positive seal between the valve face and seat.

FIGURE 10-24 Damper spring.

FIGURE 10-25 A variable-rate valve spring.

Heat is one of the major contributing factors to valve spring fatigue. Oil flowing over the valve spring is the only cooling method used. The majority of the heat generated in the valve spring is the result of friction between the coils of the spring. Some valve spring manufacturers attempt to reduce the amount of friction by coating the valve springs. Common coatings include Teflon® and SDF-1. Teflon® sheds the oil, while SDF-1 retains the oil by use of a wetable matrix. In tests, a coated spring runs about 20°F (11°C) cooler than an uncoated spring.

> **AUTHOR'S NOTE: Many years ago, when I was a novice technician, I unbolted my stock overhead camshaft and replaced it with a "hot" performance camshaft. I adjusted the valves but left the replacement springs that came with the kit in my toolbox to install later. The camshaft combined with my new headers made a very noticeable improvement in engine power. Months later, when I was on a long and fast drive, my engine started misfiring. When I pulled over, I could hear clatter from under the valve cover. When I removed the valve cover, I saw the broken spring. It would have been easier to have installed the replacement springs in the shop than to replace the broken one in the back lot of someone else's garage in the rain.**

Valve Spring Retainers and Rotators

Valve springs are held in place by retainers locked onto the valves with keepers (Figure 10-26). The keepers lock onto the valve stem and are tapered so the retainer opening cannot fit over the keepers. The retainer may be a simple piece of steel to hold the spring in place. Other designs use a rotator built into the retainer. A rotator is used to turn the valve on its seat. They look thicker than an ordinary retainer and are either ball or spring type. The ball type has small ball bearings that rotate on a ramped track in the assembly to turn the valve. The spring type has an unequal height spring to achieve the same effect. Often they are used only on the exhaust valves. The rotation helps clean carbon away from the valve and seat and eliminates hot spots on the valve (Figure 10-27).

FIGURE 10-26 Valve keepers lock the valve spring retainer in place.

FIGURE 10-27 The exhaust valves on this engine are fitted with valve rotators to help clean the valve face and seat.

SUMMARY

- The camshaft is driven by the crankshaft and timed precisely to provide valve opening at the correct times.
- The valvetrain components work together to open and close the valves. The common components include: camshaft, lifters or lash adjusters, pushrods, rocker arms or followers and valves.
- Simply stated, the camshaft is used to control valve opening, the rate of opening and closing, and how long the valve will be open.
- The lobes of the camshaft push the valves open as the camshaft is rotated. The height of the lobe determines the amount of valve lift or opening. The design of the lobe will also determine the duration of valve opening.
- Lift can be expressed in two methods. The first is cam lift, which is the measured lift of the cam lobe. The second is valve lift. This is the cam lift multiplied by the rocker arm ratio to provide a specification for the amount of valve movement.
- The camshaft lobes generally have a very slight taper, which works with offsetting the centerlines of the lifter to provide for lifter rotation.
- Lifters are mechanical (solid) or hydraulic connections between the camshaft and the valves. Lifters follow the contour of the cam lobes to lift the valve off its seat. Flat tappet lifters are designed to rotate in the lifter bore to reduce wear.
- Hydraulic lifters use oil to automatically compensate for the effects of engine temperature and valvetrain wear.
- Roller lifters create less friction than standard flat tappet lifters and use a different roller camshaft.
- OHV engines with the camshaft located in the block use pushrods to transfer motion from the lifter to the rocker arms.
- The rocker arm is a pivoting lever used to transfer the motion of the pushrod to the valve stem. It changes the direction of the camshaft lift and spring closing forces, and it provides a leverage during valve opening.
- OHC or DOHC engines have the camshafts mounted on the cylinder heads. The camshaft lobes may contact the rocker arms directly, and a lash adjuster may be positioned under one end of the rocker arm.
- There are two basic functions of the valve spring: first, it closes the valve against its seat; second, it maintains tension in the valvetrain when the valve is open, to prevent float.
- To accomplish its functions, the valve spring can be designed with different characteristics and features. The most common designs include dual springs, dampers, and variable-rate springs.

TERMS TO KNOW

Advanced camshaft
Camshaft bearings
Camshaft lift
Clearance ramp
Duration
Lifters
Lobe separation angle
Lobes
Pedestal mounted rocker arm
Pump-up
Pushrods
Retarded camshaft
Rocker arm
Rocker arm ratio
Roller lifter
Shaft-mounted rocker arms
Stud-mounted rocker arm
Valve lift
Valve overlap
Valve spring

REVIEW QUESTIONS

Short Answer Essays

1. What is the function of the valvetrain?

2. Explain the purpose and function of the camshaft.

3. Describe the relationship among the camshaft lobe design and lift, duration, and overlap.

4. Explain the purpose of the lifters.

5. Explain the purpose of pushrods.

6. List and explain the common methods of mounting the rocker arms.

7. List three types of valve springs.

8. What is meant by the term *duration* as it relates to the camshaft?

9. If the rocker arm ratio is 1.5:1 and the cam lobe lift is 0.335 inch, what is the actual valve opening?

10 What is the purpose of a valve rotator?

Fill-in-the-Blanks

1. The camshaft is used to control valve _____, the _____ of opening and closing, and how _____ the valve will be open.

2. The _____ of the lobe determines the amount of valve lift or opening.

3. _____ is the length of time, expressed in degrees of crankshaft rotation, the valve is open.

4. The camshaft lobes generally have a very slight taper. This taper, along with the _____ of the lifter and the lobe, provides for lifter rotation.

5. Lifters are mechanical or hydraulic connections between the _____ and the _____.

6. OHV engines with the camshaft located in the block use _____ to transfer motion from the lifter to the rocker arms.

7. The design of the rocker arm is such that the side to the valve stem is usually _____ than the side to the pushrod.

8. When the valve is opened 50 percent of its travel, the rocker arm-to-stem contact should be _____ on the valve stem.

9. Valve spring _____ reduces valve spring temperatures.

10. _____ lock the valve spring retainer on top of the valve spring.

Multiple Choice

1. Each of the following is part of a typical valvetrain, EXCEPT:
   A. Lifters   C. Ports
   B. Pushrods   D. Rocker arms

2. Camshaft _____ is how long the valve stays open.
   A. Lift   C. Ramp
   B. Lobe center   D. Duration

3. The lobe separation angle defines:
   A. The height of the lobe
   B. The lift of the lobe
   C. The amount of overlap
   D. The valvetrain clearance

4. *Technician A* says that overlap improves volumetric efficiency at higher rpm.
   Technician B says that overlap reduces emissions.
   Who is correct?
   A. A only   C. Both A and B
   B. B only   D. Neither A nor B

5. Hydraulic lifters can _____ at higher rpm causing _____.
   A. Collapse, clatter
   B. Pump up, valve float
   C. Leak down, low power
   D. Lose oil pressure, reduced valve lift

6. Mechanical lifters require:
   A. A specified clearance
   B. Periodic adjustment
   C. Both A and B
   D. Neither A nor B

7. A rocker arm is 0.45 in. (11.43 mm) from the pushrod to the pivot and 0.66 in. (16.76 mm) from the pivot to the valve tip. The camshaft lift is 0.36 in. (9.14 mm). How far will the valve open?
   A. 0.25 in. (6.35 mm)
   B. 0.81 in. (20.57 mm)
   C. 0.36 in. (9.14 mm)
   D. 0.53 in. (13.46 mm)

8. Which valve spring runs coolest?
   A. A coated spring
   B. A variable-rate spring
   C. A double-coil spring
   D. A quenched-coil spring

9. *Technician A* says that a weak spring can cause valve float.
   Technician B says a weak valve spring can cause the valve to burn.
   Who is correct?
   A. A only   C. Both A and B
   B. B only   D. Neither A nor B

Chapter 11

TIMING MECHANISMS

UPON COMPLETION AND REVIEW OF THIS CHAPTER, YOU SHOULD BE ABLE TO:

- Describe the function of the valve timing system.
- Describe the operation of a timing chain system.
- Explain the function and advantages of a timing belt system.
- Explain the uses of a gear-driven timing system.
- Explain the difference between an interference engine and a free-wheeling engine.
- Explain the advantages of variable valve timing systems.
- Describe the operation of the VTEC variable valve timing and lift system.

INTRODUCTION

The front end of the engine consists of the **valve timing mechanism**. This is either a belt(s), chain(s), or gears used to time the rotation of the camshaft to the crankshaft. Timing gears are more frequently found on older, heavy-duty, and some diesel engines; modern engines typically use either chains (Figure 11-1) or belts (Figure 11-2) to drive the camshaft(s). Perfect valve timing is essential for proper engine performance. Good volumetric efficiency of the engine depends on valves opening and closing at the correct time. The timing mechanism may also drive other auxiliary shafts, balance shafts, and components (such as the water pump). This chapter will expose you to the various design configurations commonly used. We will also discuss the purposes, advantages, and disadvantages of the traditional and variable valve timing and lift designs.

> The **valve timing mechanism** drives the camshaft in the correct timing with the crankshaft. It may use a belt, chain, or gears.

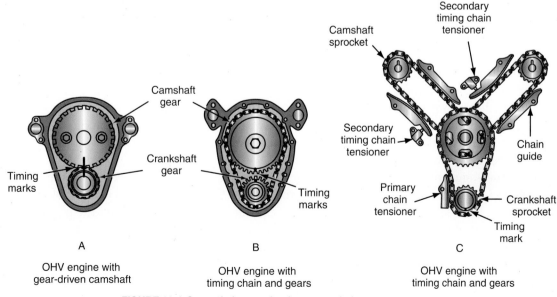

A — OHV engine with gear-driven camshaft

B — OHV engine with timing chain and gears

C — OHV engine with timing chain and gears

FIGURE 11-1 Some timing mechanisms use chains or gears.

Camshaft
timing marks
1/2 notch location

Timing
belt

Crankshaft

Crankshaft
at TDC

OHV engine with multiple
belt-driven camshafts

FIGURE 11-2 Many overhead camshaft engines use
timing belts.

VALVE TIMING SYSTEM

The timing mechanism is responsible for maintaining a perfect relationship between the rotation of the crankshaft and the camshaft, to open the valves at the right time. The intake valve must begin to open prior to the piston completing the exhaust stroke (Figure 11-3). Also, the valve must be closed at the proper time to prevent loss of compression. If the exhaust valve is not opened and closed at the precise time, the engine will not be able to breathe properly and draw in a fresh air-fuel charge.

Valve timing is a function of the camshaft, which is driven off the crankshaft. The camshaft rotates at half the speed of the crankshaft to allow the valves to open only for the desired length of time. To complete one full engine cycle, the camshaft will rotate once and the crankshaft will complete two revolutions. The camshaft sprockets will be twice the size of the crankshaft sprockets unless an intermediate shaft is used. The valve timing system must provide perfect correlation between the camshaft and crankshaft. If the cam timing is off by one tooth, the engine will likely run very poorly; emissions will increase, and power and fuel economy will decrease. Variable valve timing and lift systems must also start with the correct alignment between the crank and the cam to allow proper variations on valve timing. Some of these systems sometimes use a secondary timing chain that can allow for changes in a camshaft's timing (Figure 11-4).

The engine may be a free-wheeling or interference type. A free-wheeling or noninterference engine means that when the timing mechanism fails the valves will usually not hit the pistons. Even some **free-wheeling engines** can allow the valves to contact the pistons if the engine rpm is high enough when the mechanism breaks. On an **interference engine**, it is very likely that when the timing mechanism breaks the valves will contact the pistons (Figure 11-5). At best, valves will need to be replaced. At worst, valves will be driven into the pistons and the whole engine will need to be overhauled or replaced. Many timing belts

In a **free-wheeling engine**, even when the timing mechanism breaks, there is adequate clearance between the valves and the pistons to prevent the valves from bending or the pistons from breaking.

In an **interference engine**, the valves hit the top of the pistons when they are open and the piston is at TDC. This can occur when the timing mechanism breaks.

FIGURE 11-3 Valve opening is timed precisely by the camshaft through the valve timing mechanism.

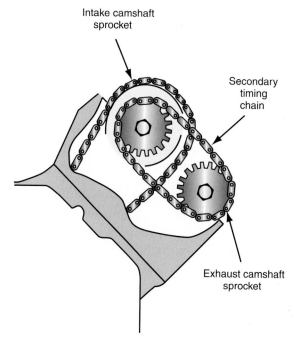

FIGURE 11-4 A secondary chain may be used for a camshaft, particularly when the engine uses variable valve timing.

require replacement at regular maintenance intervals. Timing gears and chains are generally replaced during an engine overhaul or when they begin to make noise or show signs of wear. When a timing mechanism breaks, engine damage can be severe. You can help your customers by explaining the importance of proper timing system maintenance, including routine timing belt replacement.

Every valve timing system will have alignment marks on the sprockets or gears to make sure you can properly time the engine (Figure 11-6). Camshaft or valve timing is usually established with cylinder number 1 at TDC on the compression stroke.

FIGURE 11-5 Every valve that came out of this engine was bent after the timing belt snapped.

FIGURE 11-6 Timing marks on the sprockets allow proper alignment during service.

CHAIN-DRIVEN SYSTEMS

Chain-driven systems can be used either on cam-in-the-block or on overhead-cam engines. The timing chain is a strong, long-lasting way of linking the cam to the crank. A flat-link silent chain makes much less noise than a roller-style chain. The chain system is still noisy, however, which is objectionable to some consumers. It is also more expensive for the manufacturers to fit a chain rather than a belt.

When a chain is used, steel, fiber composite, or composite plastic sprockets may be used to rotate with the chain. To drive an overhead camshaft(s), one or more **timing chain guides** and a **timing chain tensioner** must be used. The guides hold the chain against a synthetic rubber or Teflon® face. The timing chain tensioner holds proper tension on the guide to take up slack as the chain and guides wear. The tensioner is usually a spring-loaded plunger that may be fed with oil pressure to take up chain slack. The tensioner also has a spring-loaded ratcheting system. When the chain or guide is worn enough, the spring will push the plunger out one more step on the ratchet. This also prevents excessive noise at startup, before oil pressure is held behind the plunger (Figure 11-7). Timing chains receive splash oiling often with a slinger placed at an oil passage.

Engines that use balance shafts may use two chains; one will drive the camshafts and the other will drive the balance shaft (Figure 11-8). Note the guides and tensioners as well. This engine has dual balance shafts. You can see that they will spin at twice the speed of the crankshaft, which is typical of balance shafts. You can see this by comparing the size of the balance shaft sprockets to that of the crankshaft sprocket.

Chains are generally replaced during an engine overhaul or if they begin to make excessive noise. As they wear, you can often hear the chain slap during a quick deceleration.

> A **timing chain guide** is a length of metal with a soft face to hold the chain in position as it rotates.

> A **timing chain tensioner** maintains tension on the chain or belt to prevent slapping or skipping a tooth.

Primary chain   Balance shaft chain

Tensioner   Guide

FIGURE 11-8 This DOHC engine shows the tensioner, guides, and primary and balance shaft chains.

FIGURE 11-7 The timing chain tensioner uses spring pressure and oil pressure to maintain the correct tension on the chain.

AUTHOR'S NOTE: I heard the telltale rattle of my timing chain hitting the cover during deceleration for several weeks before I took a Saturday and replaced the timing chain, guides, sprockets, and tensioner. My chain and sprockets were *very* worn; I was lucky they didn't cause me to walk home one day.

BELT-DRIVEN SYSTEMS

A timing belt is frequently used on overhead camshaft engines. The timing belt is strong and very quiet. It is much more cost effective for the manufacturer than gears or chains. As belt materials have improved, belt replacement intervals have increased. This is good for the consumer because timing belt replacement can be a costly maintenance item. Some manufacturers do not even specify a replacement interval; they simply require the technician to inspect the belt periodically.

The timing belt runs from the crank sprocket to the cam sprocket(s) without a guide. Instead, an idler pulley is used to guide the belt and provide a tension adjustment. Most newer timing belt systems use an automatic tensioner that self adjusts; others require periodic adjustment (Figure 11-9). These tensioners used with timing belt systems have chambers on both sides of a piston that is filled with silicon oil. They are not fed oil externally. If belt tension increases, the piston moves downward. The piston movement causes an increase in hydraulic pressure in the lower chamber. Since the check ball is closed, the oil must seep past the piston and cylinder wall to go into the reservoir chamber. As oil seeps into the reservoir, the tension piston moves slowly down the cylinder until the load from the belt balances with spring tension (Figure 11-10). Alternately, when the belt tension is increased, spring tension forces the piston to move up in the cylinder and take up the slack.

The timing belt may also drive other components. On older engines, it frequently drove the auxiliary shaft that turned the **distributor** (Figure 11-11). Other manufacturers use the timing belt to drive the water pump. When servicing or replacing the timing belt, it is recommended that the water pump be replaced at the same time. This type of setup makes water pump removal easier when the timing belt and other components are already

FIGURE 11-9 An automatic adjusting tensioner may be used to maintain proper belt tension.

FIGURE 11-10 Hydraulic pressure flows through the tensioner when belt tension increases.

Belt tension

Reservoir chamber

Pressure chamber

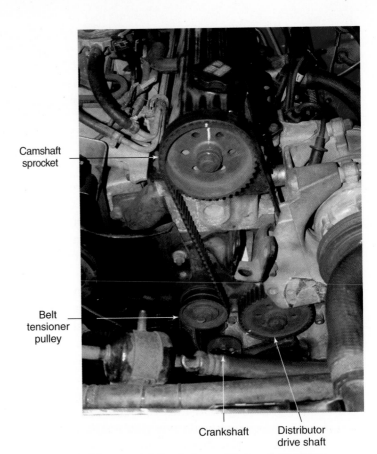

Camshaft sprocket

Belt tensioner pulley

Crankshaft

Distributor drive shaft

FIGURE 11-11 A timing belt may also be used to drive the distributor shaft.

removed. You will certainly be saving the customer a headache and their money by taking care of it now. In some dual overhead camshaft (DOHC) V6 engines, a single timing belt may be used to drive the four camshafts and idler pulleys (Figure 11-12). In other DOHC V6 engines, a timing chain surrounds the crankshaft sprocket and intermediate shaft sprocket and the chain drives the intermediate shaft sprocket. A front cover is mounted over the chain drive, and an outer sprocket is mounted on the intermediate shaft on the outside of the front cover. This allows for replacement of the belt without disassembling the

Cam gears

Coolant pump

Timing belt tensioner

Crankshaft gear

FIGURE 11-12 This DOHC timing belt drives all four camshafts.

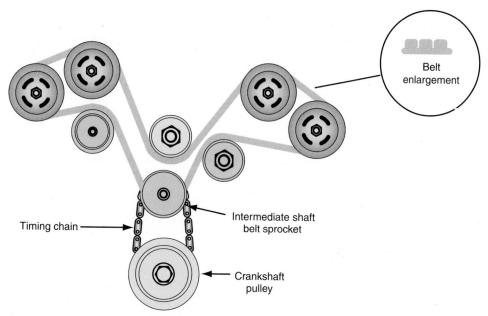

FIGURE 11-13 A DOHC engine that uses a timing chain to drive an intermediate shaft. A belt off the intermediate shaft sprocket drives the camshafts.

front chain cover. A timing belt surrounds the outer intermediate shaft sprocket to drive the overhead camshafts on both sides of the engine. The back side of the belt contacts two idler pulleys and a tensioner pulley (Figure 11-13). A timing belt may also be used on engines with a balance shaft(s). A timing belt does not need lubrication; in fact, one reason for premature failure is oil or coolant leaking onto the belt.

GEAR-DRIVEN SYSTEMS

Gear-driven systems are not as popular as they used to be, in part because they are only used on camshaft-in-the-block engines (Figure 11-14). They are a very reliable and precise system, although a little noisy. Because there is no chain or belt slack, valve timing is always very close to perfect. The gears are very durable, as long as they receive adequate lubrication. Many diesel engines use timing gears because the valve timing is very critical and it adds to the idea of the long-lasting engine (Figure 11-15).

FIGURE 11-14 The crankshaft gear drives the camshaft directly when gears are used.

A BIT OF HISTORY

Chrysler's first V-style engine was its legendary 331 c.i.d. "Hemi," introduced in 1951. It had a compression ratio of 7.5:1, with an overhead valve configuration. It put out 180 hp. Though today's new "Hemi" engine is slightly larger, at 348 c.i.d., it puts out 340 hp.

VARIABLE VALVE TIMING AND LIFT SYSTEMS

Anxious to reduce the limitations of the compromises made in camshaft design, many manufacturers are turning to variable valve timing (VVT) or variable valve lift systems. One camshaft cannot produce fabulous low end torque and also manage excellent high rpm breathing. Variable valve lift and timing systems reduce those inherent compromises. Valve lift should be increased at higher rpms when volumetric efficiency falls. The valve timing should be retarded at higher rpms to keep the valves open later in the stroke, because the stroke is occurring so quickly and it still takes significant time to get the stream of air into the cylinder. Many more manufacturers are adding variable valve timing and/or valve lift systems to their vehicles every year. Most manufacturers have at least some of their models equipped with a VVT system. Some VVT designs allow only for the intake or exhaust camshaft to be variable. Other systems control both camshafts. Honda has been using a variable timing and lift system since 1989.

There are many variations in design of the systems. Systems either change the valve timing or alter the timing and the valve lift. The systems may be designed to improve performance and/or emissions. By reducing valve overlap at idle, hydrocarbon emissions are reduced. Less of the fresh air charge is allowed out the exhaust port. By increasing the period of overlap at higher rpm, the engine can breathe better. In some systems, during cruise conditions the exhaust valves are held open longer. This allows some exhaust gases to flow in and mix with the intake charge. By adding some inert gases to the charge, the heat of combustion is reduced and oxides of nitrogen (NOx) emissions are decreased. The exhaust gas recirculation (EGR) system used on many vehicles performs this same function. It actually adds exhaust gases into the intake charge to dilute the mixture and reduce combustion temperatures. The EGR systems have had many problems over the years. This function of the VVT system used on Ford's 2.0-liter Zetec engine has allowed Ford to eliminate the EGR system and still meet the NOx standards.

Other systems focus more on the performance enhancements offered by variable valve timing. Increasing valve lift and retarding valve timing can increase the output of an engine significantly at higher rpm. This is when the engine torque traditionally falls off due to the inability to fill the chambers. The vast majority of import and domestic manufacturers are using or planning to use variable valve timing systems on some of their vehicles.

Honda's popular variable timing and lift electronic control (VTEC) system modifies both valve timing and valve lift. Honda took the lead in production variable valve timing with its VTEC system, designed to improve performance at higher rpm. It increased the engine output over a single cam non-VTEC engine by 25 percent. In the VTEC system, each pair of valves has three cam lobes and three rocker arms. During low speed operation, the low speed rockers push directly on the two regular-speed lobes. At high rpm, when the PCM determines that the third lobe will improve performance, it signals an oil control valve to engage the third rocker to the high speed lobe. A pin slides out of the low speed rockers and into the high speed rocker to engage it to use the camshaft lobe with more aggressive timing, lift, and duration. On one Acura model, the high speed lobe has an advertised duration of 290°; that is a race-quality camshaft. Figure 11-16 shows a simplified diagram of the pin actuating system (Figure 11-16). Honda's i-VTEC was introduced in 2001 and uses a more powerful PCM to control the camshaft. This system uses an adjustable oil-driven cam gear (also called a cam phasing actuator) that actually closes the intake valve well into the compression stroke when the PCM determines that it should be in "economy mode." This action is similar to what happens in the Atkinson cycle engine. Later, Honda introduced the Advanced VTEC, which uses an even faster PCM. The intake valves have less lift at low engine speeds, which gives the engine an estimated 13 percent increase in fuel economy and a reduction in emissions.

FIGURE 11-15 This 7.3-liter diesel engine uses timing gears.

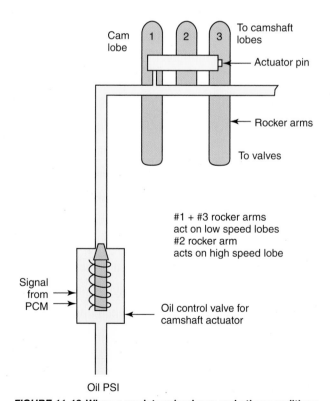

FIGURE 11-16 When a predetermined rpm and other conditions are met, the PCM signals the cam actuating control valve. This valve allows oil pressure to push a pin through the rocker arms, activating the high speed rocker arm and cam lobe.

GM's VVT system improves engine performance, fuel economy, and emissions by changing valve timing only. It also eliminates the need for an exhaust gas recirculation (EGR) system. On some applications, it allows 90 percent of the engine's torque to be available between the wide range of 1,900 and 5,600 rpm. The VVT system uses a camshaft position actuator for each of the intake and exhaust camshafts that bolt on to the end of the camshafts. The actuators lock the exhaust cam in its fully advanced position and the intake cam in its fully retarded position for startup and idle. This minimizes overlap to create a very smooth idle. Off idle, the PCM controls the camshaft actuator solenoid valves to allow the position actuators to alter intake and exhaust timing by 25 camshaft degrees. The PCM looks at information about engine coolant temperature, engine oil temperature, intake airflow, throttle position, vehicle speed, and volumetric efficiency to determine the optimum positioning of the camshafts for any given set of conditions. The solenoid valve controls oil pressure to the position actuator to effect changes in camshaft timing. The camshaft position actuator has an outer housing that is driven by a timing chain. A rotor with fixed vanes lies inside the assembly and is attached to the camshaft. The solenoid valve, as commanded by the PCM, controls the direction and pressure of oil flow against the vanes. Increasing pressure against one side of the vanes will advance the camshaft variably until the desired position is achieved. Changing the oil pressure to the opposite side of the vanes will retard the camshaft. Whenever the engine is off idle, the PCM is continuously providing electrical pulses to the solenoid valves to modify camshaft timing or to hold it in the desired position. Under light engine loads, the PCM retards the valve timing to reduce overlap and maintain stable engine operation. When the engine is under a medium load, overlap is increased to improve fuel economy and emissions. To improve mid-range torque under a heavy load and at moderate or low rpm, intake valve timing is advanced. Under a heavy load at high rpm, intake valve closing is retarded to improve engine output (Figure 11-17).

Toyota is using a VVT system that changes the phase angle of the camshaft relative to the crankshaft to advance or retard the camshaft. A hydraulic actuator controlled by the PCM sits on the end of the camshaft to effect timing changes. The camshaft timing can be varied continuously up to 60°. The system also employs a variable lift and duration system by using two different camshaft lobes for each set of intake and exhaust valves. One

FIGURE 11-17 This cam phasing actuator rotates the camshaft in relation to the crankshaft.

rocker arm opens both intake valves and another opens both exhaust valves. During low to moderate speed operation, the "high-speed" cam lobes do not contact the rocker arms. At higher speeds, a sliding pin is pushed by hydraulic pressure to take up the space between the higher-duration cam lobes. The increased rocker arm movement also increases valve lift.

The number and types of systems in use increases each year; the idea has proven to be effective in improving power and in lowering emissions. Designs in the works include valves opened by electrically operated solenoids. This system would offer infinitely variable valve lift, duration, and timing.

SUMMARY

- The front end of the engine consists of the valve timing mechanism.
- Proper valve timing is essential for proper engine performance.
- Valve timing systems may use a belt, chain, or gears to link the camshaft to the crankshaft.
- Marks are provided on the sprockets, gears, covers, block, or head to properly time the engine, usually at TDC firing cylinder number 1.
- A timing chain is a long-lasting drive mechanism that can be used on cam-in-the-block or overhead-cam engines.
- Timing chains are not generally replaced as a regular maintenance service.
- Dual timing chains may be used to drive camshafts individually on variable valve timing engines or on engines with balance shafts.
- Timing belts are a very popular and cost-effective way to drive overhead camshafts.
- Timing belts usually require periodic replacement as a maintenance service.
- Timing gears are sometimes used in heavy-duty diesel engines.
- Timing gears are very durable and provide precise control of timing.
- Manufacturers are increasingly using variable valve timing and or lift systems to overcome the limitations of the compromise of one camshaft for all engine speeds and loads.

TERMS TO KNOW

Free-wheeling engine
Interference engine
Timing chain guide
Timing chain tensioner
Valve timing mechanism

REVIEW QUESTIONS

Short Answer Essays

1. Explain the purpose of precision valve timing.
2. List the types of timing mechanisms.
3. Why do modern passenger car engine designs tend to use timing belts or chains rather than gears?
4. Describe the difference between a free-wheeling and an interference engine.
5. What can happen when the timing mechanism breaks on an interference engine?
6. Explain the maintenance requirements of the different styles of timing mechanisms.
7. Describe the advantages of a timing chain over a timing belt.
8. Describe the operation of a timing chain tensioner.
9. Explain the advantages of variable valve timing.
10. Briefly explain the operation of Honda's VTEC system.

Fill-in-the-Blanks

1. Valve timing is a function of the _____ .
2. The camshaft rotates _____ as fast as the crankshaft.

3. Proper valve timing is established with cylinder number 1 at _____ of the _____ stroke.

4. _____ _____ mechanisms may be used on OHV or OHC engines.

5. The timing chain _____ hold the chain in proper alignment.

6. The _____ takes up slack on a timing chain.

7. Balance shafts spin at _____ the speed of the crankshaft.

8. A worn timing chain will make the most noise during engine _____ .

9. By reducing the amount of overlap at idle, a VVT system can reduce _____ emissions.

10. Some VVT systems eliminate the need for the _____ emissions control system.

Multiple Choice

1. Valve timing affects each of the following EXCEPT:
   A. Engine torque
   B. Spark timing
   C. Emissions
   D. Fuel economy

2. Each of the following is a typical type valve timing mechanism EXCEPT:
   A. Chain
   B. Belt
   C. Gear
   D. Band

3. *Technician A* says that timing belts are quieter than timing chains.
   Technician B says that timing chains generally last longer than timing belts.
   Who is correct?
   A. A only
   B. B only
   C. Both A and B
   D. Neither A nor B

4. An automatic timing belt tensioner:
   A. Receives oil under pressure to tension the belt
   B. Has triple-coil springs to ensure proper tension
   C. Should be adjusted every 60,000 miles
   D. Is filled with silicon oil

5. Some _____ engines use an auxiliary shaft chain and a timing belt driven off of the auxiliary shaft sprocket.
   A. eight-cylinder OHV engine
   B. four-cylinder DOHC engine
   C. V6 DOHC engine
   D. Diesel engines

6. The quietest timing mechanism is the:
   A. Belt
   B. Chain
   C. Band
   D. Gears

7. A VVT system may be able to eliminate the _____ system.
   A. Spark timing
   B. Positive crankcase ventilation
   C. Exhaust gas recirculation
   D. Valvetrain

8. The Honda VTEC system:
   A. Modifies valve lift
   B. Changes valve timing
   C. Changes valve duration
   D. All of the above

9. Reducing valve overlap at idle reduces:
   A. Hydrocarbon emissions
   B. Oxides of nitrogen emissions
   C. Carbon monoxide emissions
   D. All of the above

10. *Technician A* says that a VVT system may increase the overlap at higher rpm.
    Technician B says that a VVT system may retard the intake timing under heavy-load, high rpm conditions to improve power.
    Who is correct?
    A. A only
    B. B only
    C. Both A and B
    D. Neither A nor B

Chapter 12

ENGINE BLOCK CONSTRUCTION

UPON COMPLETION AND REVIEW OF THIS CHAPTER, YOU SHOULD BE ABLE TO:

- Explain the purpose of the cylinder block.
- Describe the construction and components of the cylinder block.
- Explain the purpose and machining processes of the cylinder bore.
- Explain the purpose and machining process of the main bore.
- Describe a short block, long block, and crate engine and their uses.

- Name and describe the components of the crankshaft.
- Explain the relationship of crankshaft throw to stroke and firing impulses.
- Describe the forces applied to the crankshaft.
- Explain the manufacturing processes and materials used in crankshaft design.
- Explain the purpose of the harmonic balancer.
- Describe the purpose of the flywheel.

INTRODUCTION

The cylinder block and the components within it make up the bottom end of the engine. This is also called a **short block**. A **long block** is essentially a short block with cylinder head installed (Figure 12-1). Most crate engines can be ordered as short or long blocks. Engine materials and machining have changed significantly in the past 20 years. Clearances have decreased, ring and piston choices abound, and block reconditioning is often limited. A silicon- impregnated,

A **short block** is a block assembly without the cylinder heads and some other components.

A **long block** is an engine block with the cylinder head(s) and other components mounted.

FIGURE 12-1 This is a short block with a cutaway cylinder head unit.

cast aluminum alloy block with integral bores cannot be refinished, for example. The function and purpose of the flywheel and harmonic balancer are also covered in this chapter because these components affect the function of the block assembly. You must have access to thorough service information to properly service a modern engine (Figure 12-2). The service information will provide you with what you need to know about the materials used for the block, machining processes allowed for repairs, and proper repair procedures.

GENERAL MOTORS ENGINES
3.0L, 3.8L, and 3.8L "3800" V6 (Continued)

Engine Specifications (Continued)

Crankshaft Main and Connecting Rod Bearings

Engine	Main Bearings				Connecting Rod Bearings		
	Journal Diam. In. (mm)	Clearance In. (mm)	Thrust Bearing	Crankshaft End Play In. (mm)	Journal Diam. In. (mm)	Clearance In. (mm)	Side Play In. (mm)
All Models	[1] 2.4988-2.4998 (63.469-63.494)	.0003-.0018 (.008-.005)	2	.003-.009 (.08-.23)	[1] 2.2457-2.2499 (57.117-57.147)	.0003-.0028 (.008-.071)	.003-.015 (.076-.38)

[1] Maximum taper is .0003" (.008 mm).

Pistons, Pins and Rings

Engine	Pistons	Pins		Rings		
	Clearance In. (mm)	Piston Fit In. (mm)	Rod Fit In. (mm)	Ring No.	End Gap In. (mm)	Side Clearance In. (mm)
All Models	[1] .0013-.0035 (.033-.089)	.0004-.0007 (.010-.018)	.00075-.00125 (.019-.032)	1	.010-.020 (.254-.508)	.003-.005 (.08-.13)
				2	.010-.020 (.254-.508)	.003-.005 (.08-.13)
				3	.015-.035 (.381-.889)	.0035 (.09)

[1] Measured at bottom of piston skirt. Clearance for 3.8L turbo is .001-.003" (.03-.08 mm).

Valve Springs

Engine	Free Length In. (mm)	Pressure Lbs. @ In. (Kg @ mm)	
		Valve Closed	Valve Open
3.8L (VIN C)	2.03 51.6	100-110@1.73 (45-49@44)	214-136@1.30 (97-61@33)
3.0L & 3.8L (VIN 3)	2.03 51.6	85-95@1.73 (39-42@44)	175-195@1.34 (79.1-88.2@34.04)
3.8L (VIN 7)	2.03 (51.6)	74-82@173 (33-37@44)	175-195@1.34 (79.1-88.2@34.04)

Camshaft

Engine	Journal Diam. In. (mm)	Clearance In. (mm)	Lobe Lift In. (mm)
3.0L	1.785-1.786 (45.34-45.36)	.0005-.0025 (.013-.064)	Int. .210 (5.334) Exh. .240 (6.096)
3.8L (VIN C)	1.785-1.786 (45.34-45.36)	.0005-.0025 (.013-.064)	[1] .272 (6.909)
3.8L (VIN 3)	1.785-1.786 (45.34-45.36)	.0005-.0025 (.013-.064)	[1] .245 (6.223)
3.8L (VIN 7)	1.785-1.786 (45.34-45.36)	.0005-.0025 (.013-.064)	· · · ·

[1] Specification applies to both intake and exhaust

Caution: Following specifications apply only to 3.0L(VIN L) 3.8L (VIN 3), 3.8L (VIN 7) and 3.8 "3800" (VIN C) engines.

TIGHTENING SPECIFICATIONS

Application	Ft. Lbs. (N.m)
Camshaft Sprocket Bolts	20 (27)
Balance Shaft Retainer Bolts	27 (37)
Balance Shaft Gear Bolt	45 (61)
Connecting Rod Bolts	45 (61)
Cylinder Head Bolts	[1] 60 (81)
Exhaust Manifold Bolts	37 (50)
Flywheel-to-Crankshaft Bolts	60 (81)
Front Engine Cover Bolts	22 (30)
Harmonic Balancer Bolt	219 (298)
Intake Manifold Bolts	2
Main Bearing Cap Bolts	100 (136)
Oil Pan Bolts	14 (19)
Outlet Exhaust	
Elbow-to-Turbo Housing	13 (17)
Pulley-to-HarmonicBalancer Bolts	20 (27)
Outlet Exhaust	
Right Side Exhaust	
Manifold-to-Turbo Housing	20 (27)
Rocker Arm Pedestal Bolts	37 (51)
Timing Chain Damper Bolt	14 (19)
Water Pump Bolts	13 (18)

[1] Maximum torque is given. Follow specified procedure and sequence.
[2] Tighten bolts to 80 INCH lbs. (9 N.m.)

FIGURE 12-2 This service information includes critical measurement and torque specifications. You cannot service an engine properly without an engine's specific information.

BLOCK CONSTRUCTION

The **cylinder block** is the foundation of the engine. It houses all the rotating and recipro-cating parts of the bottom end and provides lubrication and cooling passages (Figure 12-3). The engine block is cast; hot, liquid metal is poured into a mold and allowed to cool and solidify. This forms the structure of the block. Sand is used inside the mold to provide water

The **cylinder block** is the main structure of the engine. Most of the other engine compo-nents are attached to the block.

Piston and connecting rod assembly

Flywheel

Oil pickup assembly

Crankshaft

Crankshaft bearings and caps

Oil pan and gasket

FIGURE 12-3 Cylinder block and related components.

jackets around the bores. Holes are drilled in the outside of the block to remove the sand after casting. These holes are then sealed with core plugs.

Blocks have traditionally been made of gray cast iron because of its strength and resistance to warpage. After it is cast, it is easily machined to provide rigid bores for the pistons, crankshaft, and lifters (and camshaft, if applicable). The block is also machined on the top, bottom, and ends to provide good sealing surfaces for the head, oil pan, transaxle, and timing cover (Figure 12-4).

Many newer designs use a cast aluminum alloy for the block. This offers a significant weight reduction that helps manufacturers meet fuel economy standards without sacrificing performance. Cast aluminum alloy blocks are more prone to cracking and warping problems when they are overheated, however. The cylinder wall may have replaceable or cast-in-place steel liners. When they are cast in place, they are called dry liners because no coolant hits the surface directly. Replaceable liners are called wet sleeves; coolant circulates all around the sleeve. The sleeve is sealed by an O-ring on the bottom and the head gasket and head on the top. Some aluminum blocks are fitted with cast iron cylinder heads and/ or thickly cast oil pans, to provide added rigidity to the block. Other newer engines are all aluminum.

In some cases, a silicon impregnated aluminum alloy is used in the casting. After the bores are machined, a special honing process removes the aluminum and uncovers a layer of hard silicon to form the thin wear wall of the bore. These cylinders cannot be bored oversize unless a sleeve is installed.

Other aluminum blocks use a special lost-foam casting process. The casting molds are made in a foam pattern that allows more complex shapes. These aluminum blocks (and heads) almost look as though they are made of silver Styrofoam (Figure 12-5). The manufacturers claim increased reliability with lower machining costs.

In recent years, some manufacturers have designed two-piece blocks, also called modular blocks, utilizing a bedplate or a one-piece main bearing cap (Figure 12-6). The two-piece design provides a stronger lower end because it ties all of the main caps together to improve block stiffness and reduce engine vibration.

Lubrication and Cooling

A cylinder block contains a series of oil passages, called **oil galleries**, that allow oil to be pumped through the block and crankshaft to reach the main bearings, and to the camshaft

The **bedplate** is the lower part of a two-piece engine block. The bedplate may also be called the main girdle.

Oil galleries are drilled passages throughout the engine that carry oil to key areas needing lubrication.

FIGURE 12-4 A bare V8 block.

FIGURE 12-5 This Saturn engine uses the lost-foam casting process.

FIGURE 12-6 A cylinder block with a bedplate.

and up into the cylinder head. The oil lubricates, cools, seals, and cleans engine components (Figure 12-7).

Some of the heat generated by the engine must be absorbed by the block and cylinder heads. This heat is wasted power and must be removed before it damages the engine. The engine has water jackets throughout the engine to help cool the engine. The area around the cylinder bores has coolant circulating and coolant flowing into the cylinder head to cool the area around the valve seats. Coolant circulates through the block jackets and coolant passages to transfer heat away from these hot areas into the rest of the cooling system.

Core Plugs

All cast cylinder blocks use **core plugs**. These may also be called expansion plugs or freeze plugs. During the manufacturing process, sand cores are used. These cores are partly broken and dissolved when the hot metal is poured into the mold; however, holes must be placed in

Core plugs are used to seal the holes machined into the block to remove the sand after casting. Core plugs are often called freeze plugs. Occasionally, when coolant is not adequately protected with antifreeze, it freezes in the block; a core plug will pop out and prevent the block from cracking. People are not always that lucky, however, and that is not the purpose of core plugs.

FIGURE 12-7 Cylinder block oil system and galleries.

FIGURE 12-8 Cylinder block core and gallery plugs.

the block to get the sand out after the block is cast. The core holes are machined, and core plugs are placed into them to seal them (Figure 12-8).

Cylinder Bores

The cylinder bores may either be integral or sleeved. Integral bores are machined out of the casting to fit the pistons. They are bored first and then honed to size to provide the proper surface finish for the rings. Bores are plateau honed from the factory to provide good cylinder wall lubrication with minimal wear. A rough stone hone is used to provide a crosshatch pattern in the cylinder. The crosshatch scratches of 50–60° on the cylinder wall hold oil to lubricate the pistons and rings. The cylinder is then finish honed with a soft stone to cut the rough edges off of the grooves in the cylinder wall. The result is a smooth plateau for the rings and pistons to contact (Figure 12-9).

Clearances between the pistons and cylinder bores are commonly very small on modern engines, as little as 0.0005 in.; the machining processes are precise. Many aluminum blocks will use cast iron sleeves either cast or dropped into the block. The block is supported at the top and bottom during the machining processes to prevent twisting. The lower portion of the block has webbing to help keep the cylinders straight during engine operation. The cylinder head helps secure the top half of the block. The top of the block must be very smooth so that the cylinder head can seal it. The base or bottom of the block is also machined to allow for proper sealing of the oil pan.

Main Bore

The **main bore** holds the crankshaft. The bore is supported by the webbing in the lower part of the block casting. Some blocks use a cast iron girdle that bolts onto the **main caps** to increase rigidity of the bore (Figure 12-10). Main bearing caps are cast separately and then machined to bolt on the bottom of the block. The main bore is then line- or align-bored to form the round hole between the bottom of the block and the main caps. The main caps are bolted in place, while a long boring bar passes through the end of the block. Because the boring bar will never turn exactly true, particularly toward the end of its reach, the bores will only be round if the same cap is installed in the same direction for each bore. The main bores must be in perfect alignment with each other in a straight line. If the main bores are misaligned, rapid and uneven main bearing wear will occur.

The **main bore** is a straight hole drilled through the block to support the crankshaft.

The **main caps** form the other half of the main bore and bolt onto the block to form round holes for the crankshaft bearings and journals.

Honed

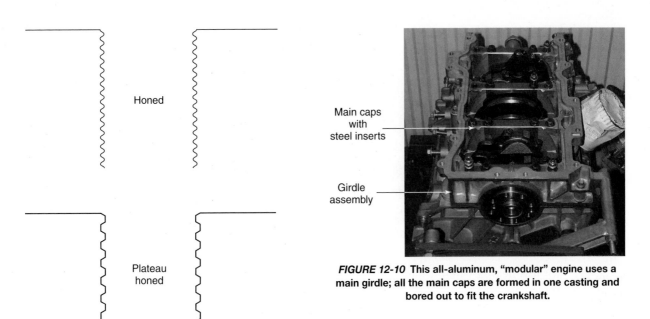

Main caps
with
steel inserts

Girdle
assembly

FIGURE 12-10 This all-aluminum, "modular" engine uses a
main girdle; all the main caps are formed in one casting and
bored out to fit the crankshaft.

Plateau
honed

FIGURE 12-9 This exaggerated view shows the rough
edges of honing, which are then cut down by plateau
honing. This process holds oil for the rings but causes
less wear to the rings than regular honing.

The main caps are held in place by two, four, or six bolts. A two-bolt main cap is the
most common (Figure 12-11), but many modern lightweight and performance engines use
a four-bolt main cap. The extra bolts may be next to each other (parallel), or they may be
bolted in from the sides (cross bolted). This design holds the crankshaft better and allows

FIGURE 12-11 Bearings are fit into the main bores to
protect the crankshaft. This is an example of a
two-bolt main block.

FIGURE 12-12 This is a four-bolt main bearing cap.

for less flexing when under higher stresses (Figure 12-12). Some engines use a six-bolt main cap.

Main **bearing inserts** snap into the halves of the main caps. They provide a soft surface to support, lubricate, and protect the crankshaft. They also prevent the main bore from wear; the crankshaft does not contact the main bores (Figure 12-11).

CRANKSHAFT

The crankshaft withstands tons of force when the piston pushes the connecting rod down on the power stroke. It converts that reciprocating motion of the piston into useful rotary motion. Crankshafts may be made of cast iron, nodular iron, or forged steel. Cast iron cranks are quite common on production vehicles. Forged crankshafts are stronger and are often found on high performance vehicles, engines with forced induction systems, and diesel engines.

The crankshaft is bolted into the main bore on its **main bearing journals**. These are machined on the centerline of the crankshaft and support the weight and forces applied to the crankshaft. Crankshaft journals are machined and polished smooth to minimize bearing wear and friction. The journals actually ride on a thin film of oil between the bearing and the journal.

The **connecting rod bearing journals** or pins are offset from the centerline (Figure 12-13). The connecting rod clamps around these journals with a bearing insert in between to protect the crankshaft and connecting rod. The offset places weight and pressure off the center of the crankshaft and provides leverage for the piston assembly to turn the crankshaft.

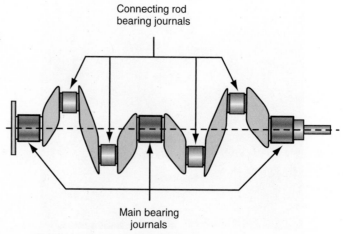

FIGURE 12-13 The rod bearing journals are offset from the centerline of the crank. The length of the offset times two is the engine's stroke.

The distance the connecting rod journals are offset from the centerline of the crankshaft is called the **crankshaft throw**. The throw determines the stroke of the engine. Rod journal offset times two equals the stroke. The throws are set so that pistons in opposite banks move in opposite directions to minimize engine vibration. This design cancels some of the forces applied to the crankshaft.

Crankshafts have many different configurations, depending on the engine. They will usually be ground so that the cylinders fire in equal intervals during one full cycle. A four-cylinder engine produces a power stroke every 180°, its rod journals are 180° off of each other. A typical four-cylinder engine with a firing order of 1–3–4–2 will fire cylinder number 1 at 180° of crank rotation, number 3 at 360°, number 4 at 540°, and complete the full cycle at 720° with the firing of cylinder number 2. This puts cylinders number 1 and number 4 and cylinders number 3 and number 2 positioned at the same point in the cylinder during rotation (Figure 12-14). The connecting rod journals from an end view form a straight line above and below the main journals.

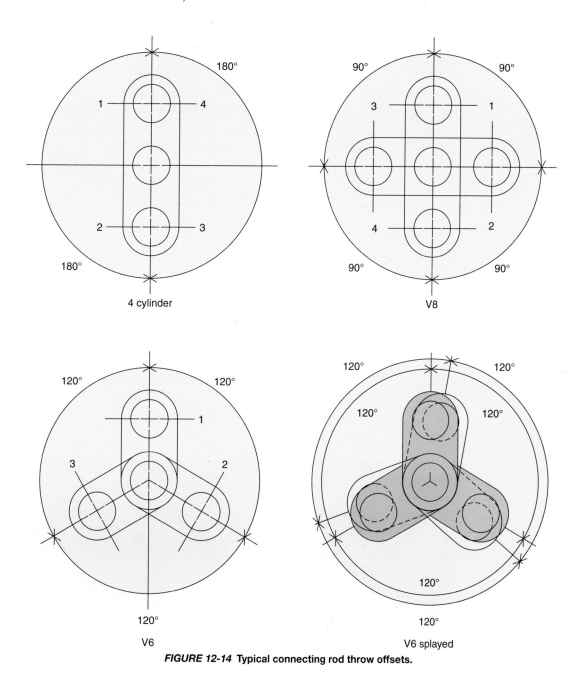

FIGURE 12-14 Typical connecting rod throw offsets.

Individual offset crankpins

FIGURE 12-15 A splayed crankshaft.

A V6 engine typically produces a power stroke every 120° of crank rotation. This is called an even-firing V6. The journals on the crankshaft are **splayed**. In this engine, on one throw, two journals are offset and separated by 30°, in this case (Figure 12-15). Other V6 engines use one throw with two equal journal surfaces. This means that two rods will rotate together. The throws are machined at 120° intervals around the crankshaft. This produces an odd-firing V6: cylinders fire at 90°, then after 150° of rotation, then again after 90° and on through the two full revolutions of the crankshaft that create a cycle.

An even firing V8 crankshaft has four throws each with two journals on them (Figure 12-16). The crankshaft produces power strokes every 90° of crank rotation. This is what makes a V8 such a smooth-running engine; the crankshaft is rotated by a connecting rod every 90°. This leaves little time for the crankshaft to speed up or slow down.

Crankshaft Oil Holes

The crankshaft is drilled to provide oil passages that feed each rod and main journal. The oil is held under pressure by the tight fit between the journal and the bearing. The bearing has an oil hole in it that lines up with the drilled hole in the crankshaft.

FIGURE 12-16 Two rods share one journal on this V8 crankshaft. Notice the oil holes to feed each of the rod bearings.

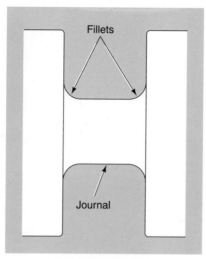

FIGURE 12-17 The fillet strengthens the crankshaft journal.

Crankshaft Fillets

The **fillet** on the journal is used to increase the strength of the journal (Figure 12-17). This joint area is subject to high stress and is a common location for cracks. Since sharp corners are weak, manufacturers call for a curved fillet of a specific radius. The fillet is increased for applications requiring severe duty. Most fillets are formed by grinding the joint. Some turbocharged or high performance engines use a process of rolling the fillet. This is done by rolling hardened steel balls under pressure on the joint. This causes the metal at the joint to become compacted and tightens the grain structure, thus strengthening the crankshaft by as much as 30 to 40 percent. The fillet is not usually a wear area unless the bearing is worn, but the fillet is an area where the crankshaft can crack and should be carefully inspected. If the journal requires machining, the fillet should be reground to the proper radius. The crankshaft is also subject to lateral forces. These come from clutch engagement and automatic transmission loads. The lateral movement of the crankshaft is controlled by a thrust bearing and a thrust surface on the crankshaft. The thrust bearing is usually attached vertically to the sides of one main bearing.

Fillets are small rounded corners machined on the edge of the journals to increase journal strength.

Crankshaft Counterweights

The crankshaft is cast or forged with large **counterweights** that oppose the rod journals. The counterweight overcomes the total weight of the rod bearing journal, bearing, connecting rod, piston pin, and piston. These help support the crankshaft and even out the speed changes as the crank rotates through firing impulses. You will often see holes drilled in the counterweights. These are drilled when the crankshaft is balanced as it spins with rods and pistons attached. Holes also lighten the crank; you may see lightening holes on the sides of the throws as well.

Crankshaft counterweights are positioned opposite the connecting rod journals and balance the crankshaft.

Crankshaft Construction

Many manufacturers construct their crankshafts from gray cast iron. This construction is strong enough for most applications. Some engine applications require additional strength above what gray cast iron can provide; in these instances, nodular iron is often used. A cast iron crankshaft is identifiable by its straight cast mold parting line (Figure 12-18).

In other applications, such as some turbocharged engines, a forged medium-carbon steel crankshaft (carbon content between 0.30 and 0.60 percent) is used. High performance engines may use a crankshaft using chromium and molybdenum (sometimes called

FIGURE 12-18 Casting lines on a cast iron crankshaft.

FIGURE 12-19 Casting lines on a forged crankshaft.

chrome-molly) alloys to increase its strength. The forging process condenses the grain of the steel and increases its strength. Because of the density of the steel, these crankshafts weigh more than a cast iron crankshaft. To lighten the weight, some manufacturers hollow the rod bearing journals.

Forged crankshafts designed for use in V8 and in-line, six-cylinder engines must be twisted to achieve the desired throw offset. This process is done immediately after the crankshaft is forged, while the meal is still hot. A forged crankshaft can be identified by the staggered die parting lines because of this twisting (Figure 12-19).

Case-hardening crankshafts protects the journals from wear and fractures. Many manufacturers use a case-hardening process called ion nitriding. The component is placed into a pressurized chamber filled with hydrogen and nitrogen gases. An electrical current is then applied through the component. This changes the molecular structure of the metal and allows the induction of the gases into the surface area of the journal.

CAMSHAFT

The camshaft may be fitted in the block. (Figure 12-20) It rides on bearings placed in the camshaft bore. A thrust plate typically limits the camshaft end play. Refer to Chapter 10 of this manual for a review of the full characteristics and functions of the camshaft.

LIFTER BORES

When a block contains the camshaft, it also has lifter bores drilled above the camshaft lobes. These bores are slightly off center from the centerline of the camshaft lobe so that the convex face of the lifter turns the lifter in its bore. Lifter bores must be clean and free of coring to provide a smooth surface for the lifters to ride on.

FIGURE 12-20 The camshaft may be housed in the block or in the head.

Pulley

Rubber damper for torsional vibration

Rubber damper for vertical vibration

Hub

FIGURE 12-22 Dual-mode harmonic balancer.

Piston

Vibration damper center area

Crank twist and snap-back action

Rubber-like material

Counterweight area

Crankshaft

* Harmonic balancer scale exaggerated.
FIGURE 12-21 Harmonic balancer action.

HARMONIC BALANCER

A two-piece **harmonic balancer** is used on the front end of the crankshaft. It is used to absorb the **torsional vibrations** of the crankshaft created as the crankshaft tries to speed up due to firing impulses on cylinders. At the same time, another cylinder is trying to sow the crankshaft down as it pushes against air pressure on the compression stroke. The inner piece is bolted to the crankshaft. An inertia ring is bonded with rubber to the inner ring. As the crankshaft winds and unwinds, the twisting of the inertia ring and the rubber ring absorb some of the vibrations (Figure 12-21). If the bonding fails, the two rings will separate and you will need to replace the harmonic balancer. A faulty harmonic balancer can cause engine vibration, resulting in a broken crankshaft.

Many of today's engines use dual-mode harmonic balancers (Figure 12-22). The first mode eliminates torsional vibration, and the other mode eliminates vertical vibration.

Many harmonic balancers are pressed onto the end of the crankshaft and a square key fits into a matching keyway in the crankshaft and harmonic balancer to prevent rotation. A bolt and washer usually retain the harmonic balancer on the end of the crankshaft.

> **AUTHOR'S NOTE: One day a customer had her vehicle towed in after the engine stopped running after developing a very serious knocking noise from the bottom end. While disassembling the engine, I noticed that the harmonic balancer had separated badly. The result: the crankshaft had broken in half!**

FLYWHEEL

The **flywheel** is a large round disc of heavy mass attached to the rear end of the crankshaft to smooth the rotation of the crankshaft (Figure 12-23). It helps the crankshaft speed remain more constant between firing pulses by storing inertial energy. The outer ring, the ring gear, is pressed on to provide teeth for the starter pinion gear to turn the engine over. The

Shop Manual
Chapter 12

Harmonic balancers are used to control and compensate for torsional vibrations.

Torsional vibration is the result of the crankshaft twisting and snapping back during each revolution.

The **flywheel** bolts to the rear end of the crankshaft and helps to maintain a steady crankshaft speed.

Flywheel

Crankshaft

FIGURE 12-23 The flywheel is attached to the rear of the crankshaft. A flywheel is used with a manual transmission. A flexplate is used with a manual transmission.

The **torque converter** mounts to the crankshaft through a flex plate. Its weight and fluid coupling do a fine job of maintaining a more constant crankshaft speed.

Crankshaft specifications, stampings, part numbers, and tolerances may be different if an engine is mated with the vehicle option of an automatic or a manual transmission.

machined center surface provides a surface for the clutch disc to grab onto to transmit the rotation of the crankshaft to the transmissions. Flywheels are only used on engines mounted to a manual transmission. An automatic transmission is mounted to the engine by a light flex plate that bolts to the **torque converter**. The torque converter is filled with transmission fluid and is as heavy as a flywheel and clutch assembly. The fluid coupling of the torque converter does an excellent job of helping to maintain a constant crankshaft speed.

SHORT BLOCKS, LONG BLOCKS, AND CRATE ENGINES

Sometimes an engine block and/or heads are damaged badly enough that it becomes more cost effective to replace the failed component(s) rather than machine or recondition it. New or rebuilt short or tall blocks are available for most engines. In addition, "crate" engines, usually supplied by an aftermarket company, are available fully assembled and rebuilt. The option of replacing a seriously damaged engine with a crate engine has become a very popular option in the past several years. Crate engines are remanufactured on assembly lines and are often cheaper that the cost of parts, machining, and labor to rebuild an engine in house (Figure 12-24).

FIGURE 12-24 This remanufactured engine comes with bored cylinders, polished or new crankshafts, refurbished and balanced connecting rods, and new rings, pistons, and bearings.

Summary

- The bottom end of the engine is made up of the block, crankshaft, pistons, and connecting rods.
- The cylinder block may be cast iron or aluminum.
- Cylinders are typically integral on a cast iron block, as are steel sleeves on aluminum blocks.
- The block's main bore holds the crankshaft.
- The main bore is line-bored with the caps in place. Main caps must be installed in their original locations and in the same direction.
- Crankshafts may be made of cast iron, nodular iron, or forged steel; forged cranks are stronger.
- The crankshaft spins on its main journals.
- The connecting rod journals are offset from the centerline, which determines the stroke.
- The connecting rod transmits combustion forces to spin the crankshaft.
- The harmonic balancer bolts to the front of the crankshaft to reduce torsional vibration.
- The flywheel or torque converter helps maintain crankshaft speed between power pulses.

Review Questions

Short Answer Essays

1. Briefly explain how a cylinder block is formed.

2. Describe the advantages and disadvantages of an aluminum block.

3. What are the advantages of two-piece blocks?

4. Why must a main cap be installed in its original location and direction?

5. What is the purpose of the crankshaft?

6. How is the stroke of an engine determined?

7. Explain the purpose of the crankshaft fillet.

8. Explain why lifter bores are off center from the center of the camshaft lobe.

9. What is the purpose of a dual-mode harmonic balancer?

10. Explain the purpose of the flywheel.

Fill-in-the-Blanks

1. The _____ _____ is the lower portion of a two-piece block.

2. _____ _____ _____ blocks are significantly lighter than cast iron blocks.

3. The passages that allow oil to flow throughout the engine are called _____ .

4. _____ _____ are used to seal holes drilled to remove sand from the block casting process.

5. Bores are _____ _____ at the factory to provide good cylinder wall lubrication with minimum wear.

6. The _____ _____ holds the crankshaft.

7. _____ crankshafts are stronger than _____ crankshafts.

8. An eight-cylinder engine produces a power stroke every _____° of engine rotation.

9. The harmonic balancer reduces _____ _____ .

10. A _____ engine is a fully assembled remanu-factured engine.

Multiple Choice

1. Aluminum blocks often use:
   A. Integral bores.   C. Steel liners.
   B. Steel housings.   D. Cast iron main girdles.

2. A harmonic balancer has separated. The response the technician should take is:
   A. Reseal the bond with rubber loctite.
   B. Replace the inertia ring.
   C. Replace the crankshaft; it is likely cracked.
   D. Replace the whole harmonic balancer.

3. An engine bedplate:
   A. Reduces vibrations.
   B. Contains the main bearing caps.
   C. Adds strength to the block.
   D. All of the above.

4. On an even-firing V6 engine with a splayed crankshaft, the throws on the splayed connecting rod journals are offset:
   A. 10°.   C. 35°.
   B. 30°.   D. 45°.

5. The crankshaft is balanced by _____ to overcome the weight of the rod journal, piston pin, connecting rod, and piston:
   A. Harmonic balancer.
   B. Fillet.
   C. Counterweights.
   D. Flywheel.

6. The lifter bores are offset from the camshaft lobe to:
   A. Allow the camshaft to turn.
   B. Spin the valve.
   C. Turn the pushrod.
   D. Rotate the lifter.

7. Each of the following is a crankshaft material, EXCEPT:
   A. Gray cast iron.
   B. Nodular iron.
   C. Chrome-moly alloys.
   D. Forged aluminum.

8. A dual-mode harmonic balancer reduces:
   A. Torsional and vertical vibrations.
   B. Crankshaft end thrust and torsional vibration.
   C. Speed fluctuations and torsional vibrations.
   D. Vertical vibrations and crankshaft end thrust.

9. The purpose of core plugs is:
   A. To fully drain coolant during engine service.
   B. To prevent the block from freezing.
   C. To remove the sand after casting.
   D. To provide cooling jackets to remove excess heat.

Chapter 13

PISTONS, RINGS, CONNECTING RODS, AND BEARINGS

UPON COMPLETION AND REVIEW OF THIS CHAPTER, YOU SHOULD BE ABLE TO:

- Describe the function and design of engine bearings.
- Describe the importance of proper bearing oil clearances.
- Describe the design and function of pistons.
- Describe the methods used in piston design to control heat expansion.
- List the advantages of cast, forged, and hypereutectic cast pistons.

- Describe the purpose of piston pin offset.
- Explain the purposes of piston rings.
- Describe the design and materials of piston rings.
- Explain the design and function of the connecting rods.

INTRODUCTION

The components within the engine block, the pistons, piston rings, connecting rods, and engine bearings, all play a critical role in transmitting the power of combustion from the top of the piston to the rod journal of the crankshaft to spin the engine. The piston is forced downward when combustion peaks just after TDC on the power stroke. The piston rings seal the small gap between the piston and the cylinder wall to use all the pressure of combustion. The connecting rods connect the piston to the crankshaft, delivering the reciprocating motion of the piston to the rotating crankshaft. Soft-coated engine bearings are held in the main, connecting rod, and camshaft bores to protect the crankshaft and camshaft. A thin film of oil is held between the bearing and the journal to reduce friction and minimize wear. Piston rings; main, connecting rod, and camshaft bearings; and timing chains are the main wear items in an engine block. Each of the bottom end components is carefully designed to match the characteristics of the engine. By knowing how these components are designed and should perform, you will be better able to evaluate their condition during engine repairs.

BEARINGS

Bearings (bearing inserts) are used to protect the main and connecting rod journals of the crankshaft and carry the rotational loads (Figure 13-1). A similar type of bearing is used on the camshaft when it is in the block. The camshaft bearings are one piece rather than two, but their materials and characteristics are nearly the same as those used to protect the crankshaft. The engine bearings are designed to wear before the crankshaft. The bearings snap into the main caps and connecting rod caps and are then bolted around the crankshaft journals. The clearance between the bearing and the journal is critical. The clearance allows a thin film of oil to lubricate the bearing and the crankshaft. That oil also absorbs

FIGURE 13-1 This is one half of a set of new aluminum-coated main bearings.

FIGURE 13-2 The crankshaft should ride on a film of oil roughly 0.001 in. thick. That is enough to reduce friction and protect the crankshaft.

the pounding of the crank as the power stroke forces it downward in the main bores. The crankshaft should spin on a film of oil and not contact the bearing (Figure 13-2). The oil pump sends oil under pressure through the holes drilled in the crankshaft through an oil hole in the bearing into this small clearance. The correct bearing clearance allows a calibrated amount of oil to leak out past the side of the bearings so the oil is constantly refreshed. When the bearing clearances become too great, more oil leaks out and engine oil pressure decreases. This loss of oil pressure results in a decrease in the oil film strength. The crankshaft will knock in its bore and cause rapid wear and distortion of the bearings and crankshaft. If the engine is not repaired promptly, the damage to the crankshaft and bearings can become so severe that the bearings spin with the crankshaft in their bores. When bearings spin in their bores, they can destroy the main bores, connecting rod bores, or camshaft bores and even cause the engine to seize. You will usually replace an engine with this type of severe damage to the block and crankshaft.

Bearing Materials

Engine bearings are made in halves that snap into the main and connecting rod bores and caps. The back of the bearings is steel, to provide strength to carry the tremendous loads impressed by the crankshaft. On top of the load bearing shell are layers of soft metal to protect the crankshaft.

There are many different materials used in the construction of bearings. The main concern of the manufacturers is to achieve a bearing construction that will meet four primary requirements:

1. *Surface action*—The ability of the bearing to withstand metal-to-metal contact without being damaged.
2. *Embedability*—The ability of the bearing material to tolerate the presence of foreign material.
3. *Fatigue resistance*—The ability of the bearing to withstand the stress placed upon it.
4. *Corrosion resistance*—The ability of the bearing to withstand the corrosive acids produced within the engine.
5. *Conformability*—The ability of the bearing material to conform to small irregularities of the journal.

Since no single material is capable of meeting all of these requirements, most bearings are constructed using multilayers of different materials. The backing is usually made of steel; then additional lining material is added. The most common materials used include:

1. Babbitt
2. Sintered copper-lead
3. Cast copper-lead
4. Aluminum

To assist the technician in the selection of bearings, most manufacturers use a part number suffix to identify the bearing lining material; for example, a Federal Mogul bearing with the part number 8-3400CP indicates the bearing is steel-backed copper-lead alloy with overplate. The following chart provides the codes used:

Part Number Suffix	Lining Material and Application
AF	Steel-backed aluminum alloy thrust washer
AP	Steel-backed aluminum alloy with overplate connecting rod or main bearings
AT	Solid aluminum alloy connecting rod or main bearings
B	Bronze-backed babbitt connecting rod or main bearings
BF	Steel-backed bronze or babbitt thrust washers
CA	Steel-backed copper alloy connecting rod or main bearings
CH	Steel-backed copper-lead alloy with overplate connecting rod and main bearings for performance applications
CP	Steel-backed copper-alloy with overplate connecting rod and main bearings
DR	Steel-backed babbitt, steel-backed copper-lead alloy, or solid aluminum camshaft bushing
F	Bronze-backed babbitt, bronze or aluminum thrust washers
RA	Steel-backed aluminum alloy connecting rod and main bearings
SA, SB, SBI, SH, SI, SO	Steel-backed babbitt connecting rod and main bearings
TM	Steel-backed copper alloy with thin overplate connecting rod and main bearings
W	Camshaft thrust washer

When choosing replacement bearings you will usually use original equipment quality components or updated designs as recommended by the parts supplier.

Bearing Characteristics

Figure 13-3 provides common nomenclature used with insert-type bearings. Most main and connecting rod bearings are designed to have a certain amount of spread. **Bearing spread** is provided by manufacturing the bearing inserts so they have a slightly larger curvature compared to the curvature of the rod bearing or cap. This design assures positive positioning of the bearing against the total bore area. When the bearing is installed into the bore, a light pressure is required to set the bearing (Figure 13-4).

In addition to bearing spread, most main and connecting rod bearings are designed to provide **bearing crush** (Figure 13-5). When the bearing half is installed in the connecting rod or cap, the bearing edges protrude slightly from the rod or cap surfaces. When the two halves are assembled, the crush exerts radial pressure on the bearing halves and forces them into the

Bearing spread means the distance across the outside parting edges is larger than the diameter of the bore.

Bearing crush refers to the extension of the bearing half beyond the seat.

FIGURE 13-3 Bearing nomenclature.

Rod and main bearings

- Length
- Locating tab
- Annular groove
- Oil hole
- End chamfer
- Distribution groove
- Gauge diameter (spread)
- Height
- Back thickness
- Total wall thickness
- Lining thickness
- Parting line chamfer
- Eccentric wall thickness

Thrust bearings

- Length
- Bar gauge
- Flange diameter
- Thrust face
- Flange oil pocket

FIGURE 13-4 Designing the bearing with a certain amount of spread assures positive positioning of the bearing against the total bore area.

- Bearing
- Housing bore

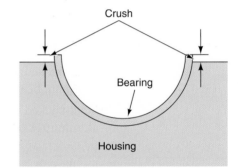

FIGURE 13-5 Bearing crush exerts radial pressure on the bearing halves and forces them securely into the bore.

- Crush
- Bearing
- Housing

The **locating tab** is a tab sticking out of bearing inserts that fits in a groove in the block, main bearing cap, connecting rod, or connecting rod cap to prevent insert rotation.

bore. Proper crush is important for heat transfer. If the bearing is not tight against the bore, it will act as a heat dam and not transfer the heat to the engine block. Bearing spread and bearing crush also help to prevent the bearing from spinning in its bore as rotational friction acts on it.

To assure proper installation of the bearing in the bore, most manufacturers use a **locating tab** (Figure 13-6). A protruding tab at the parting of the bearing halves is set into a slot in the housing bore. When properly installed, the tab will locate the bearing, preventing it from shifting sideways in the bore.

Main Bearings

Main bearings carry the load created by the movement of the crankshaft. Most of these bearings are insert design. The advantages of this design include ease of service, variety of materials, and controlled thickness.

Split bearings surround each main bearing journal. The bearings are supplied oil under pressure from the oil pump through oil passages drilled in the engine block and crankshaft (Figure 13-7). The crankshaft does not rotate directly on the main bearings; instead, it rides on a thin film of oil trapped between the bearing and the crankshaft. If the journals are worn and become out-of-round, tapered, or scored, the oil film is not formed properly.

Locating lug

Recess

FIGURE 13-6 A protruding tab at the parting of the bearing halves is mated to a slot in the housing bore. The tab locks the bearing to prevent it from spinning or shifting sideways.

Oil enters at the main bearing journals

Oil gallery

Oil comes out to lubricate the connecting rod journals

FIGURE 13-7 Internal passages in the crankshaft send lubrication to the main and connecting rod bearings.

This will result in direct contact between the crankshaft and bearing and will eventually damage the bearing and/or crankshaft. Soft materials are used to construct the bearings in an attempt to limit wear of the crankshaft.

Oil Grooves. Providing an adequate oil supply to all parts of the bearing surface, particularly in the load area, is an absolute necessity. In many cases, this is accomplished by the oil flow through the bearing oil clearance. In other cases, however, engine operating conditions are such that this oil distribution method is inadequate. When this occurs, some type of oil groove must be added to the bearing. Some oil grooves are used to assure an adequate supply of oil to adjacent engine parts by means of oil throw-off.

Oil Holes. Oil holes allow for oil flow through the engine block galleries and into the bearing oil clearance space. Connecting rod bearings receive oil from the main bearings by means of oil passages in the crankshaft. Oil holes are also used to meter the amount of oil supplied to other parts of the engine; for example, oil squirt holes in connecting rods are often used to spray oil onto the cylinder walls. When the bearing has an oil groove, the oil hole normally is in line with the groove. The size and location of oil holes is critical. Therefore, when installing bearings, you must make sure the oil holes in the block line up with holes in the bearings.

Thrust Bearings

The crankshaft is fitted with one **thrust bearing** to limit end play within the block (Figure 13-8). There must be some clearance for the crankshaft to move horizontally (end to end), to prevent it from seizing as components heat up. The thrust bearing protects the crank from slamming against the block as it moves forward and backward in its bore. Under acceleration, the crankshaft will be forced toward the back of the block. A thrust bearing allows a thin film of oil between a soft bearing and the crankshaft to absorb this force. The thrust bearing is often one of the main bearings, with flanges on the sides to control crank end play (Figure 13-9). It may be positioned at an end of the crank or in the middle. Some thrust bearings are not part of a main bearing; they can be multipiece (Figure 13-10). In this case, they are called thrust washers. Thrust bearings will typically wear faster when the engine is attached to a manual transmission because the clutch friction disk is constantly being engaged and disengaged from the flywheel while accelerating. This extra stress places horizontal loads on the crankshaft that are not found when the engine is coupled with an

The **thrust bearing** prevents the crankshaft from sliding back and forth by using flanges that rub on the side of the crankshaft journal.

Crankshaft end play is the measure of how far the crankshaft can move lengthwise in the block.

Crankshaft

Thrust bearings

FIGURE 13-8 A flanged thrust bearing.

FIGURE 13-9 This thrust bearing has a soft aluminum coated side flange to cushion crankshaft movement back and forth in the block.

FIGURE 13-10 A multiple-piece thrust bearing.

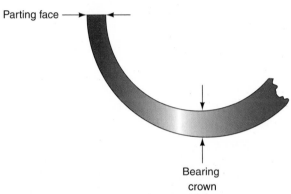

Parting face

Bearing crown

FIGURE 13-11 Bearing crown maintains close clearances at the top and bottom of the bearing, where most of the loads are applied.

automatic transmission. Some manufacturers will have a different specification for the crankshaft end play. You will measure crankshaft end play during engine disassembly and reassembly to be sure it is within specifications.

Rod Bearings

Split bearings surround each connecting rod journal. Like the main bearings, the **connecting rod bearings** do not rotate directly on the crankshaft. They allow the rod journal to rotate on a thin film of oil trapped between the bearing and the crankshaft. Journals that are worn, out-of-round, tapered, or scored will not maintain this oil film, resulting in damage to the bearing and/or crankshaft.

Most main and rod bearings are designed with a bearing crown to maintain close clearances at the top and bottom of the bearing. This is the area where most of the loads are applied. The crown also allows increased oil flow at the sides of the bearing (Figure 13-11). In addition, most bearings have a beveled edge to provide room for the journal fillet.

Bearing Oil Clearances

There must be a gap or clearance between the inside diameter of the bearing face and the outside diameter of the journal of the main, connecting rod, and camshaft (if insert bearings are used). This clearance allows engine oil pressure to hold a film of oil within the clearance and prevent too much oil from leaking out. When the clearance becomes too great, either from wear of the bearing or the crankshaft journal engine oil, pressure is reduced.

Connecting rod bearings are insert-type bearings retained in the connecting rod and mounted between the large connecting rod bore and the crankshaft journal.

The bearing oil clearances provide a calibrated "leak" of oil from between the bearing and the journal. The clearances must be within the specifications to provide good oil pressure, prevent knocking, and protect the journal from wear. Many modern oil bearing clearances fall within the range of 0.0008 in. (0.0203 mm) to 0.002 in. (0.0508 mm). These are just typical clearances; always check the manufacturer's specifications for the engine you are working on.

Bearings are a wear item. They are designed to wear and protect the crankshaft or camshaft. They do wear out after high mileage and exposure to contaminated oil. When the engine is disassembled for repairs, it is customary to replace the main, rod, and cam bearings.

When fitting new bearings you will check the clearance to be sure it is within the specified range. We will discuss this procedure in Chapter 13 of the Shop Manual.

Undersize and Oversize Bearings

When a crankshaft is worn lightly and polished to be useable, bearings that are 0.001 and 0.002 in. undersize are often available. If the crankshaft has been ground to repair serious journal wear, **undersize bearings** can be fitted. These bearings are generally available in 0.010, 0.020, and 0.030 in. undersize. Metric undersized bearings may include 0.050, 0.250, 0.500, and 0.750 mm. Undersize means that the inside diameter of the bearings is smaller, to fit the reduced diameter of the crankshaft journals.

Oversize bearings may be used when the block has been line bored to an oversized diameter. Oversize bearings are often available in 0.010, 0.020, 0.030, and 0.040 in. Available metric oversized bearings are typically 0.250, 0.500, 0.750, and 1.000 mm. These bearings have a larger outside diameter than the standard bearing. The use of oversize and undersize bearings has decreased dramatically as component replacement has become more cost-effective than many complex machining operations. Some bearings are stamped with a code that allows the technician to check the size of the bearing that is in use now. New bearings come stamped from the factory with the size correlation on it. Some engine blocks have a stamping on them that helps the technician indicate what size bearing came on the engine originally. If different sized bearings are installed in this type of engine, the technician should change this mark or remove it.

> **AUTHOR'S NOTE:** It has been my experience that most connecting rod and main bearing failures are caused by contaminated engine oil or lack of lubrication. Often the cause of failure is clear by the sludge in the oil pan, the coolant-contaminated oil, or the lack of oil in the crankcase. Try to determine the cause of the failure if it is premature. Most bearings should last the life of the engine; these days that may reach 150,000 to 200,000 miles (241,402 to 321,870 km). Be sure to discuss the importance of regular oil and filter changes with your customers.

CAMSHAFT AND BALANCE SHAFT BEARINGS

In an overhead valve (OHV) engine, the circular bearings are pressed into openings in the block (Figure 13-12). The camshaft journals are positioned in these bearings. The specified bearing clearance must be present between the camshaft journals and the bearings. Oil from the lubrication system is supplied through passages in the block to an opening in each camshaft bearing. The opening in the camshaft bearing must be properly aligned with the oil passage in the block. Excessive camshaft bearing clearance causes low oil pressure. During an engine rebuild, the camshaft bearings should be replaced. To make camshaft removal and replacement easier, the camshaft bearings on many engines are progressively larger from the back to the front of the block. Special tools are available to remove and replace the camshaft bearings. Balance shaft bearings are similar to camshaft bearings.

Shop Manual
Chapter 13

Undersize bearings have the same outside diameter as standard bearings, but the bearing material is thicker to fit an undersize crankshaft journal.

Oversize bearings are thicker than standard to increase the outside diameter of the bearing to fit an oversize bearing bore. The inside diameter is the same as standard bearings.

A BIT OF HISTORY

Isaac Babbitt invented his bearing material in 1839 using a mixture of tin, lead, and antimony. This tin-based babbitt was later replaced with a

A BIT OF HISTORY

(Continued)

lead-based material. When lead-based babbitt was used in the first automotive engines, it was melted and then poured into the block and connecting rods. After the babbitt cooled, the correct oil clearance was achieved by hand scraping the excess material. Final adjustment of the bearings was accomplished by using shims between the bearing caps. Due to the nature of these early bearings, they often smeared under heavy loads. To decrease the oil clearance as the bearing wore, the caps were refilled. This operation was considered owner maintenance, and babbitt material was carried in the vehicle to make the necessary roadside repairs.

FIGURE 13-12 Camshaft bearings in the cylinder block.

BALANCE SHAFTS

Many engine designs tend to have inherent vibrations. The in-line four-cylinder engine is a good example of this, but other engines use balance shafts for varying reasons, including customer demand for smoothness. Even some vee-style engines use balance shafts. Some engine manufacturers use a balance shaft to counteract this tendency (Figure 13-13). Four-cylinder engine vibration occurs when the piston is moving down during its power stroke. When the engine completes one revolution, two vibrations occur. The **balance shaft** rotates at twice the speed of the crankshaft. This creates a force that counteracts crankshaft vibrations.

PISTONS

The piston, when assembled to the connecting rod, is designed to transmit the power produced in the combustion chamber to the crankshaft (Figure 13-14). The piston must be able to withstand severe operating conditions. Stress and expansion problems are compounded by the extreme temperatures to which the top of the piston is exposed. In addition, the rapid movement of the piston creates stress and high pressures. To control

FIGURE 13-14 The forces of combustion push the piston down in the cylinder. The connecting rod bolts to the crankshaft and transmits this force.

FIGURE 13-13 A balance shaft in a V6 engine.

these stress conditions, most pistons are constructed from aluminum. The lightweight aluminum operates very efficiently in high rpm engines.

The piston has several parts, including (Figure 13-15):

1. *Land*—Used to confine and support the piston rings in their grooves.
2. *Heat dam*—Used on some pistons to reduce the amount of heat flow to the top ring groove. A narrow groove is cut between the top land and the top of the piston. As the engine is run, carbon fills the groove to dam the transfer of heat.
3. *Piston head or crown*—The top of the piston against which the combustion gases push. Different head shapes are used to achieve the manufacturers' desired results (Figure 13-16). Most engines use flat-top piston heads. If the piston comes close to the valves, the recessed piston head is used to provide additional clearance. Different types of domed and wedged piston heads are used to increase compression ratios.
4. *Piston pins (or wrist pins)*—Connect the piston to the connecting rod. Three basic designs are used: piston pin anchored to the piston and floating in the connecting rod; piston pin anchored to the connecting rod and floating in the piston; and piston pin full floating in the piston and connecting rod.

The **balance shaft** is a shaft driven by the crankshaft that is designed to reduce engine vibrations.

Shop Manual
Chapter 12,
page 479

FIGURE 13-15 The components of the piston and connecting rod assembly.

C.D. = Compression distance

FIGURE 13-16 Some of the common piston head designs used in automotive engines.

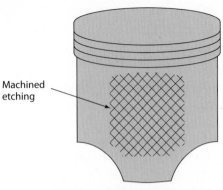

FIGURE 13-17 Some piston skirts have an etching machined into them to help trap oil to lubricate the skirt. Scuffing prevents oil from being trapped, and causing increased friction and wear.

Machined etching

Gas ports

FIGURE 13-18 Some performance engines use gas ported pistons to improve top ring sealing.

Slipper skirts allow shorter connecting rods to be used. Part of the skirt is removed to provide additional clearance between the piston and the counterweights of the crankshaft. Without this recessed area, the piston would contact the crankshaft when the shorter connecting rods are used.

Thermal expansion of aluminum is about twice as much as iron.

5. *Skirt*—The area between the bottom of the piston and the lower ring groove and at a 90-degree angle to the piston pin. The skirt forms a bearing area in contact with the cylinder wall. To reduce the weight of the piston and connecting rod assembly, many manufacturers use a slipper skirt. The piston skirt surface is etched to allow it to trap oil (Figure 13-17). This helps lubricate the piston skirt as it moves within the cylinder and prevents scuffing of the skirt.
6. *Thrust face*—The portion of the piston skirt that carries the thrust load of the piston against the cylinder wall.
7. *Compression ring grooves*—The two upper ring grooves used to hold the compression rings.
8. *Oil ring groove*—The bottom ring groove used to hold the oil ring.
9. *Gas ports*—Some pistons have holes drilled near the top compression ring to allow combustion gases behind the ring to force it out against the cylinder wall to improve sealing (Figure 13-18).

Expansion Controls

When combustion occurs, high temperatures and pressures are applied to the top of the piston. Some of this heat is transmitted to the piston body, causing the piston to expand. To control piston expansion, most pistons are cam ground. **Cam ground pistons** are made to be an oval shape when the piston is cold. The larger diameter of the piston is across the thrust surfaces. The thrust surfaces are perpendicular to the piston pin (Figure 13-19). On a cold piston, the thrust surfaces have a close fit to the cylinder wall while the surface around the piston pin has greater clearance. As the piston warms, it will expand along the piston pin. As it expands, the shape of the piston becomes more round. Pistons are also tapered from the top to the bottom. The top of the piston runs much hotter than the skirt due to its exposure to combustion gases. On **tapered pistons**, the top of the piston is manufactured with a narrower diameter than the bottom area of the skirt, to allow room for the excess expansion. When the piston is fully warm, the top is designed to fit the cylinder perfectly (Figure 13-20). To measure piston diameter, use the manufacturer's guidelines; measuring at the wrong height could give you improper readings.

Many manufacturers design their cast-aluminum pistons with steel struts located in the skirt area (Figure 13-21). The strut works to control the amount and rate of expansion. The struts are cast into the piston during manufacturing. Since steel and aluminum expand at different rates for the same temperature, the struts keep the skirt from expanding as much (Figure 13-22). This allows the manufacturer to use tighter clearances between the piston and cylinder to prevent the piston from rocking in the cylinder and generating noise.

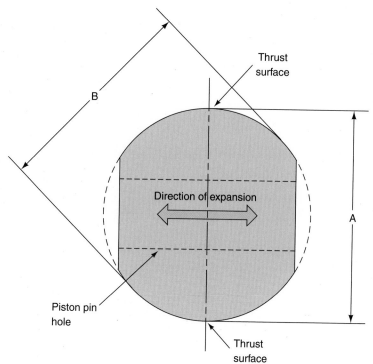

FIGURE 13-19 Cam ground pistons have the largest diameter across the thrust surfaces (A).

An exaggerated view of a cam ground piston shows that the pin side is narrower. It will expand to become round when it reaches normal operating temperature.

B
The piston is narrower at the top because it will expand more as the engine warms up.

FIGURE 13-20 The piston is cam ground (A) and tapered (B) to provide good cylinder fit when running at normal operating temperature.

FIGURE 13-21 Some manufacturers use a steel strut to help control expansion and add strength to the piston.

A graphite moly-disulfide coating is added to many modern piston skirts to reduce scuffing and increase durability (Figure 13-23). These lubrication coatings are helpful in reducing friction, to allow tighter piston-to-cylinder wall clearances. Many parts manufacturers offer replacement pistons with coatings. They are slightly more expensive but will increase the longevity of the piston and reduce cylinder wear.

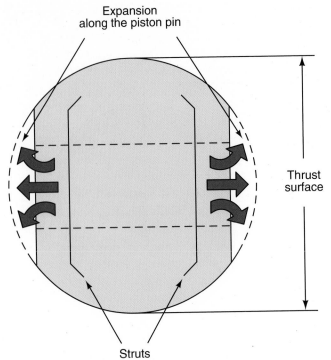

Expansion
along the piston pin

Thrust
surface

Struts

FIGURE 13-22 Steel struts in the piston help control piston expansion.

Lubrication
patch

FIGURE 13-23 Some pistons have a coating on the skirts to reduce friction and improve piston fit in the cylinder.

Piston offset is used to provide more effective downward force onto the crankshaft by increasing the leverage applied to the crankshaft.

Piston Pin

The piston is joined to the connecting rod by the wrist pin or piston pin. The piston pin bore is reinforced by a pin boss to support the huge forces transmitted to the connecting rod through the pin. The piston pin is often pressed into the piston on passenger car applications. Some higher-end engines and racing applications use full floating pins. A full floating pin is held in place by snap rings.

Piston Pin Offset

Most manufacturers offset the piston pin to reduce piston slap (Figure 13-24). The connecting rod is angled to different sides on the compression and power strokes, causing the piston to rock from one skirt to the other at TDC. During the compression stroke, not as much force is exerted on the piston skirt as during the power stroke. Offsetting the piston pin hole toward the major thrust surface about 0.062 in. (1.57 mm) reduces the tendency of the piston to slap the cylinder wall as it rocks from the minor to the **major thrust surface**. During the compression stroke, the connecting rod pushes the minor thrust surface against the cylinder wall (Figure 13-25). The **offset pin** causes more of the combustion pressure to be exerted on the larger half of the piston head (Figure 13-26). This uneven pressure application causes the piston to tilt so the top of the piston contacts the cylinder wall on the **minor thrust surface** and the bottom of the piston contacts the cylinder wall on the major thrust surface. When the piston begins its downward movement, its upper half slides into contact with the major thrust surface (Figure 13-27).

To assist in the installation of the piston, most manufacturers have a groove or other marking to indicate the side of the piston that faces the front of the block. On V-type engines, the pin offset for each bank is on opposite sides. This is because crankshaft rotation determines the major thrust surface side. If the pistons are installed in the wrong direction, it will cause excess friction and piston slap.

> **AUTHOR'S NOTE:** I had a young man bring his beautifully restored, freshly rebuilt and performance modified engine in to diagnose a knocking noise. The noise was coming from the top of the block and I suspected piston slap. After removing the cylinder head, I could see that the forward markings on the pistons were pointing toward the rear of the engine. After proper reassembly the engine was quiet and powerful; he had done fine work aside from the piston assembly problem.

FIGURE 13-24 Some piston pins are offset to reduce piston slap. A forward marking on the piston ensures the offset will be in the correct orientation during reassembly.

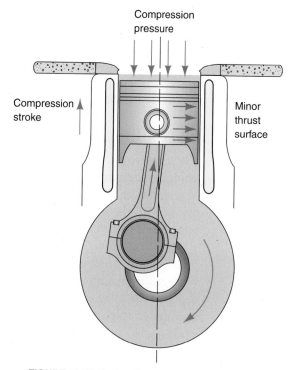

FIGURE 13-25 During the compression stroke, the connecting rod forces the piston against the minor thrust side of the cylinder wall.

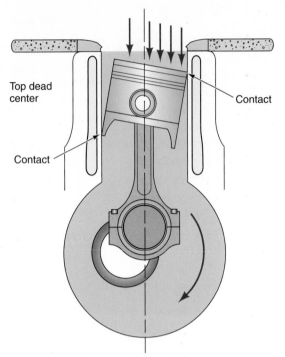

Top dead center

Contact

Contact

FIGURE 13-26 The offset pin causes more of the combustion pressure to be exerted on the larger half of the piston head, causing the piston to tilt and cushion the transition from the minor thrust side to the major thrust side. The illustration is exaggerated for clarity.

Major thrust surface

Power stroke

FIGURE 13-27 When the piston begins its downward movement, its upper half slides into contact with the major thrust surface.

Piston Speed

As the piston moves up and down in the cylinder, it is constantly changing speeds. At top and bottom dead center, the speed of the piston is zero. It then immediately accelerates to maximum speed. This action places heavy loads on the piston pin and pin boss. Today, most pistons are constructed of aluminum instead of iron. A piston made of iron will take more energy to accelerate and decelerate than one made of aluminum because of its mass. This would accelerate wear on the piston pin, connecting rod, crankshaft, and bearings. It would also decrease the power output and efficiency.

PISTON DESIGNS AND CONSTRUCTION

If it is determined that the pistons require replacement, there are several piston designs available from manufacturers and aftermarket suppliers. The technician must be capable of selecting the correct design for the engine application. This usually means replacing the piston with the same design as removed; however, if the engine is to be modified, piston selection is one aspect in increasing engine performance and durability.

It is possible to change the compression ratio of the engine by changing the head design of the piston. Most original equipment pistons have a flat head. These pistons may require valve reliefs machined into the head to prevent valve-to-piston contact (Figure 13-28). The depth of the relief is determined by camshaft timing, duration, and lift. A dished piston is designed to put most of the combustion chamber into the piston head and lower the compression ratio (Figure 13-29). When adding a turbocharger or supercharger to an engine, you may need to lower the compression ratio by installing dished pistons. A domed piston will fill the combustion area and increase the compression ratio (Figure 13-30). Other designs are variations of these three types. In addition to changing compression ratio, piston head design can increase the efficiency of the combustion process by creating a turbulence to improve the mixing of the air and fuel.

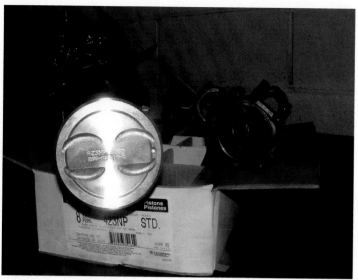

FIGURE 13-28 This piston has valve relief cutouts to prevent the valves from knocking the top of the piston. You can also see the forward marking dot on the right side of the piston head.

Dish depth

FIGURE 13-29 Dished piston design.

Dome height

FIGURE 13-30 Domed piston designs can increase the compression ratio and add turbulence to the combustion chamber.

It is important to recognize that over 90 percent of technicians will work on restoring engines to their original operating condition. Very few technicians will actually find work modifying engines to improve performance. If you understand the basic principles of engine operating and service presented in this text, you will easily be able to transfer that knowledge to high performance work if you find yourself in that often-coveted position.

AUTHOR'S NOTE: Changing the piston design can have a negative impact on emissions. For all street-application vechicles, it is illegal to raise the compression ratio of an engine by changing pistons or to increase emissions above the standard to which the vehicle was built for all street-application vehicles. While modified parts are readily available and widely used, they come with a warning that they are for off-road use only. Making modifications to an engine that has to pass an emissions test can cause test failures and costly repairs. As a technician, you are responsible for maintaining the integrity of the emissions on a customer's vehicle.

Another consideration when replacing pistons is construction type. Weight and expansion rates of the piston are important aspects in today's high rpm engines. The use of aluminum pistons reduces the weight of the assembly and the load on the crankshaft. Construction

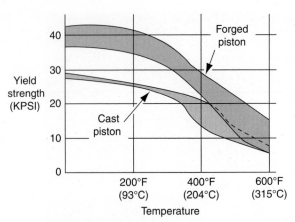

FIGURE 13-31 Yield strength comparisons between cast and forged pistons. The yield strength roughly describes the elasticity of the piston during compression and expansion before failure. Forged pistons can handle heavier loads than cast pistons and are used during engine modifications and high performance applications.

methods also affect the piston's expansion rate and amount. Most original equipment pistons are made of cast aluminum. These are very capable of handling the requirements of most original equipment engines. The advantage of cast pistons is their low expansion rate. The expansion rate is low because the grain structure of cast is not as dense as other manufacturing processes. This low expansion rate allows the engine manufacturer to tighten the piston clearance tolerances, helping to prevent piston slap and rattle when the engine is cold.

Forged pistons are stronger due to their dense grain structure (Figure 13-31). The tighter grain structure allows the piston to run about 20 percent cooler than a cast piston. Depending on the alloy used in the aluminum, the expansion rate may be greater than cast. This requires additional piston clearance and may result in cold engine piston rattle. Alloys have been developed in recent years that reduce the expansion rate of forged pistons. Consequently, it is important to follow the piston manufacturer's specifications for piston clearance.

A third alternative is hypereutectic cast pistons. Most cast pistons are constructed with about 9.5 percent silicon content, while forged pistons have between 0.10 and 10 percent, depending on the aluminum used. Silicon is added to help provide lubrication and increase strength. Hypereutectic pistons contain between 16 and 22 percent silicon. Aluminum can only dissolve about 12 percent silicon and hold it in suspension. The silicon is added when the aluminum is molten and dissolves. At the 12 percent level, the silicon remains dissolved when the piston cools. Normally, silicon added above this saturation point will not dissolve, and settles at the bottom of the mold. Hypereutectic casting keeps the undissolved silicon dispersed throughout the piston by closely monitoring and controlling the heating and cooling rates during the manufacturing process.

Hypereutectic casting increases the temperature fatigue resistance of the piston. This allows the piston to be made thinner than most cast pistons, reducing the weight of the piston. The strength of hypereutectic pistons falls between cast and forged. The drawback to this piston construction is that silicon can reduce the piston's ability to conduct heat away from the combustion chamber. Although its temperature fatigue is increased, the piston runs hotter than cast and may cause detonation.

It is possible to tell if the piston is forged or cast by looking at the underside of the piston head. Cast pistons have mold dividing lines, while forged pistons do not have these parting lines.

PISTON RINGS

For the engine to produce maximum power, compression and expansion gases must be sealed in the combustion chamber. To control **blowby**, the piston is fitted with compression rings. The rings seal the piston to the cylinder wall, sealing in combustion pressures.

Shop Manual
Chapter 13

Blowby refers to compression pressures that escape past the piston.

FIGURE 13-32 The top piston ring grooves are used to hold the compression rings. The rings seal against the cylinder wall.

FIGURE 13-33 Some compression rings are designed with a chamfer or taper. These types of rings may be directional; always look for a top marking during installation.

In addition, oil rings are used to prevent oil from reaching the top of the piston. Oil in this area will result in blue smoke being expelled from the exhaust system.

Compression rings are located on the two upper lands of the piston (Figure 13-32). The compression rings form a seal between the piston and the cylinder wall. To provide a more positive seal, many manufacturers design one or more of the compression rings with a taper (Figure 13-33). The taper provides a sharp, positive seal to the cylinder wall. This style of ring must be installed in the correct position. A dot or other marking is usually provided to indicate the top of the ring.

In addition to tapering or chamfering the ring, the ring is designed to use combustion pressures to force the ring tighter against the cylinder wall and the bottom of the ring groove (Figure 13-34). The top ring provides primary sealing of the combustion pressures. The second ring is used to control the small amount of pressures that may escape past the top ring.

The most common materials for constructing compression rings are steel and cast iron. The face of the ring is usually coated to aid in the wear-in process. This coating can be graphite, phosphate, iron oxide, molybdenum, or chromium. The bottom piston ring groove houses the oil control ring. There are two common styles of oil control rings: the cast-iron oil ring and the segmented oil ring. Both styles of oil rings have slots to allow the oil from the cylinder walls to pass through. The ring groove also has slots to provide a passage back to the oil sump.

Ring seating is accomplished by the movement of the piston in the cylinder. If the cylinder is properly conditioned, this movement laps the rings against the wall.

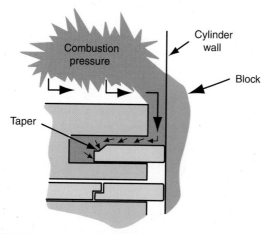

FIGURE 13-34 The rings are designed to use combustion pressures to exert additional outward forces of the rings against the cylinder walls.

FIGURE 13-35 Typical three-piece oil control rings with two thin scrapers with a spacer in between.

Many engine manufacturers now install low tension piston rings. These rings provide adequate combustion chamber sealing with reduced friction on the cylinder walls as the piston moves up and down. Since low tension rings provide a reduced internal engine friction, they also provide a slight improvement in fuel economy.

Segmented **oil control rings** have three pieces (Figure 13-35). The upper and lower portions are thin metal rails that scrape the oil. The center section of the ring is constructed of spring steel, and is called the expander or spacer. The movement of the piston causes the oil ring to scrape oil from the cylinder wall and return it to the sump.

Ring Materials

Rings are replaced whenever the engine is disassembled for repairs. Generally, you will use original equipment rings with an engine being built to original specifications. There are times, however, when you may choose an appropriate alternative. The following ring materials and their strengths and weaknesses are discussed to assist you in ring selection.

Rings have long been made from cast iron. These rings are a popular choice when overhauling an engine. The material is strong but soft. It can conform best to irregular cylinder surfaces and shapes. When re-ringing a worn engine without boring the cylinders, cast-iron rings are usually recommended. When selecting a replacement ring for a fully overhauled engine, be sure that the quality meets or exceeds the original equipment.

Other rings are made of steel, or coated with chrome or moly. Many late-model engines are fitted with steel top rings for their increased durability. The top ring suffers the most; it is exposed to the combustion flame and the highest pressures. If an engine is originally equipped with a steel top ring, it should be replaced with the same steel ring. Many modern turbocharged, supercharged, high output engines and engines with the top ring positioned very close to the piston head require steel top rings.

Moly rings are cast rings with a groove cut in the face that is filled with molybdenum. Moly rings provide excellent scuff resistance and superior durability. The moly holds a little oil to lubricate the ring and reduce the friction against the wall. Moly rings are frequently used as original equipment rings. This is an excellent replacement choice for a freshly bored engine. They seal quickly against a properly finished cylinder wall.

Chrome-coated rings are another popular replacement option. They are stronger and can last 50 percent longer than plain cast-iron rings. They do not resist scuffing as well as moly rings, but they also do not absorb contaminants the way moly rings can. When an engine is operated in dusty or sandy conditions, consider chrome rings. They will not seal an out-of-round cylinder, however, so they should only be used in freshly bored and properly finished cylinders. Chrome-coated rings are very hard and can cause excessive cylinder wall wear unless the block is cast with additional amounts of nickel to increase its strength and resistance to abrasives.

Oversize Rings

If the cylinder bore diameter is increased to correct for wear, the piston and rings must be replaced. The piston and rings are oversize to fit the new bore diameter. Piston rings are typically available in standard, 0.020, 0.030, 0.040, and 0.060 in. oversizes. If an odd size is needed, such as 0.010 or 0.050 in. oversize, use the ring one size smaller; for example, for 0.010 in. oversize, use a standard ring; for 0.050 in. oversize, use a 0.040 in. oversize. Metric oversizes are available in 0.50, 0.75, 1.0, and 1.5 mm. You cannot install oversized piston rings with a piston of standard size. An oversized piston must accompany it.

CONNECTING RODS

The connecting rod transmits the force of the combustion pressures applied to the top of the piston to the crankshaft (Figure 13-36). To provide the strength required to withstand these forces and still be light enough to prevent rough running conditions, the connecting rod is constructed from one of the following methods:

Shop Manual
Chapter 13

1. Forged from high strength steel
2. Made of nodular steel
3. Made from cast iron
4. Made of sintered powdered metal

To increase its strength, the connecting rod is constructed in the form of an I (Figure 13-37). The small bore at the upper end of the rod is fitted to the piston pin. The piston pin can be pressed into the piston and a free fit in the connecting rod. In this type of mounting, a bushing is installed in the small rod bore. The piston pin can also be pressed into the small rod bore and a free fit in the piston. The piston moves on the pin surface without the use of bushings.

The larger bore connects the rod to the crankshaft journal. The large bore is a two-piece assembly. The lower half of this assembly is called the rod cap. The rod cap and connecting

Big end refers to the end of the connecting rod that attaches to the crankshaft. Small end refers to the end of the connecting rod that accepts the piston pin.

FIGURE 13-36 The connecting rod connects the piston to the crankshaft.

Big end   I beam   Little end

FIGURE 13-37 A typical connecting rod.

FIGURE 13-38 Fracturing, instead of machining, the connecting rod cap provides a perfect concentric fit when it is bolted back together.

Shop Manual

Chapter 13

rod are constructed as a unit and must remain as a matched set. During cylinder block disassembly, the connecting rods and caps should be marked to assure proper match when installed. Most rod caps are machined from the connecting rods during the manufacturing process. Some manufacturers now use a new process of "breaking" the cap off of the connecting rod. This process uses connecting rods constructed from sintered powdered metal. The connecting rod is then shot-peened to remove any flash and to increase surface hardness. After the rod bolt holes are drilled and tapped, the bolts are installed loosely. Next, the break area is laser scribed and the cap is fractured in a special fixture. This creates a rod and cap parting surface that is a perfect fit and creates a stronger joint (Figure 13-38).

SUMMARY

TERMS TO KNOW

Balance shaft

Bearing crush

Bearing spread

Blowby

Cam ground pistons

Compression rings

Connecting rod bearings

Locating tab

Major thrust surface

Minor thrust surface

Offset pin

Oil control rings

Oversize bearings

Tapered pistons

Thrust bearing

Undersize bearings

- Bearings are used to carry the loads created by rotational forces. Main bearings carry the load created by the movement of the crankshaft. Connecting rod bearings carry the load of the power transfer from the connecting rod to the crankshaft.
- Oil is supplied to the rod bearing journals through holes in the crankshaft.
- Undersize bearings are used to maintain correct oil clearance when the crankshaft journal has been ground to provide a new journal surface.
- Oversize bearings are used if the bearing bore diameter has been increased.
- The piston, when assembled to the connecting rod, is designed to transmit the power produced in the combustion chamber to the crankshaft.
- The piston is manufactured with expansion controls to provide better cylinder fit when the engine is at normal operating temperature.
- Pistons are generally made of aluminum; they may be cast, forged, or hypereutectic cast.
- The piston pin is generally offset toward the major thrust side to reduce piston slap.
- To seal combustion pressures, the piston is fitted with compression rings. The rings seal the piston to the cylinder wall.
- Cast iron, molybdenum, chrome, and steel rings may be used, depending on the engine design and cylinder condition.
- The connecting rod links the piston to the crankshaft; it is typically formed in I-beam construction to provide the required strength to prevent bending.

REVIEW QUESTIONS

Short Answer Essays

1. Describe the function of the engine bearings.

2. Explain the bearing characteristics, spread, and crush and why they are important.

3. Explain what problems occur when the bearing oil clearances are too large.

4. List the parts of the piston and their purpose.

5. Explain the advantages and disadvantages of cast, forged, and hypereutectic pistons.

6. Describe cam ground and tapered pistons, and explain how they enhance engine performance.

7. Explain why piston pins are offset in the pistons.

8. Describe the three rings and their purposes.

9. Explain the advantages and disadvantages of cast-iron, chrome, and moly rings.

10. Describe the design and purpose of the connecting rod.

Fill-in-the-Blanks

1. Typical bearing oil clearances on modern engines are between _____ in. and _____ in.

2. _____ _____ is important for heat transfer from the bearing to the block.

3. _____ _____ pistons are narrower across the pin side to allow for expansion.

4. Connecting rods are generally made with _____ _____ construction to provide adequate strength.

5. The _____ _____ transmits the force of the combustion pressures applied to the top of the piston to the crankshaft.

6. _____ are used to carry the loads created by rotational forces. _____ _____ carry the load created by the horizontal movement of the crankshaft.

7. _____ bearings are used to maintain correct oil clearance when the crankshaft journal has been ground to provide a new journal surface. _____ bearings are used if the bearing bore diameter has been increased.

8. A _____ _____ assures proper installation of the bearing insert in the bore.

9. Steel struts help to control piston _____.

10. Some connecting rod caps are _____ off the connecting rod during the manufacturing process to provide an improved cap-to-rod fit.

Multiple Choice

1. Undersize bearings:
   A. Have a smaller outside diameter.
   B. Are used when the block has been line bored.
   C. Have a thinner lining than standard bearings.
   D. Are used when the crankshaft has been ground.

2. When bearing oil clearances are too large:
   A. Oil pressure will be reduced.
   B. Shims are used to take up the clearance.
   C. The crankshaft will bend.
   D. All of the above.

3. Each of the following is a common type of piston, EXCEPT:
   A. Forged aluminum.
   B. Forged steel.
   C. Cast aluminum.
   D. Hypereutectic.

4. The piston pin is generally offset to:
   A. Reduce bearing knock.
   B. Protect the piston pin from combustion forces.
   C. Reduce piston slap.
   D. All of the above.

5. Engine bearings may be made with each of the following as a liner EXCEPT:
   A. Babbitt.   C. Copper.
   B. Aluminum.   D. Cast iron.

6. Rings are being chosen for an engine. Which statement below is true?
   A. If the engine has been freshly bored, cast-iron rings will provide the longest service life.
   B. Moly rings should be used on slightly worn cylinders for quick wear-in.
   C. Chrome rings will last a long time but may wear the cylinder walls.
   D. Many light-duty, lower-powered engines use steel rings.

7. Piston head designs can:
    A. Increase the compression ratio.
    B. Increase turbulence.
    C. Increase emissions.
    D. All of the above.

8. While discussing connecting rod bearings:

    Technician A says the bearing curvature is slightly greater than the curvature of the connecting rod bore.

    Technician B says when half of the bearing insert is installed in the rod cap, the bearing should protrude above the cap surfaces.

    Who is correct?
    A. A only
    B. B only
    C. Both A and B
    D. Neither A nor B

9. While discussing pistons:

    Technician A says the larger diameter of a cam ground piston is across the piston pin bores.

    Technician B says pistons are manufactured with a perfectly vertical skirt.

    Who is correct?
    A. A only
    B. B only
    C. Both A and B
    D. Neither A nor B

10. Common piston ring designs include:
    A. A three-piece oil ring.
    B. A top compression ring with no chamfer or taper.
    C. A top compression ring with an expander behind the ring.
    D. An oil ring below the piston pin.

Chapter 14

ALTERNATIVE FUEL AND ADVANCED TECHNOLOGY VEHICLES

UPON COMPLETION AND REVIEW OF THIS CHAPTER, YOU SHOULD BE ABLE TO:

- Understand the reasons for manufacturers' research and development of alternate fuel and advanced technology vehicles.

- Explain the common uses for alternatively fueled vehicles.

- Describe the difference between dedicated and retrofitted alternate fuel vehicles.

- Explain the basic operation of a propane-fueled vehicle.

- Explain the benefits of E85.

- Explain the basic operation of a flexible fuel vehicle.

- Understand the benefits and limitations of compressed natural gas as an automotive fuel.

- Describe the basic operation of a CNG-fueled vehicle.

- Discuss the advantages and disadvantages of electric automobiles.

- Describe the benefits of hybrid electric vehicles.

- Differentiate between the different hybrid electric vehicle power system layouts.

- Understand the basic operation of a hybrid electric vehicle.

- Discuss some of the benefits and limitations of fuel cell vehicles.

INTRODUCTION

Automobiles contribute greatly to North America's air pollution. In many urban areas, air pollution from vehicles is the primary source of pollution. According to the U.S. EPA, mobile sources contribute 51 percent of the carbon monoxide pollution. The governments of the United States and Canada and their respective environmental protection agencies implement legislation that requires vehicle manufacturers to continuously reduce the levels of emissions their vehicles emit. They also demand increases in fuel economy. In the United States, the Clean Air Act of 1990 produced a long-range set of milestones that require manufacturers to reach lower and lower levels of emissions with a greater percentage of their fleets. In an effort to reduce emissions, increase fuel economy, and comply with the environmental laws, manufacturers have invested millions of dollars into research and design of alternatively fueled and advanced technology vehicles. Many fleet vehicles run on propane, compressed natural gas (CNG), biodiesel, or E85. Many fleets use government incentives and grants to help purchase and maintain their fleets. When crude oil prices go high, more fleets and companies may choose alternative fuels for various reasons. There are also several advanced technology hybrid vehicles now on the market that offer significantly reduced emissions and increased fuel economy.

The engines of the alternative fuel and hybrid vehicles on the market today are still very similar to the engines we have been discussing throughout the book. Changes will be noted when appropriate. The electric motors used in hybrid vehicles and their components are an entirely different type of power source. Their operation is discussed in a general fashion in this text, and safety precautions will be offered in the Shop Manual.

ALTERNATIVE FUEL VEHICLE USE

Many urban areas in the United States and in Canada have air quality that falls below safe levels on certain days or periods of the year. These areas are called non attainment areas, meaning the quality of air regularly falls below the safe standard. Individual regulations vary by state and province, but many urban areas require that fleets run a percentage of their vehicles on alternative fuels. The Clean Air Act also introduced the Clean Fuel Fleet program in the United States. The act requires that in some nonattainment areas, certain fleets meet special fuel vehicle emissions standards. Transportation money from the federal government in the United States is also tied to a state's actions to reduce mobile source emissions. Many municipal, government, and educational fleets run on alternate fuels.

Alternative fuel vehicles may be **dedicated** or **retrofitted**. A dedicated propane vehicle, for example, is one that the manufacturer designed from the ground up to run on propane. Changes to the body, suspension, and fuel system were all designed to optimize the vehicle to run on propane. The vehicle's oxygen sensors, catalytic converters, and powertrain control modules have all been designed to run on the alternative fuel. Many alternative fuel fleet vehicles have been retrofitted to run on an alternative fuel. This means that a particular gas vehicle was chosen and then conversions were made to allow the vehicle to run on the alternative fuel. The dedicated fuel vehicles are more efficient, but they are not always available from the manufacturer to serve the needs of the consumer.

PROPANE VEHICLES

The first widely used alternative fuel vehicle has been the propane vehicle. Propane is liquefied petroleum gas (LPG), but often simply called propane or LPG. Automotive engine grade LPG is **HD-5**. HD-5 consists of a minimum of 90 percent propane and a maximum of 5 percent propylene. Many fleets operate propane trucks and buses. In the United States, there are an estimated 350,000 trucks, cars, and industrial vehicles operating on propane. Propane's popularity as an alternative fuel is growing because retrofitters are now starting to produce both throttle body and multiport injection systems. There are also retrofit kits available for diesel engines. The propane fuel is blended with the diesel fuel.

Propane cannot start the engine alone because of the high compression pressure in the engine. Therefore the ratio of propane to diesel is lower during engine starting and continues to increase as the engine temperature rises. This continues to a point where there is only enough diesel fuel being injected to cause ignition and then the propane is allowed to complete the combustion. A few manufacturers have offered dedicated propane vehicles or vehicles with factory conversion kits with modern fuel injection systems in the past several years.

Vehicles running on propane produce lower hydrocarbon (HC) and carbon monoxide (CO) emissions. The reductions can be as high as 50–80 percent. These vehicles also produce up to 30 percent lower oxides of nitrogen (NOx) emissions. Propane is also a sealed system, so it has no evaporative emissions. It is a clean-burning fuel with low particulate and sulfur emissions. This makes it an attractive fuel for fleet operators in urban locations who are required to use an alternate fuel. LPG is a byproduct of natural gas and crude oil refining; 65 percent of propane comes from natural gas. This is advantageous to both the United States and Canada, which have the largest stores of natural gas in the world. Natural gas is plentiful and cheaper to refine.

Dedicated alternative fuel vehicles were designed to run on alternative fuel.

A vehicle that was originally designed to run on gasoline but has been changed to run on an alternate fuel is often called **retrofitted** or converted.

HD-5 LPG is a "consumer grade" propane.

FIGURE 14-1 Liquid propane storage tanks may be mounted under the vehicle or in the trunk.

Most propane supply companies run all their vehicles on propane. The United States has over 10,000 refilling stations available, while Canada has about 5,000 refilling facilities.

Unfortunately, fuel consumption, by volume, is slightly higher with propane than with gasoline. Propane actually has more energy available in it but it is not as dense, so by volume it has less available energy. Because it is roughly 30 percent less dense than gasoline, it is also difficult to carry as much fuel as would be necessary to maintain the range of a gas vehicle. Propane tanks are specially made sealed units because the fuel is stored at pressures that may be over 200 psi (Figure 14-1).

Propane vehicles run very similarly to gas-fueled vehicles. With a few modifications to the engine, virtually any vehicle can be converted to run on propane. Because of the size of the storage tanks, it is more common to convert trucks or larger vehicles. Vehicles can be equipped to run solely on propane or to be dual fuel vehicles and able to run on either gasoline or propane. The tanks are either fitted in the trunk or under the vehicle (Figure 14-2).

Propane is stored in the tank(s) at pressures roughly between 100 and 200°F (55–110°C). A conversion system includes a fuel controller, valves, actuators, electronics, and software modifications. The liquid fuel travels from the tank to a vaporizer/pressure regulator (also called a converter), which converts the propane to a gaseous state. From the vaporizer the fuel goes to the throttle body injector, the carburetor, or the fuel injectors for delivery to the intake manifold.

There are federal, and often state and local, tax incentives for converting vehicles to cleaner burning alternative fuels that help make up the cost of conversion. Other factors include:

- Original fuel economy of the vehicle
- Number of miles traveled per year
- Initial cost of conversion
- Engine life expectancy

Injector

Fuel rails

Fuel lines

Fill fitting

Fill filter

Pressure regulator

Overfill prevention and stop-fill valve

Pump and filter

FIGURE 14-2 Fuel delivery components of the liquid propane injection system.

An **open loop system** delivers fuel based on demand and has no way of detecting whether the amount sent was ideal in terms of performance, emissions, or fuel economy.

A **closed loop system** uses sensors (typically an oxygen sensor) that inform the PCM about the air-fuel ratio leaving the engine, so it can adjust accordingly.

Factory conversions of passenger cars and light-duty trucks can cost about $2500 above the cost of the conventional vehicle. Aftermarket conversion kits may cost $1500–$3,000 depending on the amount of components needed and whether the system is an **open loop system** or a **closed loop system**. California requires warranties on conversion kit components, and Canada has regulations on component safety as well as installation and installers.

Dedicated propane fuel vehicles differ very little from gas vehicles aside from the fuel tank. A pump in the tank delivers liquid fuel at the correct pressure to the fuel injectors. Unused vaporized fuel returns to the tank. A single hose with two sections is used to deliver and return fuel (Figure 14-3).

When the system is a bifuel system, a fuel selector switch is used to inform the PCM which fuel is being used. The operator can change the fuel setting, or, if the propane tank runs low enough, the PCM will automatically change to gasoline. This changes the fuel delivery strategy to optimize performance, emissions, and fuel economy (Figure 14-4).

The engines in propane vehicles may be identical to those in gasoline engines. Using propane reduces the wear on the engine because it is a cleaner fuel and forms less carbon and corrosive acids. Some fleet operators claim that propane engines last two to three times longer than their gasoline-powered counterparts. The National Propane Gas Association claims that spark plugs last 80,000–100,000 miles. The same mechanical failures and maintenance procedures exist; the occurrences are often extended.

FIGURE 14-3 A single hose performs the functions of delivery and return.

Fuel rail

Injectors

Return valve

Tank - 150psi (1035 kPa) at 90°F

Supply valve

Pump

FIGURE 14-4 A manual selector switch allows a change from propane to gasoline.

PCM

From fuel sender

Fuel selector switch

Vaporizer

Coolant circuit

Filter

From propane fuel tank

E85 AND FLEXIBLE FUEL VEHICLES

E85 is made from 85 percent ethanol and 15 percent petroleum. Ethanol can be made from virtually any starch feedstock such as corn, sugar cane, or wheat. Most of the U.S. production of ethanol today comes from corn. A bushel of field corn can be converted into 2.7 gallons of fuel ethanol. Some ethanol producers are reaching even greater efficiencies. E85 is a domestically renewable energy source that helps reduce dependency on imported oil. This has a positive influence on the economy. It also helps reduce tailpipe and greenhouse gas emissions (carbon dioxide). E85 has the highest oxygen content of any fuel currently available. This makes it burn much more completely and cleaner than gasoline. EPA studies have shown that high-blend ethanol fuels can reduce harmful exhaust emissions by over 50 percent. E85 also reduces carbon dioxide (CO_2), the emission contributing to global warming, by nearly 30 percent.

Flexible fuel vehicles are designed to run on E85, or gasoline. M85 shares many of the attributes and benefits of E85. Older flexible fuel vehicles could run on E85, regular gasoline, or M85. **M85** uses a mix of methanol and gasoline. Methanol is sometimes used by race cars but is quickly being replaced by E85. The 2007 Indianapolis 500 race cars were all powered by ethanol. The use of M85 as an automotive fuel has decreased in popularity because of its toxicity, corrosiveness, and groundwater contamination issues. Flexible fuel vehicles are dedicated flexible fuel vehicles available from many manufacturers. They typically use a fuel sensor that determines the level of alcohol in the fuel, and the PCM modifies its fuel delivery, ignition timing, and emissions strategies accordingly (Figure 14-5). The sensor can be located in the tank or in the fuel line after the tank (Figure 14-6). Instead of a special sensor, some manufacturers use wide-band oxygen sensors to determine the fuel content and adjust the fuel curve accordingly. The engine is essentially the same as a gasoline model, though different engine oil may be specified when

E85 is an automotive fuel made of 85 percent ethanol and 15 percent petroleum.

A **flexible fuel vehicle** is designed to be able to run on either gasoline or a methanol or ethanol blend, M85 or E85.

M85 is an automotive fuel that contains 85 percent methanol and 15 percent gasoline. It is not commonly used anymore.

FIGURE 14-5 A flexible fuel sensor.

FIGURE 14-6 This E85 flexible fuel system mounts the fuel sensor in line between the tank and the fuel rail.

running exclusively on E85. The fuel system components have been modified to resist the added corrosive effects of the alcohol in M85 and E85. A typical layout of an E85 vehicle is shown in Figure 14-7. At the end of 2008, the following manufacturers had flexible fuel vehicles available:

Chrysler
   Selected 2007–2009 4.7L Chrysler Aspen
   Selected 2003 & 2009 2.7L Chrysler Sebring Convertible
   Selected 2003–2009 2.7L Chrysler Sebring Sedan
   Selected 1998–2003 3.3L Chrysler/Plymouth Voyager Minivan
   Selected 1998–2003 & 2008–2009 3.3L Chrysler Town & Country Minivan
   Selected 2007–2009 2.7L Dodge Avenger
   Selected 1998–2003 & 2007 3.3L Dodge Caravan Minivan
   Selected 2007–2009 3.3L Dodge Grand Caravan Minivan

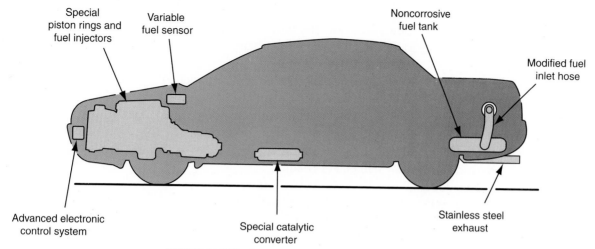

FIGURE 14-7 The layout of an E85 flexible fuel vehicle.

Selected 2003 & 2007–2009 4.7L Dodge Dakota
Selected 2007–2009 4.7L Dodge Durango
Selected 2004–2009 4.7L Dodge Ram 1500
Selected 1993 2.5L Dodge Spirit
Selected 2003–2006 2.7L Dodge Stratus
Selected 2007–2009 4.7L Jeep Commander
Selected 2007–2009 4.7L Jeep Grand Cherokee
Selected 1993 2.5L Plymouth Accliam

Ford
Selected 2006–2008 4.6L Ford Crown Victoria (including taxi/police units)
Selected 2009 5.4L Ford E-series
Selected 3.0L Ford Escape
Selected 2009 5.4L Ford Expedition
Selected 2002–2005 4.0L Ford Explorer
Selected 2004–2005 4.0L Ford Explorer Sport Trac
Selected 2006–2009 5.4L Ford F-150
Selected 1999–2000 3.0L Ford Ranger
Selected 2001–2003 3.0L Ford Ranger Supercab
Selected 1995–2004 & 2006 3.0L Ford Taurus
Selected 2009 5.4L Lincoln Navigator
Selected 2006–2007 &2009 4.6L Lincoln Town Car
Selected 2007–2008 4.6L Mercury Grand Marquis
Selected 2002–2005 4.0L Mercury Mountaineer
Selected 2002–2004 3.0L Mercury Sable

General Motors
Selected 2009 3.9L Buick Lucerne
Selected 2007 3.9L Buick Terraza
Selected 2009 5.3L Cadillac Escalade
Selected 2009 5.3L Cadillac Escalade ESV & EXT
Selected 2005–2009 5.3L Chevrolet Avalanche
Selected 2007–2009 5.3L Chevrolet Express
Selected 2009 2.2L & 2.4L Chevrolet HHR & Panel
Selected 2006–2009 3.5L Chevrolet Impala
Selected 2009 3.9L Chevrolet Impala
Selected 2009 3.5L Chevrolet Impala (police package)
Selected 2006–2007 3.5 L Chevrolet Monte Carlo

Selected 2000–2002 2.2L Chevrolet S10
Selected 2003–2009 5.3L Chevrolet Silverado 1/2 Ton Pickup
Selected 2002–2009 5.3L Chevrolet Suburban
Selected 2002–2009 5.3L Chevrolet Tahoe
Selected 2005–2008 5.3L Chevrolet Tahoe (police package)
Selected 2007–2008 3.9L Chevrolet Uplander
Selected 2007–2009 5.3L Chevrolet Savanna
Selected 2003–2009 5.3L GMC Sierra 1/2 Ton Pickup
Selected 2000–2002 2.2L GMC Sonoma
Selected 2002–2009 5.3L GMC Yukon
Selected 2002–2009 5.3L GMC Yukon XL
Selected 2009 5.3L Hummer H2 & H2 SUT
Selected 2007–2008 3.9L Pontiac Montana
Selected 2007 3.9L Saturn Relay

Isuzu
Selected 2000–2001 2.2L Isuzu Hombre

Mazda
Selected 2001–2003 3.0L Mazda B3000

Mercedes
Selected 2007 2.5L Mercedes C230 Sedan
Selected 2005 2.6L Mercedes C240 Luxury Sedan & Wagon
Selected 2008–2009 3.0L Mercedes C300 Luxury & Sport
Selected 2003–2005 3.2L Mercedes C320 Sedan & Wagon

Nissan
Selected 2007–2009 5.6L Nissan Armada
Selected 2005–2009 5.6L Nissan Titan

COMPRESSED NATURAL GAS VEHICLES

Compressed Natural Gas (CNG) is lightly refined natural gas stored under high pressure.

Compressed natural gas (CNG) is a promising alternative fuel for the future. It is a clean-burning fuel, is available in abundance domestically (both in the U.S. and Canada), and it is commercially available to end users at refilling stations. Natural gas is the cleanest burning fuel used today. CNG vehicles can emit 70 percent less CO, 89 percent less HC, and 87 percent less NOx. CNG vehicles also produce significantly lower (20 percent) greenhouse gas emissions. Dedicated CNG vehicles produce little or no evaporative emissions during fueling or use. This is a huge advantage over conventional gas vehicles and gas filling stations. Gas vehicle evaporative emissions and fueling contribute over 50 percent of the vehicles' HC emissions.

Natural gas is minimally refined and widely available as a resource in the United States and Canada. Prices for CNG should be low once production is increased and refueling stations are abundant. There are already many CNG refilling stations available, which is another benefit of CNG as an alternative fuel choice (Figure 14-8).

CNG has an octane rating of between 110 and 130. This allows dedicated CNG vehicles to run higher compression ratios and advanced ignition timing. The potential power from a CNG vehicle is increased, even though CNG has a lower energy content than gasoline. Vehicles can be converted to CNG, or they can be dedicated CNG or bifuel vehicles. Several manufacturers produce CNG vehicles, including:

- Chevrolet Cavalier
- Dodge Caravan and B Van
- Ford Crown Victoria and Contour
- Select GM trucks
- Honda Civic GX
- Toyota Camry
- Volvo S70 and V70

FIGURE 14-8 A compressed natural gas filling station.

Many of these models are offered as bifuel vehicles, which can run on CNG or unleaded gasoline.

Cylinder Safety

The layout of a typical CNG vehicle is shown in Figure 14-9. The CNG is stored in tanks under pressures between 3,000 and 3,600 lbs. Some of the tanks are going as high as 100 bar. These tanks are typically aluminum cylinders wrapped in fiberglass because they cost less and weigh less than cylinders made of steel or composites. All tanks

FIGURE 14-9 The layout of a CNG vehicle.

are sealed with epoxy paint and clear polyurethane to further protect them from environmental factors.

All cylinders are certified by the U.S. Department of Transportation and labeled with the psi capacity, serial number, manufacturer's name, and date of manufacture. The U.S DOT requires that tanks be inspected every 3 years. The inspection must be performed by a qualified service person in accordance with the manufacturer's inspection process. Retest marking labels must be affixed to the tank near the original test date label and coated with epoxy. No repairs are allowed to the CNG cylinders. Cylinders must be removed from service 15 years after the date of manufacture.

In Canada, cylinders must be inspected and also hydrostatically tested every 3 years, as regulated by the Canadian Standards Association. A retest label must be placed near the original label and sealed with epoxy. Repairs are not allowed, and all cylinders must be replaced 15 years after the date of manufacture. A label may be attached to the fuel filler door stating the retest and expiration date.

Vehicle Safety

There are two safety devices designed to stop fuel flow if leaks or service need to be performed. The fuel control valve is a manually operated ball valve located on each tank (Figure 14-10). Full clockwise rotation turns the fuel flow off. The control valve also houses a pressure relief valve. The cylinder releases excess pressure when temperatures reach approximately 215°F (100°C). Remember, temperature increases as pressure increases. The relief valve will vent all the CNG from the cylinder. This does not pose a fire hazard, as CNG rises into the atmosphere rather than hangs at ground level like propane or gasoline. Once a relief valve has popped, it must be replaced.

There is also a quarter turn manual shut-off valve (Figure 14-11). It is accessed from the outside of the vehicle and is labeled "Manual Shut-Off Valve" on a body panel. When the valve is parallel to the line, fuel can flow. When the handle is perpendicular to the fuel line, fuel flow is stopped from the tanks to the fuel delivery system.

CNG vehicles use high-pressure stainless steel seamless fuel tubes. Fittings on the high-pressure side of the system are double-ferrule, compression-type fittings made of special stainless steel (Figure 14-12). Fittings on the low pressure side of the system use national pipe thread (NPT) and 45-degree flare fittings.

FIGURE 14-10 A fuel control valve at the tank with a pressure relief valve.

Gas flow   No gas flow

FIGURE 14-11 Typical manual valve operation.

FIGURE 14-12 CNG fuel line fittings.

Vehicle Operation

Fuel flows through either a single dual stage regulator or a primary and then a secondary regulator. The regulator reduces the pressure to between 90 and 140 psi. The regulator may contain a built-in pressure relief valve and fuel filter (Figure 14-13). The relief valve is on the low side and relieves pressure above roughly 200 psi, depending on the vehicle model. The pressure regulator has engine coolant running to it because, as the

FIGURE 14-13 CNG fuel pressure regulator.

FIGURE 14-14 Engine coolant hoses routed to the regulator prevent icing as the pressure drops at the regulator.

FIGURE 14-15 The fuel low pressure sensor and fuel temperature sensor help the PCM determine the injector on time.

pressure drops rapidly, so does the fuel temperature. The coolant prevents fuel from freezing (Figure 14-14). The PCM controls a high pressure fuel shut-off solenoid and a low pressure fuel solenoid. The solenoid is an on/off valve that controls fuel flow. The PCM operates a relay to turn the solenoid on and off. The PCM also uses a fuel low pressure sensor to help with its injector timing strategy (Figure 14-15). It also uses a fuel temperature sensor as an input to help calculate fuel delivery and ignition timing. The PCM uses these sensors along with other typical engine sensors to optimize performance, emissions, and fuel economy.

Refilling a CNG vehicle is done through a traditional fuel door (Figure 14-16). The connector to which you attach the quick connect fitting contains a one-way check valve (Figure 14-17). Another one-way valve is installed between the fuel fill line and the fuel pressure regulator to prevent any release of gas to the atmosphere through the filling connection.

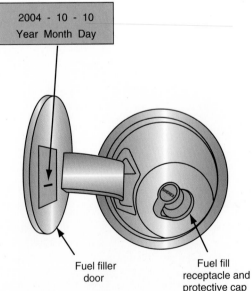

FIGURE 14-16 CNG tank filling receptacle and cylinder test date sticker.

FIGURE 14-17 A one-way check valve prevents fuel from escaping the system during refueling.

THE HONDA CIVIC GX CNG VEHICLE

As an example of a specific dedicated CNG vehicle, let's look at Honda's Civic CNG vehicle. It was the first vehicle in the United States to meet the California Air Resources Board (CARB) "Advanced Technology Partial Zero–Emissions Vehicle" (AT-PZEV) standard. This gives users emissions credits toward highway funding and EPA regulations. It also easily meets the Super-Ultra-Low-Emission Vehicle (**SULEV**) standard that is required of certain vehicles in some states in 2005.

Its engine is an aluminum alloy in-line 4 cylinder with 1.7L (1,590 cc) displacement. It runs a compression ratio of 12.5 to 1, and the CNG octane is rated at 130. It has a SOHC 16-valve head with variable valve timing (VTEC-E). The valves and valve seats have been modified. The pistons and connecting rods have been strengthened (Figure 14-18). It produces 98 lbs-ft. of torque at 4,100 rpm. It is connected to a continuously variable transmission (CVT). It holds the equivalent of 7.5–8 gallons of gasoline and has a "real world" driving range of 200 miles (320 km). It has a fuel economy rating of 31 mpg city and 34 mpg highway.

The emissions from the vehicle are particularly impressive (Figure 14-19). The CO, NMOG (hydrocarbon), and NOx emissions compared to the Civic gasoline Ultra-Low-Emission Vehicle (ULEV) shows the significant reductions. The CNG Civic uses two specially coated oxygen sensors to reduce the effects of hydrogen on their operation. The catalyst system includes a closed couple catalyst and an under-floor catalyst to reduce emissions (Figure 14-20). Fleets are widely using Honda CNG vehicles (Figure 14-21).

Piston and connecting rod
Ratio: 12.5
Piston material
Rod material

Pressure regulator
Integrated type
1st stage unit
2nd stage unit

Cylinder head
CNG special
valves and valve
seat materials

Catalyst
Close-coupled
under floor converter
2 oxygen sensors

FIGURE 14-18 Engine modifications are made to the dedicated CNG Civic.

FIGURE 14-20 The CNG Civic uses two catalysts and two oxygen sensors.

FIGURE 14-19 The CNG Civic easily meets the ultra-low-emissions vehicle (ULEV) and super-ultra-low-emission vehicle (SULEV) standards. It also reduces emissions dramatically compared to its very clean Civic gasoline model.

FIGURE 14-21 This urban cab company uses CNG vehicles in its fleet.

ELECTRIC VEHICLES

Electric vehicles are driven by an electric motor powered by batteries. General Motors introduced the EV1 electric car in 1996. Many manufacturers have produced some version of an electric vehicle in the 1990s for fleet use, in part to meet California's regulation that a

FIGURE 14-22 The batteries supply power to the electric motor that drives the vehicle.

Battery pack

Electric motor

Transmission

Electric wires

Battery pack

certain percentage of fleet vehicles be zero-emission vehicles (ZEV). These vehicles are ideal for urban driving where range is short. The driving range of GM's EV1 is roughly 75 miles, depending on the type of driving. This limits the use of electric vehicles dramatically.

Battery technology has also limited the further development of electric vehicles as an alternative for the general public. The EV1 used 26 12-volt batteries to power the 102-kilowatt AC electric motor that drives the front wheels. The basic layout of the electric vehicle is shown in Figure 14-22. Fully charging a discharged set of batteries through a 220-volt outlet can take 14 hours. A Magne Charge inductive charger was developed by AC Delco to reduce the time to 3 to 4 hours. Nonetheless, until batteries that can provide longer range are developed, the production of electric vehicles for use by the general public is diminished. GM discontinued the production of the EV1 in 1999 and removed all of them from the road by 2003 because they were all originally leased. Some of the technology and experiences learned from the EV1 led to the development of today's hybrid electric vehicles (HEV). Electric vehicles and transport mechanisms are widely used in industrial applications, such as in warehouses, where emissions cannot be easily vented.

HYBRID ELECTRIC VEHICLES

Hybrid electric vehicles (HEV) use two power sources to propel the vehicle. Versions sold in the United States and Canada use an electric motor(s) and a gas or diesel engine to reduce emissions, improve fuel economy, and, in some cases, increase power. The first mass produced commercially available hybrid vehicles were the Honda Insight and the Toyota Prius (Figure 14-23). The problem of vehicle range has been eliminated, as the gas engine and brake system can be used to recharge the battery pack. The Honda Insight has a range of over 600 miles on its full 10.6-gallon gas tank, depending on driving conditions. Tailpipe emissions are also greatly reduced; the 2008 Toyota Prius earned the AT PZEV government emissions rating. The Toyota Prius continues to be a well-selling vehicle; as of May 2008, total sales of the car reached over 1,000,000. Many people believe that until fuel cell and hydrogen-powered vehicles become a mass-produced affordable reality, HEVs will have a significant impact on market product and emissions. A partial list of HEV models sold in the U.S. market up to the 2009 model year include:

- Chevrolet Equinox
- Chevrolet Malibu
- Chevrolet/GMC 1/2 Ton Pickup
- Chevrolet Tahoe/GMC Yukon
- Ford Escape
- Honda Accord
- Honda Insight
- Honda Civic

A **hybrid electric vehicle (HEV)** uses an electric motor and a gas, diesel, or alternate fuel engine to power the vehicle.

FIGURE 14-23 The Toyota Prius has become a very popular hybrid vehicle.

- Lexus LS600hl
- Lexus GH450h
- Lexus RX 400h
- Mazda Tribute
- Mercury Mariner
- Nissan Altima
- Saturn Aura
- Saturn Vue
- Toyota Camry
- Toyota Prius
- Toyota Highlander

A **series system** HEV uses the electric motor to drive the vehicle; the gas motor recharges the batteries to power the electric motor.

A **parallel system** HEV can use the electric motor and the engine to power the vehicle. It can also use just one power source at a time.

In a **dual parallel system** HEV, two electric motors and an engine are used to drive the vehicle.

AUTHOR'S NOTE: I see more and more HEVs commuting on the road with me as I regularly trek through three states in the northeastern U.S. Friends report that they are very pleased with the performance of their Prius and the Honda Civic hybrids. I was impressed with the pickup of the second-generation Prius, and I regularly drive a turbocharged vehicle. A few of these folks with these cars have, however, complained of lower fuel mileage than the posted estimates. These fuel economy ratings are provided by the Environmental Protection Agency, not the vehicle manufacturer. Most vehicles achieve lower actual fuel economy than the posted rating. Another consideration is that HEVs are designed to get better fuel economy during city driving, when the electric motor gets more use, than during extended travel on the highway.

HEV OPERATION

The vehicle is powered by either the electric motor, the engine, or in some cases, both. An HEV system can be characterized as a **series system,** a **parallel system,** or a **dual parallel system**. In a series system, just one of the power sources can be used to drive the vehicle. The electric motor drives the vehicle, while the gasoline or diesel engine powers a generator that charges the batteries to power the motor. A parallel system, which most current models employ, can use the power of both the electric motor and the engine simultaneously to propel the vehicle. The parallel system is named as such because both power sources can apply

FIGURE 14-24 The motor is used to generate electricity as well as drive the vehicle. A rotor and stator are also used in a traditional generator.

torque to move the vehicle. The electric motor can also act in reverse, as a generator to recharge the batteries. A parallel system can also use either the electric motor or the engine exclusively to power the vehicle (Figure 14-24). A system is called **dual parallel** when it has two electric motors and one engine, all of which can drive the vehicle at the same time. This is employed on pickup and sport utility vehicles to date.

An AC electric motor is typically used to start the vehicle from a stop. Electric motors generate maximum torque at 0 rpm, so acceleration is acceptable even when a smaller gas engine is used. The electric motor is charged by a battery pack. During light acceleration or cruise, the electric motor may work alone to power the vehicle. On some vehicles when power is in high demand, both the electric motor and the gasoline engine are used to maximize performance. During moderate acceleration, the gasoline engine may run alone to provide the torque and horsepower required at higher speeds. The engine may also run without the electric motor when the batteries need to be charged. The gasoline engine powers the generator to charge the batteries while it is running. When the vehicle comes to a stop, both the gas engine and the electric motors are turned off to reduce emissions and consumption to zero. At the instant the throttle is pressed, the electric motor begins to spin and accelerate the vehicle. Some trucks and SUVs using a dual parallel hybrid electric vehicle have a traction motor at each axle to retain the four-wheel-drive capacity (Figure 14-25).

Some vehicles also employ **regenerative braking**. On these systems, the kinetic energy typically wasted during braking is converted into electrical energy to recharge the batteries. It also helps to stop the vehicle, as a noticeable drag is felt while the vehicle momentum turns the motor or generator to charge the batteries.

Regenerative braking uses the lost kinetic energy during braking to help recharge the batteries.

FIGURE 14-25 This dual parallel hybrid has a traction motor at each axle to drive the vehicle, plus a traditional gas engine.

On vehicles that use the gasoline motor and regenerative braking to recharge the batteries, external charging is never needed. This makes these vehicles much more appealing to the consumer. The consumer only needs to fill up the gas tank, like any traditional vehicle owner.

Batteries

The batteries currently being used or in prototype are our familiar lead acid batteries, nickel-metal hydride (NiMH) batteries, and lithium ion batteries. Each of these batteries has advantages and disadvantages over the other. Considerations for use include longevity, recharging ability, weight, and cost. The battery packs must be maintained at a relatively constant temperature, so they require a heating and cooling system (Figure 14-26). Manufacturers estimate that battery packs may need to be replaced after 10 years of use. The current price of a battery pack runs between $1,800 and $3,000. Some of the other electronic components can cost several thousand dollars; an inverter for a new Prius is close to $5,000. Manufacturers are offering excellent warranty programs for their batteries and hybrid components to keep the new technology popular and customers satisfied. Battery packs and HEV electronic components are generally warrantied for 10 years.

Voltages on hybrid vehicles can run up to 500 volts, so a technician must thoroughly understand the safety precautions before servicing the hybrid vehicle.

Electric Motors

Modern HEVs typically use one or more alternating current (AC) permanent magnet (PM) motors to drive the vehicle. The operation of the motor is similar in theory to that of the operation of the permanent magnet starter motors used on many of today's vehicles (Figure 14-27). As a matter of fact, one of the electric motors may be used to start the engine and take the place of a traditional starter. The HEV motors use high AC voltage to deliver the high power demands of moving the vehicle. They may also be called traction motors. The motor shown in Figure 14-28 is mounted between the engine and transmission on the flywheel.

Other Hybrid Technology Components

A **power splitting device** is used to change the power from one source to the other and to combine the forces of both when necessary (Figure 14-29). A planetary gear set may be used to allow the electric motor to drive one member and the gasoline motor to drive another

A **power splitting device** changes the mode of power delivery from one source to the other or to a combination of both sources.

FIGURE 14-26 A typical 360-volt battery pack layout with a heating and cooling system.

FIGURE 14-27 The components of a permanent magnet motor.

FIGURE 14-28 A flywheel mounted PM motor.

3-phase
power cable
connections

Stator

Rotor

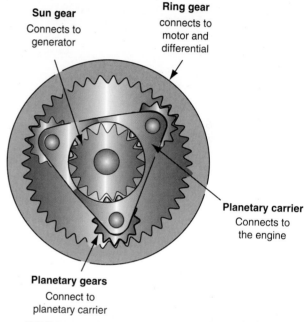

Sun gear
Connects to
generator

Ring gear
connects to
motor and
differential

Planetary carrier
Connects to
the engine

Planetary gears
Connect to
planetary carrier

FIGURE 14-30 A planetary gear set uses different gear combinations to distribute power from the motor and engine.

Electric motor
/generator

Fuel
tank

Electrical
storage

Combustion
engine

FIGURE 14-29 The power splitting device allows the vehicle to run on a combination of power sources or either one individually.

(Figure 14-30). Some HEVs use a continuously variable transmission (CVT) that houses some of the components of, and performs some of the functions of, the power splitting device.

An **inverter** is used to convert the DC voltage stored by the battery pack to the AC voltage required by the electric motors. A **converter** is used to change the high voltage battery output of 270–500 volts down to 12 volts, which can be used by the conventional electrical system. In most cases, there is no longer a traditional 12-volt generator because the battery pack can supply plenty of power for the vehicle accessories.

Gasoline Engine

The gasoline engine used in an HEV vehicle may be the same as the one used in the traditional variant of the model or it may be specially fitted to be optimized for hybrid use (Figure 14-31). The changes to the gasoline engine are minimal in terms of operation and diagnosis. GM's HEV truck offers the same engine in its traditional model as well. Dodge says that the electric motor used in its hybrid version of the Durango will give the 3.9L V6 engine the power of a V8 engine while offering a 20 percent savings in fuel economy. Many of the HEV gasoline engines are newer technology engines and employ variable valve timing and some changes to the intake system.

An **inverter** is used to convert DC volts to AC volts.

A **converter** is used to change the very high battery pack voltage to 12 volts for the conventional electrical system.

FIGURE 14-31 This Honda Civic Hybrid uses a 1.3L engine in its Integrated Motor Assist system.

The Toyota Prius, the Ford Escape, and perhaps others have modified the engine to an **Atkinson cycle engine**. The Atkinson cycle engine actually sacrifices a little bit of power but gains 7–10 percent fuel mileage and further reduces emissions. The Atkinson cycle allows the intake valve to be open longer during the compression stroke, which reduces the engine's **pumping losses**. The engine actually loses power as it tries to pump the piston up against pressurized air during the compression stroke; these are the pumping losses that the Atkinson cycle engine reduces. The intake camshaft lobe is modified in an Atkinson cycle engine to hold the intake valve open longer (Figure 14-32).

The **Atkinson Cycle engine** holds the intake valve open longer during the compression stroke to increase fuel economy.

Pumping losses are wasted power from the pistons pushing up against compressed air during the compression stroke.

FIGURE 14-32 Lobe #1 shows the Atkinson style camshaft, compared to the traditional lobes shown in #2.

TOYOTA PRIUS HEV

The second-generation Toyota Prius with its Hybrid Synergy Drive system is a very popular vehicle in many parts of North America (Figure 14-33). It replaced the first-generation Prius in 2004. The Prius is a true hybrid, engineered from the ground up as an HEV. It uses a 1.5L inline 4 cylinder Atkinson cycle engine. Its DOHC cylinder head is fitted with 16 valves and variable valve timing, VVT-i. The engine itself delivers 78 hp at 5,000 rpm and 85 lbs.-ft. of torque at 4,200 rpm. The 500-volt/67 hp permanent magnet drive motor delivers 295 lbs.-ft. of torque between 0 and 1,200 rpm. The estimated total horsepower available from the power plant is 106 hp. Acceleration from 0 to 60 mph (0–100 km) is comparable to that of the Toyota Camry LE four cylinder, in the ten-second range. Its drag coefficient is the lowest in the industry, helping to contribute to its best-in-class (mid-size car) fuel economy rating of 48 mpg city/45 mpg highway. The fuel tank holds 11.9 gallons (45 L) and has a range of approximately 600 miles (1,000 km). The batteries never need external charging and do not have a facility for it. The Prius is also the best vehicle on the North American market in terms of emissions performance. The vehicle is classified as a SULEV vehicle and meets CARB's AT-PZEV rating. It is rated as one of the cleanest production cars on Earth, producing 90 percent fewer emissions than the traditional conventional internal combustion engine. Toyota has increased production for worldwide sales from one vehicle every 8–10 minutes to one vehicle per minute.

The Prius is coupled to an Electronically Controlled Continuously Variable Transmission (ECCVT). A small joystick mounted on the dash controls shifting. It also has a power split device within the transmission to allow the vehicle to run on either the electric motor or the gasoline engine exclusively. One of the motors (also called generators or MG1 and MG2) works as a starter and a generator, eliminating those traditional components (Figure 14-34). Vehicle acceleration is provided by "drive-by-wire" technology. Three throttle pedal position sensors inform the powertrain management system about the requested power, and, depending on other inputs from the engine and motor circuitry, the electronic control units respond with acceleration from an appropriate power source. The throttle plate on the gasoline engine is driven by an electric motor to the degree requested by the control module. Braking is also managed in a similar manner. The regenerative braking system captures about 30 percent of the energy normally lost in traditional braking systems.

FIGURE 14-33 The second generation Toyota Prius uses its Hybrid Synergy Drive system.

FIGURE 14-34 The layout of the Toyota Prius power system.

The Prius gas engine stops when the vehicle is at a stop or at very low speeds. This further reduces fuel consumption. It is started automatically by the powertrain management system when accelerator input and other inputs indicate the need. The battery pack is a newly developed NiMH comprised of six 1.2-volt DC cells connected in series. Twenty-eight of these modules connected in series gives the Prius motors their operating voltage of 201 volts and reduces the weight of the previous battery packs used on the Generation I Prius. The Prius is equipped with a dashboard that displays available battery power, instantaneous fuel economy, and a mileage bar for each 5-minute segment of driving (Figure 14-35).

FIGURE 14-35 The dash board information offers the driver an opportunity to drive with fuel economy in mind.

FUEL CELL VEHICLES

There are no commercially available fuel cell powered vehicles on the North American market for the 2009 model year. Many experts believe, however, that eventually the fuel cell will become the primary power plant for the modern automobile, or at least capture the advanced technology market. Toyota has already developed a hydrogen fuel cell RAV-4 that can achieve 900 miles on one tank of hydrogen. The reality of mass marketed fuel cell vehicles may be at least a decade off, however, while research and development into hydrogen conversion, fuel cell technology, and refilling infrastructure concerns are resolved. Interest and research investment is high because hydrogen is a regenerative fuel and its use could drastically reduce our consumption of nonrenewable fuel sources.

A fuel cell vehicle operates very much like an electric vehicle except that hydrogen (or another type of fuel cell) supplies electrical power to the motor(s) rather than batteries. Fuel cells electrochemically combine oxygen (from the air) and hydrogen to produce electricity. A special membrane is used to allow interaction between hydrogen and oxygen. Hydrogen will normally "explode" when introduced into the atmosphere with any electrical charge in it. Safety issues related to using hydrogen as a fuel have slowed development of the hydrogen fuel cell vehicle. Obviously, all development will place safety as the foremost issue, considering hydrogen's potential hazards.

Many other fuels are being considered to power fuel cells. Hydrogen is at the forefront because it can be regenerated, but methanol, ethanol, and gasoline are also being tested. A reformer can be used to separate the hydrogen out of the fuels, and the hydrogen feeds the fuel cell. The fuel cell produces electricity to run the electric motor that powers the vehicle. Additionally, solid fuel cells are being researched.

SUMMARY

- Federal and provincial regulations are driving manufacturers to develop more fuel-efficient, environmentally friendly vehicles.
- Automobile sources are a primary source of air pollution.
- Many areas of North America have air quality that at times is below the safe standard.
- Alternate fuel and hybrid electric vehicles can produce fewer emissions.
- Vehicles can be designed to run on alternate fuels or retrofitted to run on alternate fuels.
- Many fleet vehicles choose to run on alternative fuels.
- Propane vehicles produce lower HC and CO emissions.
- E85 is a renewable fuel source that could reduce dependence on imported oil.
- Flexible fuel vehicles can run on either E85, gasoline, or a combination of both.
- Natural gas is a plentiful resource in North America.
- Compressed natural gas vehicles store their fuel in tanks under pressures between 3,000 and 3,600 psi.
- CNG vehicles reduce emissions and can run higher compression ratios.
- CNG tank inspections are regulated by the U.S. and Canadian governments to ensure safety.
- Electric vehicles produce zero emissions but have a short driving range between recharging.
- Hybrid electric vehicles (HEVs) use an electric motor and a gasoline or alternate fuel engine to power the vehicle.
- HEVs can dramatically increase fuel economy and decrease emissions.
- HEVs offer adequate power and excellent range between fill-ups.

TERMS TO KNOW

Atkinson cycle engine

Closed loop system

Compressed natural gas (CNG)

Converter

Dedicated alternative fuel vehicle

Dual parallel hybrid system

E85

Flexible fuel vehicle

HD-5 LPG

Hybrid electric vehicle (HEV)

Inverter

Open loop system

TERMS TO KNOW
(continued)

Parallel hybrid system

Power splitting device

Pumping losses

Regenerative braking

Retrofitted alternative
fuel vehicle

Series hybrid system

- No external battery recharging is required on modern HEVs.
- Fuel cell vehicles may be the next big development in reducing vehicle emissions and increasing fuel economy.

REVIEW QUESTIONS

Short Answer Essays

1. Describe the reasons that manufacturers are looking for ways to reduce emissions and fuel consumption.

2. Explain the difference between a dedicated and a retrofitted alternate fuel vehicle.

3. Explain two benefits of propane as an alternative fuel.

4. What are the benefits of E85 as an automotive fuel?

5. What are some of the added or changed components on a flexible fuel vehicle?

6. Explain the basic fuel system of a CNG vehicle.

7. List three safety precautions incorporated into CNG vehicles.

8. What are the emission reduction percentages when using CNG as a fuel?

9. Describe the power flows in a parallel system HEV.

10. Explain the basic operation of a typical parallel system HEV from a stop sign to full acceleration.

Fill-in-the-Blanks

1. The _____ _____ _____ _____ produced regulations in the U.S. that required manufacturers to produce vehicles with lower and lower emissions.

2. _____ is a mix of ethanol and petroleum.

3. A _____ alternative fuel vehicle has been originally designed to run on the designated fuel.

4. CNG is typically stored at pressures between _____ and _____.

5. AT-PZEV stands for _____ _____.

6. The Honda Civic CNG vehicle engine has a compression ratio of _____:1.

7. A system that uses the wasted kinetic energy from braking is called a _____ _____ _____.

8. An HEV that uses a gasoline engine strictly to power an electric motor is a _____ system HEV.

9. The Atkinson cycle engine increases fuel economy by reducing the engine's _____ _____.

10. The second-generation Toyota Prius can accelerate from 0 to 60 mph (0–100 km) in roughly _____ seconds.

Multiple Choice

1. According to the U.S. EPA, mobile sources contribute roughly _____ percent of the carbon monoxide emissions to the air pollution problem.
   A. 10   C. 50
   B. 25   D. 75

2. When a conversion is made to a gas vehicle to allow it to run on an alternative fuel, it is called a _____ vehicle.
   A. HEV   C. Flexible fuel
   B. Dedicated   D. Retrofitted

3. When discussing propane, which statement is true?
   A. Propane is denser than gasoline.
   B. Propane has more available energy.
   C. Propane is stored at 2,000 psi.
   D. Propane has lower fuel consumption than gasoline.

4. When running a vehicle on propane, engine wear is greatly reduced because:
   A. Propane is a cleaner fuel.
   B. The engine runs higher compression pressures.
   C. It produces lower HC emissions.
   D. The engine runs on higher fuel pressures.

5. E85 is typically made from all of the following, EXCEPT:
   A. Corn.
   B. Wheat.
   C. Sugar cane.
   D. Peanuts.

6. The "greenhouse gas" that contributes to global warming is:
   A. CO
   B. CO_2
   C. HC
   D. NOx

7. Flexible fuel vehicles use a fuel sensor that detects _____ so the PCM can better calculate spark timing and fuel delivery.
   A. Fuel temperature
   B. Fuel density
   C. Alcohol content
   D. Oxygen content

8. The use of CNG vehicles would reduce _____ emissions more than the use of any other alternate fuel or HEV other than the electric vehicle.
   A. HC
   B. CO
   C. NOx
   D. CO_2

9. Natural gas has an octane rating of:
   A. 60–80
   B. 80–105
   C. 110–130
   D. 120–150

10. *Technician A* says that an HEV uses a smaller gas engine than the conventional model vehicle.
    Technician B says that some parallel HEVs can use the electric motor(s) to boost the power of the gasoline engine.
    Who is correct?
    A. A only
    B. B only
    C. Both A and B
    D. Neither A nor B

Automotive Engine Rebuilders Association
330 Lexington Drive
Buffalo Grove, IL 60089
www.aera.org

Engine Repower Council
4600 East-West Highway
Suite 300
Bethesda, MD 20814
www.carcare.org/Repower/ERC_index.shtml

Hiperformer
Remanufactured automotive and marine engines
www.hiperformer.com

International Automotive Technicians Network
www.iatn.com

National Automotive Technicians Education Foundation
13505 Dulles Technology Drive
Herndon, VA 22071
www.ase.com

National Institute for Automotive Service Excellence
13505 Dulles Technology Drive
Herndon, VA 22071
www.natef.org

Production Engine Remanufacturers Association
415 West Golf Road
Arlington Heights, IL 60005
www.pera.org

APPENDIX B

Engine Specifications Chart

General Engine Specifications

Manufacture of Engine			
Model Year			
Engine Block Casting Number(s)			
Cylinder Head Casting Number(s)			
Cylinder Block Casting Material			
Cylinder Head Casting Material			
Combustion Chamber Design			
# of Cylinders		Cyl. Arrangement	
Bore		Firing Order	
Stroke		Valve Arrangement	
Rated Horsepower		Rated Torque	
Compression Pressure		Oil Capacity	
Compression Ratio		Oil Pressure	
Cooling Capacity		Oil Type	

Engine Torque Specifications

Crankshaft Main Bearing Bolts		Rocker Arm Shaft/Stud	
Connecting Rod Bolts/Nuts		Vibration Damper	
Oil Pump Housing		Flywheel/Flexplate	
Water Pump		Flywheel	
Timing Cover		Clutch Pressure Plate	
Thermostat Housing		Camshaft Caps	
Spark Plugs		Timing Tensioner	
Cylinder Head Bolts		Camshaft Sprocket	
Intake Manifold		Oil Pan	
Exhaust Manifold		Oil Pump Bolts/Nuts	

Engine Dimensions

Camshaft			
Bearing Journal Diameter			
Base Circle Runout			
Cam Lobe lift			
Valve Lift			
Intake Valve Timing	Opens @		Closes @
Exhaust Valve Timing	Opens @		Closes @
Valve Overlap			
Duration	Intake		Exhaust

Crankshaft	
End Play	
Main Bearing Journal Diameter	
Main Bearing Journal Width	
Main Bearing Clearance	
Connecting Rod Journal Diameter	
Connecting Rod Journal Width	
Maximum Out-of-Round (All Journals)	
Maximum Taper (All Journals)	

Cylinder Block	
Deck Height	
Deck Clearance	
Cylinder Bore Diameter (Standard)	
Maximum Taper	
Maximum Out-of-Round	
Maximum Cylinder Block Flatness	

Connecting Rods	
Total Length (Center-to-Center)	
Piston Pin Bore Diameter	
Connecting Rod Bore	
Bearing Clearance (Standard)	
Side Clearance	

Pistons	
Piston-to-Bore Clearance	
Piston Ring Gap Clearance	
Piston Ring Side Clearance	
Piston Ring Groove Height	
Piston Ring Groove Diameter	
Piston Pin Diameter	
Piston-to-Pin Clearance	

Cylinder Head	
Combustion Chamber Volume	
Valve Arrangement	
Valve Guide Inside Diameter	
Valve Stem-to-Guide Clearance	
Intake Valve Seat Angle	
Exhaust Valve Seat Angle	
Intake Valve Seat Width	

Exhaust Valve Seat Width	
Valve Seat Runout	
Cylinder Head Flatness	
Cylinder Head Thickness	

Rocker Arms & Push Rods	
Rocker Arm Ratio	
Push Rod Length	
Push Rod Diameter	
Lifter/Tappet Diameter	

Valves		
Intake Valve Length		
Exhaust Valve Length		
Intake Valve Stem Diameter		
Exhaust Valve Stem Diameter		
Intake Valve Head Diameter		
Exhaust Valve Head Diameter		
Intake Valve Face Diameter		
Exhaust Valve Face Diameter		
Intake Valve Face Angle		
Exhaust Valve Face Angle		
Valve Interference Angle		
Valve Margin		
Valve Lash	Cold	Hot

Valve Springs	
Intake Valve Spring Free Length	
Exhaust Valve Spring Free Length	
Intake Spring Tension (force@length)	
Exhaust Spring Tension (force@length)	
Installed Height	

Oil System		
Oil Pressure	@ Idle RPM	@ 2000 RPM
Oil Pump Gear-to-Body Clearance		
Oil Pump Gear End Clearance		

GLOSSARY
GLOSARIO

Note: **Terms are highlighted in color,** followed by **Spanish translation in bold.**

Advanced camshaft Camshaft design that has the intake valves opened more than the exhaust valves at TDC. This type of camshaft is used to increase output at low rpm.

Arbol de levas de avance Un diseño del árbol de levas que tiene las válvulas de admisión abiertas más que las válvulas de escape en PMS. Este tipo de árbol de levas se emplea para aumentar la producción del par en bajas rpm.

Aerobic sealants Sealants that require the presence of oxygen to cure.

Sellantes aeróbicos Los sellantes que requieren la presencia del oxígeno para curarse.

Air filter The air filter cleans all air entering the engine.

Filtro de aire El filtro de aire limpia el aire que entra al motor.

Airflow restriction indicator An airflow restriction indicator displays the amount of air cleaner element restriction.

Indicador de la restricción del flujo de aire Un indicador de la restricción del flujo del aire muestra la cantidad de restricción del elemento de aire limpio.

Alloys Mixtures of two or more metals; for example, brass is an alloy of copper and zinc.

Aleación La mezcla de dos o más metales; por ejemplo, el latón es una aleación de cobre con zinc.

Anaerobic sealants Sealants that will only cure in the absence of oxygen.

Sellantes anaeróbicos Los sellantes que sólo se curan en la ausencia del oxígeno.

Annealing Annealing is a heat-treatment process to reduce metal hardness or brittleness, relieve stresses, improve machinability, or facilitate cold working of the metal.

Esmaltación La esmaltación es un proceso de tratamiento por calor para reducir el peligro que los metales se pongan duros o agrios, relevar las tensiones, mejorar el torneo, o facilitar el trabajo en frío del metal.

Armature The armature is the movable component of the motor that consists of a conductor wound around a laminated iron core and is used to create a magnetic field.

Armadura La armadura es un componente movible del motor que consiste en un conductor enredado alrededor de un núcleo de hierro la-minado y se usa para crear un campo magnético.

Atkinson cycle engine The Atkinson cycle engine holds the intake valve open longer during the compression stroke to reduce pumping losses and increase fuel economy.

Motor de ciclo Atkinson El motor de ciclo Atkinson mantiene la válvula de entrada abierta más tiempo durante el golpe de compresión para reducir las pérdidas de bombeo e incrementar la economía del combustible.

Backpressure The pressure created within the engine cylinder as a result of a restricted exhaust system. With the creation of backpressure, vacuum cannot be produced as efficiently.

Contrapresión La presión creada dentro del cilindro del motor causada por una restricción en el sistema de escape. Al crear esta presión, el vacío no se puede producir eficazmente.

Balance shaft The balance shaft is a shaft driven by the crankshaft that is designed to reduce engine vibrations.

Eje de balance El eje de balance es un eje accionado por el cigüeñal que está diseñado para reducir las vibraciones del motor.

Battery terminals Battery terminals provide a means of connecting the battery plates to the vehicle's electrical system.

Bornes de la batería Los bornes de la batería proporcionan una manera de conectar las placas de la batería al sistema eléctrico del vehículo.

Bearing crush The extension of the bearing half beyond the seat which is crushed into place when the seat is tightened.

Aplastamiento del cojinete La parte de extensión del cojinete debajo del asiento que se aplasta en su lugar al apretar el asiento.

Bearing inserts Bearing inserts are soft-faced bearings used to protect the crankshaft main and rod journals as well as the camshaft journals.

Insertos de casquillos Los insertos de casquillos son casquillos de cara lisa que se usan para proteger los gorrones principales y los de bielas del cigüeñal, como también los gorrones del árbol de levas.

Bearing spread The distance between the outside parting edges is larger than the diameter of the bore.

Envergadura del cojinete La distancia al través de los rebordes exteriores es más grande que el diámetro del taladro.

Bedplate The bedplate is the lower part of a two-piece engine block and holds the crankshaft in place.

Placa de fundación La placa de fundación es la parte más inferior de un bloque motor de dos piezas y mantiene al cigüeñal en su lugar.

Body control module (BCM) A body control module (BCM) is a computer used to control systems such as the instrument panel, climate control, and driver information.

Módulo de control de la carrocería El módulo de control de la ca-rrocería es una computadora que se usa para controlar los sistemas tales como el tablero de control, el control del aire acondicionado y la información para el conductor.

Bolt diameter Bolt diameter indicates the diameter of an imaginary line running through the center of the bolt threaded area.

Diámetro del tornillo El diámetro del tornillo indica el diámetro de una línea imaginaria que atraviesa el centro del área enroscada del tornillo.

Bolt length Bolt length indicates the distance from the underside of the bolt head to the tip of the bolt threaded end.

Longitud del tornillo La longitud del tornillo indica la distancia de la parte de abajo de la cabeza del tornillo hasta la punta del enroscado del tornillo.

Bore The diameter of a hole.

Taladro El diámetro de un agujero.

Bottom Dead Center (BDC) Term used to indicate the piston is at the very bottom of its stroke.

Punto Muerto Inferior (PMI) El término uqe indica que el pistón esta en el punto más inferior de su carrera.

Brake horsepower Brake horsepower is the usable power produced by an engine. It is measured on a dynamometer using a brake to load the engine.

Caballos de fuerza del freno Los caballos de fuerza del freno es la potencia de uso que produce un motor. Se mide sobre un dinamómetro usando un freno para cargar el motor.

Burn time The amount of time from the instant the mixture is ignited until the combustion is complete.

Tiempo de encendido La cantidad del tiempo desde el instante que se enciende la mezcla hasta que se haya completado la combustión.

Burned valves Valves that have warped and melted, leaving a groove across the valve head.

Válvulas quemadas Las válvulas que se han deformado y fundido, dejándo una ranura al través de la cabeza de la válvula.

Cam ground pistons Pistons cast or forged into a slight oval or cam shape to allow for expansion. As the piston warms, it will become round.

Pistones rectificados por leva Los pistones colados o forjados en una forma lijeramente ovulada o en forma de la leva para permitir la expansión. Al calentarse el pistón se redondea.

Camshaft The shaft containing lobes to operate the engine valves.

Arbol de levas Un eje que tiene lóbulos que operan las válvulas del motor.

Camshaft lift Camshaft lift is the valve lifter movement created by the rotating action of the camshaft lobe. Typical lobe lift of original equipment in the block camshafts is between 0.240 and 0.280 in. (6 and 6.6 mm).

Elevación del árbol de levas La elevación del árbol de levas es el movimiento de la carrera de la válvula que crea la acción giratoria del lóbulo del árbol de levas. La elevación típica del lóbulo del equipo original en los árboles de levas de garrucha está entre 6 y 6.6 mm (0.240 y 0.280 pulg.).

Carbon steel Carbon steel has a specific carbon content.

Acero semiduro El acero semiduro tiene un contenido específico de carbono.

Case hardening Case hardening is a heating and cooling process for hardening metal. Because the case hardening is only on the surface of the component, it is usually removed during machining procedures performed at the time the engine is being rebuilt. It is not necessary to replace the case hardening in most applications.

Cimentación en caja La cimentación en caja es un proceso de calentamiento y enfriamiento para carburar o cimentar el metal. Ya que la cimentación en caja sólo es en la superficie del componente, generalmente se quita durante las gamas de mecanizado que se realizan al tiempo que se reconstruye el motor. No es necesario reemplazar la cimentación en caja en la mayoría de las aplicaciones.

Catalytic converter A catalytic converter reduces tailpipe emissions of carbon monoxide, unburned hydrocarbons, and nitrogen oxides.

Convertidor catalítico Un convertidor catalítico reduce las emisiones de monóxido de carbono del tubo de escape, hidrocarburos sin quemar y óxidos de nitrógeno.

Ceramic Ceramic is a combination of nonmetallic powdered materials fired in special kilns. The end product is a new product.

Cerámica La cerámica es una combinación de materiales en polvo no metálicos que se cuecen en hornos especiales. El producto final es un producto nuevo.

Chamber-in-piston A chamber-in-piston is a piston with a dish or depression in the top of the piston.

Pistón en cámara Un pistón en cámara es un pistón con un disco o depresión en la parte superior del pistón.

Channeling Local leakage around a valve head caused by extreme temperatures developing at isolated locations on the valve face and head.

Acanalado Una fuga local alrededor de una cabeza de válvula causada por las temperaturas severas que ocurren en puntos aislados en la cara y en la cabeza de la válvula.

Clearance ramp The area of the mechanical lifter camshaft from base circle to the edge comprises the clearance ramp.

Rampa de desahogo El área de la elevación mecánica del árbol de levas desde el círculo base hasta la orilla incluyen la rampa de desahogo.

Closed loop system A closed loop system uses sensors (typically an oxygen sensor) that inform the PCM about the air-fuel ratio, leaving the engine so it can adjust accordingly.

Sistema de circuito cerrado Un sistema de circuito cerrado utiliza sensores (típicamente un sensor de oxígeno) que informa al MCM sobre la relación combustible aire, dejando al motor para que pueda ajustarse en consecuencia.

Coil A coil is an electrical device that converts low voltage from the battery into a very high voltage that is able to push a spark across the gap of the spark plug.

Bobina La bobina es un dispositivo eléctrico que convierte el bajo voltaje de la batería a un muy alto voltaje que puede empujar una chispa a través del huelgo de la bujía.

Coke Coke is a very hot-burning fuel formed when coal is heated in the absence of air. Coal becomes coke above 1022° F (550° C). It is produced in special coke furnaces.

Coque El coque es un combustible hirviendo que se forma cuando se calienta el carbón sin aire. El carbón se convierte en coque arriba de los 550°C (1022°F). Se produce en hornos especiales para coque.

Combustion The process of burning.

Combustión El proceso de quemar.

Combustion chamber The volume of the cylinder above the piston with the piston at TDC.

Cámara de combustión El volumen del cilindro arriba del pistón cuando el pistón está en PMS.

Composites Man-made materials using two or more different components tightly bound together. The result is a material consisting of characteristics that neither component possesses on its own.

Compuestas Las materiales artificiales que usan dos o más componentes distinctos combinados estructuralmente. El resultado es una material que tiene las características que ninguno de los componentes posee indivi-dualmente.

Compressed natural gas (CNG) Compressed Natural Gas (CNG) is lightly refined natural gas stored under high pressure.

Gas natural comprimido El gas natural comprimido (GNC) es un gas natural poco refinado que se guarda bajo alta presión.

Compression-ignition (CI) engine A compression ignition (CI) engine ignites the air-fuel mixture in the cylinders from the heat of compression.

Motor de combustión interna de encendido por compresión Un motor de combustión interna de encendido por compresión enciende la mezcla combustible aire en los cilindros por el calor de la compresión.

Compression ratio A comparison between the volume above the piston at BDC and the volume above the piston at TDC.

Indice de compresión Una comparación entre el volumen arriba del pistón en el PMI y el volumen arriba del pistón en el PMS.

Compression rings The upper rings of the piston designed to hold the compression in the cylinder.

Anillos (aros) de compresión Los anillos superiores de un pistón di-señados a mantener la compresión dentro del cilindro.

Compressor wheel A compressor wheel is a vaned wheel mounted on the turbocharger shaft that forces air into the intake manifold.

Rueda compresora Una rueda compresora es un difusor de paletas montado en un eje turboalimentador que fuerza aire dentro del colector de entrada.

Connecting rod The link between the piston and crankshaft.

Biela La conexión entre el pistón y la cigüeñal.

Connecting rod bearings Connecting rod bearings are insert-type bearings retained in the connecting rod and mounted between the large connecting rod bore and the crankshaft journal.

Cojinetes de la biela Los cojinetes de la biela son cojinetes tipo encarte que se retienen en la biela y se montan entre el orificio grande de la biela y el gorrón del cigüeñal.

Connecting rod bearing journals Connecting rod bearing journals are offset from the crankshaft centerline to provide leverage for the piston to turn the crankshaft.

Gorrones de los cojinetes de la biela Los gorrones de los cojinetes de la biela están desajustados desde la línea central del cigüeñal para proporcionar apalancamiento para que el pistón gire el cigüeñal.

Converter A converter is used to change the high DC battery volts to 12 volts for the conventional electrical system.

Convertidor Un convertidor se usa para cambiar los voltios de una batería de alta DC a 12 voltios para el sistema eléctrico convencional.

Coolant recovery system The coolant recovery system prevents loss of coolant if the engine overheats, and it keeps air from entering the system.

Sistema de recobro de refrigerante El sistema de recobro de refrige-rante previene la pérdida de fluido refrigerador si se sobrecalienta el motor, y no permite que entre aire al sistema.

Cooling system The cooling system is used to disperse the heat of engine operation to the atmosphere.

Sistema de enfriamiento El sistema de enfriamiento se usa para dispersar el calor de la operación del motor a la atmósfera.

Core plugs Metal plugs screwed or pressed into the block or cylinder head at locations where drilling was required or sand cores were removed during casting.

Tapones del núcleo Los tapones metálicos fileteados o prensados en el monoblock o en la cabeza de los cilindros ubicados en donde se había requerido taladrar o quitar los machos de arena durante la fundición.

Crankshaft A mechanical device that converts the reciprocating motions of the pistons into rotary motion.

Cigüeñal Un dispositivo mecánico que convierte los movimientos recíprocos de los pistones a un movimiento giratorio.

Crankshaft counterweights Crankshaft counterweights are positioned opposite to the connecting rod journals and balance the crankshaft.

Contrapesos del cigüeñal Los contrapesos del cigüeñal se colocan opuestos a los gorrones de la biela y balancean el cigüeñal.

Crankshaft throw The measured distance between the centerline of the rod bearing journal and the centerline of the crankshaft.

Codo de cigüeñal Una distancia medida entre el centro del muñón del cojinete de biela y el centro del cigüeñal.

Crate engine A crate engine is a new or remanufactured engine that is built with fresh components, such as bearings, rings, and lifters. You can purchase crate engines with or without cylinder heads.

Motor en enrejado Un motor en enrejado es un motor Nuevo o refabricado que está construido con nuevos componentes, tales como cojinetes, anillos y taqués.

Cycle A cycle is a complete sequence of strokes in an engine.

Ciclo Un ciclo es una secuencia completa de golpes en un motor.

Cylinder A hole bored into the engine block that the piston travels in, and that makes up part of the combustion chamber where the air/fuel mixture is compressed.

Cilindro Un agujero taladrado en el monoblock en el cual viaja el pistón, y que forma parte de la cámara de combustión en donde se comprime la mezcla de aire/combustible.

Cylinder block The main structure of the engine. Most of the other engine components are attached to the block.

Monoblock La estructura principal del motor. La mayoría de los otros componentes del motor se conectan al monoblock.

Cylinder head On most engines, the cylinder head contains the valves, valve seats, valve guiders,. valve springs, and the upper portion of the combustion chamber.

Cabeza del cilindro En la mayoría de los motores la cabeza del cilindro contiene las válvulas con sus asientos, sus guías, sus resortes y la parte superior de la cámera de combustión.

Cylinder head gasket The cylinder head gasket prevents compres-sion pressures, gases, and fluids from leaking. It is located on the connection between the cylinder head and the engine block.

Guarnición de la cabeza del cilindro La guarnición de la cabeza del cilindro previene que se fuguen las presiones de compresión, gases y líquidos. Se encuentra en la conexión entre la cabeza del cilindro y el bloque motor.

Dedicated alternative fuel vehicle Dedicated alternative fuel vehicles were designed to run on the alternative fuel.

Vehículo de combustible alternativo dedicado Los vehículos de combustible alternativo dedicado se diseñaron para correr con combustible alternativo.

Detonation A defect that occurs if the air/fuel mixture in the cylinder is burned too fast.

Detonación Un defecto que ocurre si la mezcla aire/combustible en el cilindro se quema demasiado rápido.

Differential The differential allows the wheels to turn at different speeds as the vehicle turns corners. It is internal to a transaxle and external to a transmission.

Diferencial El diferencial permite que den vuelta las ruedas a diferentes velocidades mientras el vehículo da vueltas en esquinas. Es interno hacia un tras-eje y externo hacia una transmisión.

Displacement A measure of engine volume.

Deplazamiento Una medida del volumen del motor.

Distributor The mechanism within the ignition system that controls the primary circuit and directs the secondary voltage to the correct spark plug.

Distribuidor El mecanismo dentro del sistema de arranque que controla al circuito primario y dirige el voltaje secundario o la bujía correcta.

Dual overhead cam (DOHC) A dual overhead cam (DOHC) engine has two camshafts mounted above each cylinder head.

Leva de disco superior dual El motor de leva de disco superior dual tiene dos árboles de levas montados sobre cada cabeza del cilindro.

Dual parallel hybrid system In a dual parallel system HEV, two electric motors and an engine can be used to drive the vehicle.

Sistema paralelo dual En un sistema paralelo dual HEV dos motores eléctricos y un motor pueden usarse para impulsar el vehículo.

Duration The length of time, expressed in degrees of crankshaft rotation, the valve is open.

Duración La cantidad del tiempo, representado por los grados de rotación del cigüeñal, que esta abierta la válvula.

E85 E85 is an automotive fuel made of 85 percent ethanol and 15 percent petroleum.

E85 El E85 es un combustible para automóviles hecho con 85 por ciento de etanol y 15 por ciento de petróleo.

Eddy currents A current that runs against the main current.

Interrupción Un corriente que fluye contra el corriente principal.

Efficiency A ratio of the amount of energy put into an engine as compared to the amount of energy produced by the engine.

Rendimiento Un índice de la cantidad de energía introducido en el motor comparado a la cantidad de energía que produce el motor.

Electronic ignition (EI) An electronic ignition system uses an electronic module to turn the coils on and off.

Encendido electrónico Un sistema de encendido electrónico usa un modulo electrónico para encender y apagar las bobinas.

Electrolysis The result of two different metals in contact with each other. The lesser of the two metals is eaten way.

Electrólisis El resultado de dos metales distinctos que se ponen en contacto. El menor de los dos metales se consume.

Emission control system The emission control system includes various devices connected to the engine or exhaust system to reduce harmful emissions of hydrocarbons (HC), carbon monoxide (CO), and oxides of nitrogen (NOx).

Sistema de control de emisiones El sistema de control de emisiones incluye varios dispositivos conectados al motor o al sistema de escape para reducir las emisiones dañinas e hidrocarburos (HC), monóxido de carbono (CO) y óxidos de nitrógeno (NOx).

End gases The unburned gases in the combustion chamber are called the end gases.

Gases finales Los gases sin quemarse en la cámara de combustión se llaman gases finales.

Engine bearings Engine bearings are two bearing halves that form a circle to fit around the crankshaft journals.

Cojinetes de eje motor Los cojinetes de eje motor son dos mitades de cojinete que forman un círculo para ajustarse alrededor de los gorrones del cigüeñal.

Engine block The main structure of the engine that houses the pistons and crankshaft. Most other engine components attach to the engine block.

Monoblock La estructura principal del motor que contiene los pistones y el cigüeñal. La mayoría de los otros componentes del motor se conectan al monoblock.

Engine cradle The engine cradle, which may also be called the suspension cradle, is the front-frame member(s) in a unibody body design. The engine is mounted to this frame, as is the suspension.

Soporte del motor El soporte del motor, al que también se le llama soporte de la suspensión neumática, es el miembro(s) del bastidor delantero en un diseño de una sola carrocería. El motor está montado en su alojamiento, tomo también la suspensión.

Engine mount An engine mount attaches the engine to the chassis and absorbs the engine vibrations.

Bancada del motor La bancada del motor se sujeta del motor al chasis y absorbe las vibraciones del motor.

Exhaust manifold A component that collects and then directs engine exhaust gases from the cylinders.

Múltiple de escape Un componente que colecciona y luego dirige los gases de escape del motor desde los cilindros.

Exhaust manifold gasket A part that seals the connection between the cylinder head and the exhaust manifold.

Junta del múltiple de escape Una parte que sella la conexión entre la cabeza del cilindro y el múltiple de escape.

Face The face is the machined part of the section of the valve head that makes a seal with the valve seat.

Cara La cara es la parte mecanizada de la sección de la cabeza de la válvula que forma un sello con el asiento de la válvula.

Ferrous metal Metal containing iron or steel.

Metal férreo Un metal que contiene el hierro o el acero.

Fillets Small, rounded corners machined on the edges of journals to increase strength.

Cantos redondeados Las esquinas pequeñas, redondeadas labradas por máquina en los bordes de un muñón para reinforzarla.

Fire ring A fire ring is a metal ring surrounding the cylinder opening in a head gasket.

Anillo de encendido Un anillo de encendido es un anillo metálico circuncidante a la abertura del cilindro en la guarnición de la cabeza.

Flexible fuel vehicle A flexible fuel vehicle is designed to be able to run on either gasoline or a methanol or ethanol blend, M85 or E85.

Vehículo de combustible flexible Un vehículo de combustible flexible está diseñado para poder correr ya sea con gasolina o con una mezcla de metanol o de etanol, M85 o E85.

Flywheel A heavy circular component located on the rear of the crankshaft that keeps the crankshaft rotating during non-productive strokes.

Volante Un componente pesado circular ubicado en la parte trasera del cigüeñal que procure que el cigüeñal sigue girando durante las carreras no productivas.

Freewheeling engine An engine in which valve lift and angle prevent valve-to-piston contact if the timing belt or chain breaks.

Motor de piñón libre Un motor en el cual el levantamiento y el ángulo de las válvulas previenen el contacto entre la válvula y el pistón si se quiebra la correa o la cadena de sincronización.

Front main seal A front main seal is a lip-type seal that prevents leaks between the timing gear cover and the front of the crankshaft.

Sello principal delantero Un sello principal delantero es una junta redonda de tipo escarpia de vía que previene las fugas entre la cubierta del engranaje de distribución y el frente del cigüeñal.

Fuel injection Fuel injection systems use the PCM to control electromechanical devices, fuel injectors that deliver a closely controlled amount of fuel to the cylinders.

Por inyección Los sistemas por inyección usan el MCM para controlar los dispositivos electromecánicos, los inyectores que mandan una cantidad de combustible muy controlada a los cilindros.

Fuel system A system that includes the intake system which brings air into the engine and the components that deliver the fuel to the engine.

Sistema de combustible Un sistema que incluye al sistema de admisión que introduce el aire dentro del motor y a los componentes que entregan el combustible al motor.

Full filtration system A full filtration system means all of the oil is filtered before it enters the oil galleries.

Sistema de filtración máxima Un sistema de filtración máxima significa que todo el aceite ese filtra antes de entrar a las galeras de aceite.

Gasket A rubber, felt, cork, or metallic material used to seal surfaces of stationary parts.

Empaque Una material de caucho, fieltro, corcho o metal que sirve para sellar las superficies de las partes fijas.

Glow plugs Electrical heating elements threaded into the combustion chamber of a diesel engine to assist in cold temperature starts by heating the intake air.

Bujías del precalentamiento Los elementos eléctricos incandescentes fileteados dentro de la cámara de combustión de un motor diesel que calientan al aire de admisión para asistir en los encendidos fríos.

Grade Grade is a classification of material.

Grado El grado es una clasificación de material.

Grade marks Radial lines on the bold head that indicate the strength of the bolt.

Marcos de grado Las lineas radiales en la cabeza del perno que indican su fuerza.

Gray cast iron Gray cast iron is easy to cast and machine and absorbs vibration and resists corrosion.

Fundición con grafito laminar La fundición con grafito laminar es fácil de moldear y tornear, absorbe la vibración y resiste la corrosión.

Gross horsepower Gross horsepower is the maximum engine output as measured on a dynamometer.

Potencia global La potencia global es la salida máxima del motor tal como lo mide un dinamómetro.

Harmonic balancer (vibration dampener) A component attached to the front of the crankshaft used to reduce the torsional or twisting vibration that occurs along the length of the crankshaft.

Equilibrador armónico (amortiguador de vibraciones) Un componente conectado a la parte delantera de un cigüeñal que sirve para reducir las vibraciones torsionales que ocurren por la longitud del cigüeñal.

HD-5 LPG HD-5 LPG is 95 percent propane and 5 percent butane.

HD-5 Gas LP HD-5 Gas LP es 95 por ciento de propano y 5 por ciento de butano.

Head gasket Gasket used to prevent compression pressures, gases, and fluids from leaking. It is located on the connection between the cylinder head and engine block.

Junta de la cabeza Una junta que se emplea en prevenir que se escapen las presiones, los gases y los fluidos de compresión. Se ubica en la conexión entre la cabeza de los cilindros y el monoblock.

Head (valve) The head is the enlarged part of the valve.

Cabeza (válvula) La cabeza es la parte agrandada de la válvula.

Heater core The heater core dissipates heat like a radiator. The radiated heat is used to warm the passenger compartment.

Núcleo calentador El núcleo calentador disipa el calor como un ra-diador. El calor del radiador se usa para calentar el compartimiento del pasajero.

Hemispherical chamber A nonturbulence, half circle-shaped combustion chamber used on many older high-performance engines. The valves are on either side with the spark plug in the center.

Cámara hemisférica Una cámara de combustión sin turbulencia, en forma de una semiesfera que se empleaban en los viejos modelos de motores de alta producción. Las válvulas se ubican da cada lado con la bujía de chispa en el centro.

Hybrid electric vehicle (HEV) A hybrid electric vehicle (HEV) uses an electric motor and a gas, diesel, or alternate fuel engine to power the vehicle.

Vehículo eléctrico híbrido Un vehículo eléctrico híbrido usa un motor eléctrico y un motor de combustible alternativo, de diesel o de gas para impulsar el vehículo.

Ignition system System that delivers the spark used to ignite the compressed air/fuel mixture.

Sistema de encendido Un sistema que entrega la chispa que sirve para encender la mezcla comprimida de aire/combustible.

Indicated horsepower Indicated horsepower is the amount of horsepower the engine can theoretically produce.

Potencia disponible La potencia disponible es la cantidad de caballos de fuerza que teóricamente puede producir el motor.

Inlet check valve Keeps the oil filter filled at all times so when the engine is started an instantaneous supply of oil is available.

Válvula de retención Mantiene siempre lleno al fitro de aceite para que cuando se arranque el motor dispone de un suministro instantáneo de aceite.

Intake air temperature (IAT) sensor The intake air temperature (IAT) sensor sends an analog voltage signal to the PCM in relation to air intake temperature.

Sensor de la temperatura del aire de entrada El sensor de la temperatura del aire de entrada envía una señal e voltaje análogo al MCM en relación con la temperatura de entrada.

Intake manifold Component that delivers the air or air/fuel mixture to each engine cylinder.

Múltiple de admisión Un componente que entrega el aire o la mezcla del aire/combustible a cada cilindro del motor.

Intake manifold gasket Gasket that fits between the manifold and cylinder head to seal the air/fuel mixture or intake air.

Junta del múltiple de admisión Una junta que queda entre el múltiple y la cabeza del cilindro para sellar la mezcla de aire/combustible o el aire de admisión.

Integral valve seats Integral valve seats are part of the cylinder head.

Asientos de la válvula integral Los asientos de la válvula integral son parte de la cabeza del cilindro.

Intercooler The intercooler cools the intake air temperature to increase the density of the air entering the cylinders on turbocharged or supercharged vehicles. It is also referred to as a charge air cooler or an aftercooler.

Intercooler El intercooler enfría la temperatura del aire de entrada para aumentar la densidad del aire que entra a los cilindros en los vehículos turboalimentados o en los sobrealimentados. También se refiere como un refrigerante de aire de cambio o un postenfriador.

Interference engine An engine design in which, if the timing belt or chain breaks or is out of phase, the valves will contact the pistons. In addition, in multivalve interference engines, valve-to-valve contact is possible if the belt or chain is out of phase.

Interferencia del motor Un diseño del motor en el cual, si se quiebra la correa o la cadena de sincronización, o si esta fuera de fase, las válvulas se pondrán en contacto con los pistones. Además, en los motores de interferencia de múltiples válvulas, es posible el contacto de válvula a válvula si la correa o la cadena esta fuera de fase.

Internal combustion engine An engine that burns its fuels within the engine.

Motor de combustión interna Un motor que quema su combustible dentro del motor.

Inverter An inverter is used to convert DC volts to AC volts and to change AC volts to DC volts.

Invertidor Un invertidor se usa para convertir los voltios de DC a voltios de AC y cambiar los voltios de AC a voltios de DC.

Journal An inner bearing operated by a shaft.

Muñón Un cojinete interior operado por una flecha.

Kinetic energy Energy that is working.

Energía cinética La energía trabajando.

Knock sensor (KS) A knock sensor (KS) screws into the engine and creates an electrical input for the PCM when abnormal knocking is detected.

Sensor de detonación Un sensor de detonación se atornilla dentro del motor y crea una entrada eléctrica para el MCM cuando se detecta una detonación anormal.

Lifters Mechanical (solid) or hydraulic connections between the camshaft and the valves. Lifters follow the contour of the camshaft lobes to lift the valve off its seat.

Levantaválvulas Las conexiones mecánicas (sólidas) o hidráulicas entre el árbol de levas y las válvulas. Los levantaválvulas siguen el contorno de los lóbulos del árbol de levas para levantar la válvula de su asiento.

Lip seal Molded synthetic rubber seal with a slight raise (lip) that is the actual sealing point. The lip provides a positive seal while allowing for some lateral movement of the shaft.

Junta con reborde Las juntas de caucho sintético moldeado que tienen un reborde ligero (un labio) que es el punto efectivo del sello. El reborde provee un sello positivo mientras que permite algún movimiento lateral de la flecha.

Lobe The part of the camshaft that raises the lifter.

Lóbulo La parte del árbol de levas que alza al levantaválvulas.

Lobe separation angle The lobe separation angle is the average of the intake and exhaust lobe centers. It defines the amount of valve overlap. The tighter the angle (lower number of degrees), the more overlap the camshaft provides.

Ángulo de separación del lóbulo El ángulo de separación del lóbulo es el promedio de los centros de los lóbulos de entrada y de escape. Se define como la cantidad de superposición de la válvula. Entre más cerrado el ángulo (menor número de grados), mayor superposición proporcionará el árbol de levas.

Locating tab The locating tab is a tab sticking out of bearing inserts that fits in a groove in the block, main bearing cap, connecting rod, or connecting rod cap to prevent insert rotation.

Aleta de posicionamiento La aleta de posicionamiento es una aleta que sale de los insertos del cojinete que cabe en una ranura del bloque, la tapa del cojinete principal, la biela o la tapa de la biela para prevenir que gire el inserto.

Long block A long block is an engine block with the cylinder head(s) and other components mounted.

Bloque largo Un bloque largo en un bloque motor con las cabezas del cilindro y otros componentes montados.

Longitudinally mounted engine A longitudinally mounted engine is placed in the vehicle lengthwise, usually front to rear.

Motor montado longitudinalmente Un motor montado longitudinalmente se coloca a lo largo del vehículo, generalmente del frente a la parte posterior.

Lubrication system System that supplies oil to high friction and wear locations.

Sistema de lubricación Un sistema que proporciona el aceite a las áreas da alta fricción y desgaste.

M85 M85 is an automotive fuel made of 85 percent methanol and 15 percent petroleum.

M85vEl M85 es un combustible para automóviles hecho con el 85 por ciento de metanol y el 15 por ciento de petróleo.

Main bearing journal The crankshaft journal that is supported by the main bearing.

Muñón del cojinete principal El muñón del cigüeñal apoyado por el cojinete del cigüeñal.

Main bore The main bore is a straight hole drilled through the block to support the crankshaft.

Agujero principal El agujero principal es un hoyo derecho taladrado a través del bloque para apoyar el cigüeñal.

Main caps The main caps form the other half of the main bore and bolt onto the block to form round holes for the crankshaft bearings and journals.

Tapas principales Las tapas principales forman la otra mitad del agujero principal y se sujetan con tornillos al bloque para formar agujeros redondos en los gorrones y en los cojinetes del cigüeñal.

Major thrust surface The side of the piston skirt that pushes against the cylinder wall during the power stroke.

Superficie de empuje principal El lado de la faldilla que empuja contra el muro del cilindro durante la carrera de fuerza es la superficie de empuje principal.

Margin The margin is the material between the valve face and the valve head.

Margen El margen es el material entre la cara de la válvula y la cabeza de la válvula.

Mass airflow (MAF) sensor The mass airflow sensor (MAF) measures the mass of airflow entering the engine and sends an input to the PCM reflecting this value. This is a primary input for the PCM's control of fuel delivery.

Sensor del flujo de aire en masa El sensor de flujo de aire en masa mide la masa del flujo de aire que entra al motor y envía una entrada al MCM que refleja su valor. Esto es una entrada primaria para el control de entrega de combustible del MCM.

Mechanical efficiency A comparison of the power actually delivered by the crankshaft to the power developed within the cylinders at the same rpm.

Rendimiento mecánica Una comparación entre la energía que actualmente entrega el cigüeñal y la energía desarrollado dentro de los cilindros en la misma rpm.

Metallurgy Metallurgy is the science of extracting metals from their ores and refining them for various uses.

Metalurgia La metalurgia es la ciencia de extraer metales de sus mi-nerales y refinarlos para sus varios usos.

Micron A thousandth of a millimeter or about .0008 inch.

Micrón La milésima parte de un milímetro o aproximadamente un .0008 de una pulgada.

Mid-mounted engine A mid-mounted engine is located between the rear of the passenger compartment and the rear suspension. It may also be called a mid-engine.

Motor central Un motor central se localiza entre la parte posterior del compartimiento de pasajeros y la suspensión posterior. También se le llama motor de en medio.

Minor thrust surface The areas of the piston skirt that pushes against the cylinder wall during the compression stroke.

Superficie de empuje menor El faldón del pistón que empuja contra el muro del cilindro durante la carrera de compresión.

Misfire A misfire means that combustion is either incomplete or does not occur at all.

Fallo de encendido Un fallo de encendido significa que la combustión fue incompleta o no sucedió del todo.

Muffler The muffler is a device mounted in the exhaust system behind the catalytic converter that reduces engine noise.

Mofle El mofle es un dispositivo montado en el sistema de escape atrás del convertidor catalítico que reduce el ruido del motor.

Multi-point fuel injection (MPFI) Multi-point fuel injection systems use one injector for each cylinder to deliver fuel.

Inyección del combustible entre múltiples puntos Los sistemas de inyección del combustible entre múltiples puntos utilizan un inyector por cada cilindro para entregar el combustible.

Net horsepower Net horsepower is the maximum engine horsepower as measured on a dynamometer.

Potencia neta La potencia neta es la potencia máxima del motor tal como se mide en un dinamómetro.

Nodular iron Nodular iron contains a specific carbon content plus magnesium and other additives.

Hierro nodular El hierro nodular tiene un contenido específico de carbono más magnesio y otros aditivos.

Nonferrous metals Nonferrous metals contain no iron. Aluminum, magnesium, and titanium are examples of nonferrous metals. These metals will not attract a magnet.

Metales no ferrosos Los metales no ferrosos no contienen hierro. El aluminio, el magnesio y el titanio son ejemplos de metales no fe-rrosos. Estos metales no atraerán un imán.

Nucleate boiling The process of maintaining the overall temperature of a coolant to a level below its boiling point, but allowing the portions of the coolant actually contacting the surfaces (the nuclei) to boil into a gas.

Ebullición nucleido El proceso de mantener un nivel de temperatura general de un fluido refrigerante menos de la de su punto de ebullición, pero permitiendo que las porciones del fluido refrigerante que actualmente estan en contacto con las superficies (el nucleido) hiervan para formar un gas.

Octane rating The octane rating of a fuel describes its ability to resist knock. The lower the octane number, the easier it detonates.

Grado de octanaje El grado de octanaje de un combustible describe su habilidad para resistir el golpe. Entre más bajo el número de octano, detona más fácilmente.

Offset pin An offset pin is a piston pin that is not mounted on the vertical center of the piston.

Garrucha de desvío Una garrucha de desvío es una garrucha de pistón que no está montada en el centro vertical del pistón.

Off-square When the seat and valve stem are not properly aligned, they are off-square. This condition causes the valve stem to flex as the valve face is forced into the seat by spring and combustion pressures.

Fuera de escuadra Refiere a que el asiento y el vástago de la válvula no estén alineadas correctamente. Esta condición causa que el vástago de la válvula se dobla cuando la cara de la válvula se asienta bajo la fuerza de las presiones del resorte y la combustión.

Oil control rings Oil control rings are piston rings that control the amount of oil on the cylinder walls.

Anillos de control del aceite Los anillos de control del aceite son ani-llos de pistón que controlan la cantidad de aceite en las paredes del cilindro.

Oil cooler An oil cooler is a heat exchanger that looks much like a small radiator.

Refrigerador del aceite El refrigerador del aceite es un intercambiador de calor que se parece mucho a un radiador pequeño.

Oil breakdown Condition of oil that results from exposure to high temperatures for extended periods of time. The oil will combine with oxygen and can cause carbon deposits in the engine.

Deterioro del aceite Un resultado de las temperaturas elevadas por largos períodos de tiempo. El aceite combinará con el oxígeno y puede causar los depósitos del carbono en el motor.

Oil gallery The main oil supply line in the engine block.

Canalización de aceite La linea principal del suministro de aceite en el monoblock.

Oil pan gaskets Gaskets used to prevent leakage from the crankcase at the connection between the oil pan and the engine block.

Empaque de la tapa del cárter Los empaques que se emplean para prevenir las fugas del cárter en la conexión entre el colector de aceite y el monoblock.

Oil pump A device that pulls oil from the sump and pressurizes, then delivers it to throughout the engine by use of galleries.

Bomba de aceite Un dispositivo que toma el aceite del suministro, lo presurisa, y lo entrega por el motor por medio de las canalizaciones.

Open loop system An open loop system delivers fuel based on demand and has no way of detecting whether the amount sent was ideal in terms of performance, emissions, or fuel economy.

Mecanismo de bucle abierto El mecanismo de bucle abierto entrega combustible basado en la demanda y no tiene manera de detectar si la cantidad que mandó era la ideal en términos de rendimiento, emisiones o economía del combustible.

Oversize bearings Bearings that are thicker than standard to increase the outside diameter of the bearing to fit an oversize bearing bore. The inside diameter is the same as standard bearings.

Cojinete de medidas superiores Los cojinetes que son de un espesor más grueso de lo normal para aumentar el diámetro exterior del cojinete para quedarse en un taladro que rebasa la medida. El diámetro interior es lo mismo del cojinete normal.

Overhead camshaft engine (OHC) An overhead camshaft engine (OHC) has the camshaft held in the cylinder head.

Motor de árbol de levas en cabeza Un motor de árbol de levas en cabeza tiene el árbol de levas pegado en la cabeza del cilindro.

Oxidation Oxidation occurs when hot engine oil combines with oxygen and forms carbon.

Oxidación La oxidación ocurre cuando el aceite caliente del motor se combina con oxígeno y forma carbono.

Parallel hybrid system A parallel system HEV can use both power systems simultaneously or individually to power the vehicle; the electric motor is strong enough to power the vehicle on its own.

Sistema paralelo Un sistema paralelo HEV puede usar tanto sistemas de potencia simultáneamente o individualmente para impulsar el vehículo; el motor eléctrico es lo suficientemente potente para impulsar el vehículo por sí mismo.

Pedestal-mounted rocker arm A pedestal-mounted rocker arm is mounted on a threaded pedestal that is an integral part of the cylinder head.

Palanca de basculador montada en pedestal La palanca de basculador montada en pedestal está montada en un pedestal enroscado que es una parte integral de la cabeza del cilindro.

Pentroof combustion chambers Pentroof combustion chambers are a common design for multivalve engines using two intake valves with one or two exhaust valves and provide good surface-to-volume ratio.

Cámaras de combustion de cubierta a una pendiente Las cámaras de combustión de cubierta a una pendiente son un diseño común para los motores con multiválvulas que usan dos válvulas de entrada con una o dos válvulas de escape y proporcionan una buena proporción entre la superficie y el volumen.

Piezoresistive sensor A piezoresistive sensor is sensitive to pressure changes. The most common use of this type of sensor is to measure the engine oil pressure.

Sensor piezorresistente El sensor piezorresistente es sensible a los cambios de presión. El uso más común para este tipo de sensor es el de medir la presión del aceite del motor.

Piston An engine component in the form of a hollow cylinder that is enclosed at the top and open at the bottom. Combustion forces are applied to the top of the piston to force it down. The piston, when assembled to the connecting rod, is designed to transmit the power produced in the combustion chamber to the crankshaft.

Pistón Un componente del motor que consiste de un cilindro hueco cerrado en la parte de arriba y abierto en la parte de abajo. Las fuerzas de combustión se aplican en la parte superior del pistón para forzarlo hacia abajo. El pistón, al conectarse a la biela, es diseñado para transmitir la fuerza producida en la cámara de combustión al cigüeñal.

Piston pin Component that connects the piston to the connecting rod. There are three basic designs used: a piston pin anchored to the piston and floating in the connecting rod, a piston pin anchored to the connecting rod and floating in the piston, and a piston pin full floating in the piston and connecting rod.

Eje del pistón Un componente que conecta el pistón a la biela. Se usan tres diseños básicos: un eje de pistón fijo al pistón y libre en la biela, un eje de pistón fijo a la biela y libre en el pistón, y un eje de pistón completamente libre en el pistón y la biela.

Piston rings Components that seal the compression and expansion gases, and prevent oil from entering the combustion chamber.

Anillos (aros) del pistón Un componente que sella los gases de compresión y expansión, y previene que entra el aceite en la cámara de combustión.

Pitch Pitch is the number of threads per inch on a bolt in the USC system.

Declive El declive es el número de roscas por pulgada en un tornillo en el sistema de medidas acostumbradas en EU.

Plateau honing A cylinder honing process that uses a coarse stone to produce the finished cylinder size and a very fine stone to remove an immeasurable amount and plateau the cut.

Esmerilado de nivelación Un proceso de esmerilar el cilindro usando una piedra tosca para producir el tamaño del cilindro acabado y una piedra muy fina para quitar una cantidad inmensurable y nivelar el corte.

Pole shoes Pole shoes are made of high-magnetic permeability material to help concentrate and direct the lines of force in the field assembly.

Expansiones polares Las expansiones polares están hechas con un material de alta impermeabilidad magnética que ayudan a concentrar y dirigir las isostáticas en el ensamblado de campo.

Poppet valve A valve design consisting of a circular head with a stem attached in the center. Poppet valves are used to control the opening or closing of a passage by linear movement.

Válvula champiñón Un diseño de una válvula que consiste de una cabeza redonda con un vástago conectado en el centro. Las válvulas champiñones controlan la apertura o cerradura de un pasaje por medio de un movimiento linear.

Positive displacement pumps Pumps that deliver the same amount of oil with every revolution, regardless of speed.

Bombas de desplazamiento positivo Las bombas que entregan la misma cantidad del aceite con cada revolución, sin que importa la velocidad.

Potential energy Potential energy is energy that is not being used at a given time, but which can be used.

Energía potencial La energía potencial es la energía que no está usándose en un tiempo específico, pero que puede usarse.

Powder metallurgy Powder metallurgy is the manufacture of metal parts by compacting and sintering metal powders.

Pulvimetalúrgico El pulvimetalúrgico es la manufactura de las partes de un metal al compactar y sinterizar polvos metálicos.

Power splitting device A power splitting device changes the mode of power delivery from one source to the other or to a combination of both sources.

Dispositivo de división de la potencia Un dispositivo de división de la potencia cambia el modo de entregar la potencia de una fuente a otra o a una combinación de ambas fuentes.

Powertrain A powertrain includes the engine and all the components that deliver that force to the road, including the transmission and axles.

Motor El motor incluye a la máquina y todos los componentes que entregan esa fuerza en el camino, incluyendo la transmisión y los ejes.

Powertrain control module (PCM) The powertrain control module is an onboard computer that controls functions related to the powertrain, such as fuel delivery and spark timing.

Módulo de control del motor El modulo de control del motor es una computadora incorporada que controla las funciones relacionadas al motor, tales como la entrega de combustible y la chispa de distribución.

Preignition Defect that is the result of spark occurring too soon.

Autoencendido Un defecto que resulta de una chispa que ocurre demasiado temprano.

Prove-out circuit A prove out circuit completes the warning light circuit to ground through the ignition switch when it is in the start position. The warning light will be on during engine cranking to indicate to the driver that the bulb is working properly.

Circuito de comprobación Un circuito de comprobación completa el circuito de luz de alerta a tierra mediante el interruptor de encendido cuando está en la posición de iniciar. La luz de alerta se encenderá durante el arranque del motor para indicarle al conductor que el foco está funcionando apropiadamente.

Pumping losses Pumping losses are wasted power from the pistons pushing up against compressed air during the compression stroke.

Saldo de bombeo Los saldos de bombeo son potencia perdida de los pistones empujando hacia el aire comprimido durante los golpes de compresión.

Pump-up Lifter pump-up occurs when excessive clearance in the valvetrain allows a valve to float. The lifter attempts to compensate for the clearance by filling with oil. The valve is unable to close since the lifter does not leak down fast enough.

Bombear Los bombeos del empujador suceden cuando la holgura excesiva en el tren de la válvula permite que flote la válvula. El levantador intenta compensar por la holgura al llenarla de aceite. La válvula no puede cerrarse ya que el levantador no deja que gotee lo suficientemente rápido.

Pushrod A connecting link between the lifter and rocker arm. Engines designed with the camshaft located in the block use pushrods to transfer motion from the lifters to the rocker arms.

Varilla de presión Una conexión entre la levantaválvulas y el balancín empuja válvulas. Los motores diseñados con el árbol de levas en el bloque usan las varillas de presión para transferir el movimiento de la levantaválvulas a los balancines.

Quenching The cooling of gases as a result of compressing them into a thin area. The quench area has a few thousandths of an inch clearance between the piston and combustion chamber. Placing the crown of the piston this close to the cooler cylinder head prevents the gases in this area from igniting prematurely.

Amortiguamiento El enfriamiento de los gases que resulta de su compresión en una área pequeña. El área de amortiguamiento es una holgura de algunas milésimas partes de una pulgada entre el pistón y la cámara de combustión. Al poner el punto superior de la cabeza del pistón tan cerca a la cabeza del cilindro enfriada previene que los gases en esta área se enciendan prematuramente.

Radiator A heat exchanger used to transfer heat from the engine to the air passing through it.

Radiador Un cambiador de calor que sirve para transferir el calor del motor al aire que lo atraviesa.

Rear main seal The rear main seal is a seal installed behind the rear main bearing on the rear main bearing journal to prevent oil leaks.

Junta principal posterior La junta principal posterior es un sello insta-lado detrás del cojinete principal posterior en el gorrón del cojinete principal posterior para prevenir fugas de aceite.

Reciprocating An up-and-down or back-and-forth motion.

Alternativo Un movimiento oscilante de arriba a abajo o de un lado a otro.

Reed valve A one-way check valve. The reed opens to allow the air/fuel mixture to enter from one direction, while closing to prevent movement in the other direction.

Válvula de lengüeta Una válvula de una vía. La lengüeta se abre para permitir entrar la mezcla de aire/combustible de una dirección, mientras que se cierre para prevenir el movimiento de la otra dirección.

Regenerative braking Regenerative braking uses the lost kinetic energy during braking to help recharge the batteries.

Frenado regenerativo El frenado regenerativo utiliza la energía cinética perdida durante el frenado para ayudar a recargar las baterías.

Reid vapor pressure (RVP) Reid Vapor Pressure (RVP) is a method of describing the volatility of the fuel. You can test the RVP to rate the fuel's volatility.

Presión de vapor Reid La presión de vapor Reid es un método para describir la volatilidad del combustible. Se puede probar la PVR para estimar la volatilidad del combustible.

Remanufactured engine A remanufactured engine is one that has been professionally rebuilt by a manufacturer or engine supplier.

Motor re-fabricado Un motor re-fabricado es el que el fabricante o un proveedor de motores volvió a fabricar profesionalmente.

Resonator A resonator is a device mounted on the tailpipe to reduce exhaust noise further.

Resonador El resonador es un dispositivo montado en el tubo de escape para reducir más el ruido de escape.

Retarded camshaft A camshaft design that has the exhaust valves open more than the intake valves at TDC.

Arbol de levas retrasado Un diseño del árbol de levas cuyas válvulas de escape abren más que las válvulas de admisión en el PMS.

Retrofitted alternative fuel vehicle A vehicle that was originally designed to run on gasoline but has been changed to run on an alternate fuel is often called retrofitted or converted.

Vehículo modificado de combustible alternativo Un vehículo que originalmente se diseño para correr con gasolina pero se cambió para correr con un combustible alternativo generalmente se le llama modificado o convertido.

Reverse flow cooling system In a reverse flow cooling system, coolant flows from the water pump through the cylinder heads and then through the engine block.

Sistema de reverso del flujo de refrigerante En un sistema de reverso del flujo de refrigerante el refrigerante fluye de la bomba del agua a través de las cabezas del cilindro y luego a través del bloque motor.

Ring gear The ring gear around its circumference allows the starter motor drive gear to turn the engine.

Engranaje del anillo El engranaje del anillo alrededor de su circunfe-rencia permite que el engranaje de mando del motor de arranque encienda el motor.

Rocker arm Pivots that transfer the motion of the pushrods or followers to the valve stem.

Balancín Un punto pivote que transfiere el movimiento de las levantaválvulas o de los seguidores al vástago de la válvula.

Rocker arm ratio A mathematical comparison of rocker arm dimensions. The rocker arm ratio compares the center-to-valve-stem measurement against the center-to-pushrod measurement.

Indice del balancín Una comparación de las dimensiones del ba-lancín. El índice del balancín compara las dimensiones del balancín del centro-al-vástago de la válvula con las dimensiones del centro-al-levantaválvulas.

Room-temperature vulcanizing (RTV) Room-temperature vulcanizing (RTV) sealant is a type of engine sealer that may be used in place of a gasket on some applications.

Vulcanización a temperatura ambiente La junta de vulcanización a temperatura ambiente es un tipo de sellador del motor que puede usarse en lugar de una junta plana en algunas aplicaciones.

Roots The roots-type supercharger is a positive displacement supercharger that uses a pair of lobed vanes to pump and compress the air.

Roots El compresor tipo Roots es un compresor de desplazamiento positivo que usa un par de aletas lobadas para bombear y comprimir el aire.

Rotary valve A valve that rotates to cover and uncover the intake port. A rotary valve is usually designed as a flat disc that is driven from the crankshaft.

Válvula rotativa Una válvula que gira para cubrir y descubrir la puerta de admisión. Suelen ser diseñadas como un disco plano impulsado por el cigüeñal.

Saddle The portion of the crankcase bore that holds the bearing half in place.

Asiento del cojinete (silleta) La parte del taladro del cigüeñal que mantiene en su lugar a la mitad con el cojinete.

S-class oil S-class oil is used for automotive gasoline engines. C-class is designed for commercial or diesel engines. The "S" stands for spark ignition; the "C" stands for compression ignition.

Aceite tipo S El aceite tipo S se usa en motores para automóviles de gasolina. La clase C se diseña para los motores comerciales o de diesel. La "S" significa encendido por chispa; la "C" significa encendido por compresión.

Seal Component used to seal between a stationary part and a moving one.

Junta Un componente que se emplea para sellar entre una parte fija y una que mueva.

Sealant A special liquid material commonly used to fill irregularities between the gasket and its mating surface. Some sealants are designed to be used in place of a gasket.

Compuesto obturador Un material líquido especial que normalmente se emlpea para rellenar las irregularidades entre un empaque y su superficie de contacto. Algunos compuestos son diseñados de uso sin empaque.

Series hybrid system A series system HEV uses the electric motor to drive the vehicle; the gas motor supports the electric motor.

Sistema en serie Un sistema en serie HEV usa un motor eléctrico para conducir el vehículo; el motor de gas apoya al motor eléctrico.

Shaft-mounted rocker arms Shaft-mounted rocker arms are used in a mounting system that positions all rocker arms on a common shaft mounted above the cylinder head.

Conjunto de levanta válvulas montadas en el eje El conjunto de le-vanta válvulas montadas en el eje se usan en un sistema de montaje que coloca todo el conjunto de levanta válvulas en un eje común montado encima de la cabeza del cilindro.

Short block A short block is a block assembly without the cylinder heads and some other components.

Bloque corto Un bloque corto es el ensamblado de bloque sin las cabezas del cilindro y algunos otros componentes.

Shot-peening A tempering process that uses shot under pressure to tighten the outside surface of the metal. Shot-peening is used to help strengthen the component and reduce chances of stress or surface cracks.

Chorreo con granalla Un proceso de templado que usa la granalla bajo presión para estrechar la superficie exterior del metal. El chorreo con granalla fortalece al componente y disminuye la aparencia del fatiga o de grietas en la superficie.

Single-pattern camshaft lobes Single-pattern lobes are ground the same on both opening and closing ramps. Lobes ground differently on the two ramps are asymmetrical.

Lóbulos del árbol de levas en patrón sencillo Los lóbulos en patrón sencillo están bruñidos de igual manera en las rampas de abertura y cierre. Los lóbulos bruñidos de diferente manera en las dos rampas son asimétricos.

Sintering Sintering is done by heating the metal to a temperature below its melting point in a controlled atmosphere. The metal is then pressed to increase its density.

Sinterizado El sinterizado se hace al calentar el metal a una temperatura debajo de su punto de fusión en una atmósfera controlada. Luego el metal se presiona para incrementar su densidad.

Skin effect Skin effect is a small layer of unburned gases formed around the walls of the combustion chamber.

Efecto skin El efecto skin es una capa pequeña de gases sin quemar que se forma alrededor de las paredes de la cámara de combustión.

Solenoid A solenoid is an electromechanical device. It is typically operated by the PCM through an electrical signal and mechanically controls flow of air, oil, fuel or other substances. A fuel injector is a type of solenoid.

Solenoide Un solenoide es un dispositivo electromecánico. Típicamente lo maneja el MCM mediante una señal eléctrica y mecánicamente controla el flujo de aire, aceite, combustible u otras sustancias. Un inyector es un tipo de solenoide.

Spark-ignition engine (SI) A spark-ignition (SI) engine ignites the air-fuel mixture in the combustion chamber by a spark arcing across the spark plug electrodes.

Motor de encendido por chispa Un motor de encendido por chispa enciende la mezcla combustible aire en la cámara de combustible por medio de una chispa que hace un arco a través de los electrodos de la bujía.

Splayed crankshaft A splayed crankshaft is a crankshaft in which the connecting rod journals of two adjacent cylinders are offset on the same throw.

Cigüeñal ensanchado Un cigüeñal ensanchado es un cigüeñal en el que los gorrones de la biela de dos cilindros adyacentes no están paralelos en el mismo alcance.

Split overlap Term describing when the intake and exhaust valves are equally open at TDC.

Apertura compratida Un término que describe que las válvulas de admisión y escape estan igualmente abiertas en el PMS.

Squish area The area of the combustion chamber where the piston is very close to the cylinder head. The air/fuel mixture is rapidly pushed out of this area as the piston approaches TDC, causing turbulence and forcing the mixture toward the spark plug. The squish area can also double as the quench area.

Area de compresión El área de la cámara de combustión en donde el pistón esta muy cerca a la cabeza del cilindro. La mezcla del aire/combustible se expulsa rapidamente de esta área al aproximarse el pistón al PMS, causando una turbulencia y empujando la mezcla hacia la bujía. El área de compresión tambien puede servir de área de extinción.

Stainless steel Stainless steel is an alloy that is highly resistant to rust and corrosion.

Acero inoxidable El acero inoxidable es una aleación que es altamente resistente a la oxidación y la corrosión.

Steel Iron containing very low carbon (between 0.05 and 1.7 percent) is called steel.

Acero El hierro que contiene muy bajo carbono (entre 0.05 y 1.7 por ciento) se llama acero.

Stellite A hard facing material made from a cobalt-based material with a high chromium content.

Estelita Una material de recarga compuesta de una material de base cobáltico con un contenido muy alto del cromo.

Stem necking Stem necking is a condition of the valve stem in which the stem narrows near the bottom.

Rebajamiento del vástago El rebajamiento del vástago es una condición del vástago de la válvula en el que el vástago se estrecha cerca del fondo.

Stratified charge A stratified-charge engine has two combustion chambers. A rich air-fuel mixture is supplied to a small auxiliary chamber, and the very lean air-fuel mixture is supplied to the main combustion chamber to improve fuel economy.

Carga estratificada Un motor de carga estratificada tiene dos cámaras de combustión. Se abastece una rica mezcla combustible aire a la pequeña cámara auxiliar, y una mezcla combustible aire muy pobre se abastece a la cámara de combustión principal para mejorar la economía del combustible.

Stroke The distance traveled by the piston from TDC to BDC.

Carrera La distancia que viaja el pistón dle PMS al PMI.

Strip seal A strip seal provides a seal between the flat surfaces on the front and back of the engine block and the intake manifold.

Junta de impermeabilización Una junta de impermeabilización proporciona un sello entre las superficies planas del frente y la parte posterior del bloque motor y del colector de entrada.

Stud-mounted rocker arm A stud-mounted rocker arm is mounted on a stud that is pressed or threaded into the cylinder head.

Balancín montado en espárrago roscado El balancín montado en espárrago roscado está montado en un espárrago roscado que está prensado o enroscado en la cabeza del cilindro.

Supercharger The supercharger is a belt-driven air pump used to increase the compression pressures in the cylinders.

Superalimentador El superalimentador es una bomba de aire operada por banda que se usa para incrementar las presiones de compresión en los cilindros.

Surface-to-volume ratio A mathematical comparison between the surface area of the combustion chamber and the volume of the combustion chamber. The greater the surface area, the more area the mixture can cling to, and mixture that clings to the metal will not burn completely because the metal cools it. Typical surface-to-volume ratio is 7.5:1.

Indice superficie-volumen Una comparación matemática entre el área de la superficie de la cámara de combustión y el volumen de la cámara de combustión. Lo mayor el área de la superficie, lo mayor el área en donde puede pegarse la mezcla y la mezcla pegada al metal no se quemará completamente puesto que la enfría el metal. El índice del superficie-volumen es el 7.5:1.

Swirl chamber Combustion chamber that creates an airflow in a horizontal direction.

Cámara vórtice Cámara de combustión que crea un flujo de aire de una dirección horizontal.

Synchronization timing Timing design used on vehicles with computer-controlled fuel-injection systems. A pickup in the distributor is used to synchronize the crankshaft position to the camshaft position.

Regulación (tiempo) sincronizado Un diseño de regulación empleado en los vehículos equipados con sistemas de inyección de combustible de control computerizado. Un captador en el distribuidor sirve para sincronizar la posición del cigüeñal con la posición del árbol de levas.

Tailpipe The tailpipe conducts exhaust gases from the muffler to the rear of the vehicle.

Tubo de escape El tubo de escape conduce los gases de emisión desde el mofle a la parte posterior del vehículo.

Tapered pistons Tapered pistons have a narrower diameter at the top of the piston than at the bottom of the piston skirt, to allow for greater expansion at the top as the piston heats up.

Pistones cónicos Los pistones cónicos tienen un diámetro más estrecho en la parte superior del pistón que en la parte inferior de la falda del pistón, para permitir una mayor expansión en la parte superior mientras se va calentando el pistón.

Tensile strength The metal's resistance to be pulled apart.

Resistencia a la tracción La resistencia de un metal a ser estirado.

Tempering In tempering, metal is heated to a specific temperature to reduce the brittleness of hardened carbon steel.

Templado En templado el metal se calienta a una temperatura específica para reducir la fragilidad del acero semiduro mejorado.

Thermal efficiency Thermal efficiency is the difference between potential and actual energy developed in a fuel measured in BTUs per pound or gallon.

Eficiencia térmica La eficiencia térmica es la diferencia entre la energía potencial y la actual desarrollada en un combustible medido en UTB por libra o por galón.

Thermistor The thermistor is a resistor whose resistance changes in relation to changes in temperature. It is often used as a coolant temperature sensor.

Termostato El termostato es una resistencia cuya resistencia cambia en relación con los cambios de temperatura. Generalmente se usa como un sensor de la temperatura del refrigerante.

Thermodynamics The study of relationship between heat energy and mechanical energy.

Termodinámica El estudio de la relación entre la energía del calor y la energía mecánica.

Thermostat A control device that allows the engine to reach normal operating temperatures quickly and maintains the desired temperatures.

Termostato Un dispositivo de control que permite que le motor llegue rápidamente a las temperaturas de funcionamiento normales y mantenga las temperaturas deseadas.

Thread depth The height of the thread of a bolt from its base to the top of its peak.

Profundidad de la rosca La longitud de la rosca de un perno desde su fondo a la parte superior.

Thrust bearing A double-flanged bearing used to prevent the crankshaft from sliding back and forth.

Cojinete de empuje Un cojinete con doble brida que sirve para prevenir que el cigüeñal desliza de un lado a otro.

Timing chain The chain that drives the camshaft off of the crankshaft.

Cadena de sincronización La cadena procedente del cigüeñal que acciona al árbol de levas.

Timing chain guide A timing chain guide is a length of metal or plastic with a soft face to hold the chain in position as it rotates.

Guía de la cadena de tiempo La guía de la cadena de tiempo es una longitud de metal o plástico con una cara suave que detiene la cadena en su posición mientras gira.

Timing chain tensioner A timing chain tensioner maintains tension on the chain or belt to prevent slapping or skipping a tooth.

Tensor de la cadena de tiempo El tensor de la cadena de tiempo mantiene la tensión en la cadena o banda para prevenir el ruido de impulso o saltearse un diente.

Top dead center (TDC) Term used to indicate the piston is at the very top of its stroke.

Punto muerto superior (PMS) El término que indica que el pistón esta en la cima de su carrera.

Torque Twisting force applied to a shaft or bolt.

Torsión Una fuerza de torsión aplicada en un eje o en un perno.

Torque convertor A series of components that work together to reduce slip and multiply torque. It provides a fluid coupling between the engine and the automatic transmission.

Convertidor del par Un serie de componentes que trabajan juntos para disminuir el deslizamiento y multiplicar el par. Provee un acoplamiento flúido entre el motor y la transmisión automática.

Torque-to-yield Torque-to-yield is a stretch-type bolt that must be tightened to a specific torque and then rotated a certain number of degrees.

De tensión a rendimiento De tensión a rendimiento es un tornillo de tipo alargado que debe estar apretado a una tensión específica y luego girarlo a cierto número de grados.

Torsional vibration Vibration that is the result of the crankshaft twisting and snapping back during each revolution.

Vibración torsional Una vibración que resulta cuando el cigüeñal se tuerce y regresa repentinamente en cada revolución.

Transverse-mounted engine Engine placement in which the block faces from side to side instead of front to back within the vehicle.

Motor de motaje transversal La ubicación del motor en la cual la longitud del monoblock queda de un lado a otro dentro del vehículo en vez de estar colocado de frente a atrás.

Tumble port Intake port design that creates an airflow in a vertical direction.

Puerta de remolino Un diseño de la puerta de entrada que crea un flujo del aire en una dirección vertical.

Turbine wheel A turbine wheel is a vaned wheel in a turbocharger to which exhaust pressure is supplied to provide shaft rotation.

Rodete Un rodete es una rueda de paletas en un turboalimentador en la cual la presión de salida se suministra para proporcionar la rotación del eje.

Turbo boost Turbo boost is a term used to describe the amount of positive pressure increase of the turbocharger; for example, 10 psi (69 kPa) of boost means the air is being induced into the engine at 24.7 psi (170 kPa). Normal atmospheric pressure is 14.7 psi (101 kPa); add the 10 psi (69kPa) to get 24.7 psi (170 kPa).

Turbosoplante Turbosoplante es un término que se usa para describir la cantidad de aumento de presión positiva del turboalimentador; por ejemplo, 10 psi (69 kPa) de soplo significa el aire que se introduce en el motor a 24.7 psi (170 kPa). La presión atmosférica normal es de 14.7 psi (101 kPa); se añaden los 10 psi (69 kPa) para obtener 24.7 psi (170 kPa).

Turbocharger A turbocharger uses the expansion of exhaust gases to rotate a fan-type wheel, which increases the pressure inside of the intake manifold.

Turboalimentador Un turboalimentador usa la expansión de los gases de escape para girar la rueda en forma de abanico, la cual aumentará la presión adentro del colector de entrada.

Turbo lag Turbo lag is a short delay period as the turbocharger winds up to develop sufficient boost pressures.

Turboarrastre El turboarrastre es un corto período de retrazo mientras el turboalimentador se levanta con torno a desarrollar suficiente presión de empuje.

Turbulence A rapid movement of the air/fuel mixture. Turbulence results in improved combustion because of better mixing.

Turbulencia Un movimiento rápido de la mezcla del aire/combustible. La turbulencia resulta en una combustión mejorada porque se mezcla mejor.

Undersize bearing Bearing with the same outside diameter as standard bearings but constructed of thicker bearing material in order to fit an undersize crankshaft journal.

Cojinete de dimensión inferior Un cojinete que tiene el diámetro exterior del mismo tamaño que los cojinetes normales pero que se ha fabricado de una material más gruesa para que puede quedar en un muñón de cigüeñal más pequeño.

Valley pans A pan located under the intake manifold in the valley of a "V"-type engine used to prevent the formation of deposits on the underside of the intake manifold.

Cárter en V Un cárter ubicado en la parte inferior del múltiple de admisión en la parte más baja de un motor tipo "V" que previene la formación de los depósitos en la parte inferior del múltiple de admisión.

Valve cover gaskets Components that seal the connection between the valve cover and cylinder head. The valve cover gasket is not subject to pressures, but must be able to seal hot, thinning oil.

Juntas de la tapa de válvula Los componentes que sellan la conexión entre la tapa de las válvulas y la cabeza del cilindro. La junta de la tapa de válvula no se sujeta a la presión, pero si tiene que sellar el aceite caliente muy fluido.

Valve fillet The valve fillet is the curved area between the stem and the head.

Chaflán de la válvula El chaflán de la válvula es un área curveada entre el vástago y la cabeza.

Valve float A condition that allows the valve to remain open longer than it is intended. Valve float is the effect of inertia on the valve.

Flotación de la válvula Una condición que permite que queda abierta la válvula por más tiempo de lo designado. Es el efecto que tiene la inercia en la válvula.

Valve guide A part of the cylinder head that supports and guides the valve stem.

Guía de válvula Una parte de la cabeza del cilindro que apoya y guía al vástago de la válvula.

Valve lift Valve lift is the total movement of the valve off of its seat, expressed in inches or millimeters.

Carrera de la válvula La carrera de la válvula es el movimiento total de la válvula fuera de su asiento, y se expresa en milímetros o en pulgadas.

Valve overlap The length of time, measured in degrees of crankshaft revolution, that the intake and exhaust valves of the same combustion chamber are open simultaneously.

Períodos de abertura de las válvulas El período del tiempo, que se mide en los grados de las revoluciones del cigüeñal, en el cual las válvulas de admisión y de escape de la misma cámara de combustión estan abiertas simultáneamente.

Valve seat Machined surface of the cylinder head that provides the matting surface for the valve face. The valve seat can be either machined into the cylinder head or a separate component that is pressed into the cylinder head.

Asiento de la válvula La superficie acabada a máquina de la cabeza del cilindro que provee una superficie de contacto para la cara de la válvula. El asiento puede ser labrado a máquina en la cabeza del cilindro o puede ser un componente aparte para prensar en la cabeza de la válvula.

Valve seat inserts Valve seat inserts are pressed into a machined recess in the cylinder head.

Insertos del asiento de la válvula Los insertos del asiento de la válvula están prensadas dentro de la ranura torneada de la cabeza del cilindro.

Valve seat recession The loss of metal from the valve seat which causes the seat to recede into the cylinder head.

Encastre del asiento de la válvula La pérdida del metal del asiento de la válvula, lo que causa que el asiento retroceda de la cabeza del cilindro.

Valve spring A coil of specially constructed metal used to force the valve closed, providing a positive seal between the valve face and seat.

Resorte de válvula Un rollo de metal de construcción especial que provee la fuerza para mantener cerrada la válvula, así proveyendo en sello positivo entre la cara de la válvula y el asiento.

Valvetrain The series of components that work together to open and close the valves.

Tren de válvulas Un serie de componentes que trabajan juntos para abrir y cerrar las válvulas.

Valve stem The valve stem guides the valve through its linear movement.

Vástago de la válvula El vástago de la válvula guía la válvula a través de su movimiento linear.

Valve timing Valve timing refers to the relationship between the rotation of the crankshaft and camshaft. That correct relationship ensures that the valves open at the correct time.

Reglaje de válvula El reglaje de válvula se refiere a la relación entre el giro del cigüeñal y del árbol de levas. Esa correcta relación asegura que las válvulas se abren al tiempo correcto.

Valve timing mechanism The valve timing mechanism drives the camshaft in the correct timing with the crankshaft. It may use a belt, chain or gears.

Mecanismo de reglaje de válvula El mecanismo de reglaje de válvula corre el árbol de levas en la puesta a punto correcta con el cigüeñal. Puede hacer uso de una banda, una cadena o ruedas dentadas.

Variable-length intake manifold The variable-length intake manifold has two runners of different lengths connected to each cylinder head intake port.

Colector de entrada de longitud variable El colector de entrada de longitud variable tiene dos muelas de diferentes longitudes conectadas a cada orificio de admisión de la cabeza del cilindro.

Viscosity The measure of oil thickness.

Viscosidad La medida de lo espeso de un aceite.

olatility Fuel volatility is the ability of the fuel to vaporize. The higher the volatility, the easier it is for the fuel to evaporate.

Volatilidad La volatilidad del combustible es la habilidad de un combustible para evaporarse. A mayor volatilidad, es más fácil que se evapore el combustible.

Volumetric efficiency A measurement of the amount of air/fuel mixture that actually enters the combustion chamber compared to the amount that could be drawn in.

Rendimiento volumétrico Una medida de la cantidad de la mezcla del aire/combustible que actualmente entra en la cámara de combustión comparada con la cantidad que podría entrar.

Warning light A warning light is a lamp that is illuminated to warn the driver of a possible problem or hazardous condition.

Luz de alerta La luz de alerta es una lámpara que está prendida para avisarle al conductor de un posible problema o de una condición peligrosa.

Wastegate A wastegate limits the maximum amount of turbocharger boost by directing the exhaust gases away from the turbine wheel. The wastegate valve is also referred to as a bypass valve.

Válvula de expulsión La válvula de expulsión limita la máxima cantidad de empuje del turboalimentador al desviar los gases de escape de la rueda de la turbina. La válvula de expulsión también se conoce como válvula de desviación.

Wedge chamber Combustion chamber employing a wedge design used to create turbulence to mix the air and fuel.

Cámara biselada Una cámara de combustión que emplea un diseño biselado para crear una turbulencia que mezcla el aire y el combustible.

Wrist pin A wrist pin is a hardened steel pin that connects the piston to the connecting rod and allows the rod to rock back and forth as it travels with the crankshaft. It is also called the piston pin.

Rótula de muñeca Una rótula de muñeca es una rótula o eje de acero mejorado que conecta el pistón a la biela y permite que el eje se mesa de un lado a otro mientras se mueve con el cigüeñal. También se le llama eje del pistón.

Zero lash Zero lash is the point where there is no clearance or interference between components of the valve train.

Tira nula La tira nula es el punto en donde no hay huelgo o interfe-rencia entre los componentes del tren de la válvula.

INDEX

Note: Page references with *f* notation refer to a figure on that page.

A

Additives, gasoline, 98
  detergents, 98
  oxidation inhibitors, 98
  tetraethyl lead (TEL), 98
Adhesives, 115
Advanced camshafts, 203
"Advanced Technology Partial Zero-Emissions Vehicle" (AT-PZEV) standard, 277
Aerobic sealants, 122
Air, 21, 22
  flow of, through supercharger system components, 141–143
Air cleaners, 128, 129–130
Air filters
  design of, 127–129
  function of, 127
Airflow restriction indicator, 127
Air induction systems. *See* Intake system
Air pollution, 1, 95, 265
Alloys, 103, 179–180
Aluminum, 77, 104, 176
American Petroleum Institute (API), 65
  oil ratings, 66, 68
  Starburst symbol, 67
Anaerobic sealants, 122
Annealing, 108
Antifoaming agents, 65
Antifreeze, first permanent, 75
Anti-oxidation agents, 65
Armature, 63
Atkinson cycle engine, 284
Atmospheric pressure, 22–23, 38
  normal, at sea level, 133
  on pressure gauges, 133
Austenitic steel, 180
Automotive engines. *See* Engine(s)
Automotive fuels, 95–99
  ethanol, 97
  gasoline, 95
    additives, 98
  octane number, 95
    detonation, 96
    knock sensor, 96

B

Babbitt, Isaac, 249
Back pressure, 148
Balance shafts, 28–30, 31, 250
  bearings, 249
  rotation of, 29
Batteries, 60–62, 282
  technology for, 279
Battery Council International (BCI), 62

Battery terminals, 60
  common types of, 61
Bearing crush, 245–246
Bearing inserts, 234, 243. *See also* Bearings
Bearing journals, 30
Bearings, 243–244. *See also* Bearing inserts
  camshaft, 204
    and balance shaft, 249
  characteristics of, 245–248
  connecting rod, 247
  construction requirements for, 244–245
  main, 246–247
  materials, 244–245
    codes used for, 245
  oil clearances, 248–249
  rod, 248
  thrust, 247–248
  undersize and oversize, 249
Bearing spread, 245–246
Bedplate, 230, 231
Bentley Motors, 238
Biomass-to-liquid (BTL), 98
Blowby, 258–259
Blowers, 135, 140. *See also* Superchargers
Body control module (BCM), 88
Bolt diameter, 110
Bolt length, 110
Bolts, 110
  terminology for, 110
  torque specifications
    metric, 113
    standard, 113
    typical, 111
Bores, 31–32, 232–234
  lifter, 238
  main, 232–234
Bottom dead center (BDC), 23, 201–202
Brackets, 148–138
Brake horsepower, 36
Breakdown, oil, 68
Buick, changes since 1954, 212
Burned valves, 181–182
Burn time, 93

C

California Air Resources Board (CARB), 277
Cam ground pistons, 252–253, 253*f*
Camshaft bore, 178
Camshaft lift, 200
Camshafts, 3, 18, 175, 200–204, 221*f*, 224
  advanced, 203
  bearings, 204, 243–244
  duration of, 201

journals, 73*f*
lobe separation angle, 202*f*
lobes of, 200
retarded, 203
valvetrains and, 206–208, 207*f*
Carbon monoxide (CO), 145, 146
Carbon steel, 103
Carburetor, Pogue, 148
Case hardening, 108, 238
Cast aluminum, 106
Casting, 106
Cast iron, 103, 106
Catalytic converters, 145–147
  problems with, 147
  types of, 146*f*, 147*f*
C-class oil, 65
Central port fuel injection (CPFI), 91
Ceramics, 105–106, 179
Chamber-in-piston combustion chambers, 190
Channeling, 182
Chemical energy, conversion of, 20
Chrysler, "Hemi" engine, 221
Clamps, 148
Clean Air Act of 1990, 265, 266
Clean Fuel Fleet program, 266
Clearance ramps, 202
Closed loop system, 268
Clutch, 5*f*, 21
Cohesion agents, 65
Coil, 27*f*
Coil-on-plug (COP) ignition systems, 26, 93, 94*f*
Coke, 103
Cold cranking amperage (CCA), 62
Combustion, 2
  abnormal, 158, 159*f*, 160–161, 162–163, 194*f*, 195, 196
    support systems' contribution to, 162–163
  energy conversion during, 19–20
  expansion controls and, 252–253
  fuel and, 158–162
  heat generated by, 5
  issues, 160
  normal, 192
  process, 192–195
Combustion chamber, 2–3, 17*f*, 18–19
  designs, 188–191. *See also specific types of combustion chambers*
  heat generated within, 18–19
  leaks, 154*f*
  sealing, 154–155. *See also* Seals
  squish area of, 193
Composites, 105
Compressed natural gas (CNG), 265
Compressed natural gas (CNG) vehicles, 272–273
Compressed natural gas vehicles
  cylinder safety, 273–274
  Honda Civic GX, 277–278
  layout of, 273*f*

manufacturers of, 272
operation of, 275–276
safety, 274
Compression-ignition (CI) engine, 26
Compression ratio, 34–35, 34*f*
  calculating, 35
Compression rings, 259, 259*f*
Compressor wheel, 135–136, 135*f*
Connecting rod bearing journals, 234, 234*f*
Connecting rod bearings, 248
Connecting rods, 10, 15, 261–262
  methods of construction for, 261–262
Continuously variable transmission (CVT), 277, 283
Converter, 283
Coolant, 5, 75–77
  flow, 81*f*, 82*f*, 83–84, 84*f*
  leaks, 182
Coolant recovery system, 78–79, 79*f*
Cooling fans, 85–87, 86*f*
Cooling systems, 5, 74–87
  components of, 74–75
  coolant, 75–77. *See also* Coolant
  flow, 83–84
  cooling fans, 85–87
  heater core, 79
  hoses, 80
  radiators, 77–78
  reverse flow, 84–85
  thermostat, 80, 82*f*
  warning indicators, 87–91. *See also* Warning indicators
  water pump, 80
Core plugs, 231–232
Corrosion
  inhibitors, 65, 75
  plugging of coolant passages due to, 84
Counterweights, crankshaft, 237
Cracks, 109
  common causes for, 109
  detecting, 109–110
Crankshaft end play, 247, 248*f*
Crankshafts, 3*f*, 5*f*, 15–16, 16*f*, 234–236, 236*f*
  configurations of, 235
  construction of, 237–238
  counterweights, 237
  fillets, 237
  journals, 10, 15
  oil holes, 236
  rotation, 16, 23, 24, 29
    direction of, 33
  splayed, 236, 236*f*
  throws, 16
  twisting force on, 19. *See also* Torque
Crankshaft throw, 235
Crate engines, 166, 171, 172, 240
Crossflow radiators, 77, 78*f*
Cycle, 23

Cylinder block, 15. *See also* Engine block
  with bedplate, 230
  construction, 229–230
  core and gallery plugs, 231–232
  lubrication and cooling, 230–231
  and related components, 229*f*
Cylinder head gaskets, 115
Cylinder heads, 8, 9*f*, 15, 17*f*, 18, 72, 175–177
  component relationships, 187–188
  cracked, 109, 177
  intake and exhaust ports, 177
  reconditioning process for, 188
  valves, 178–181. *See also* Valves
Cylinders
  arrangement of, 25
  bore of, 31
  number of, 25
  wear on walls of, 156–157

D

Daimler-Chrysler, 67
Daimler, Gottlieb, 11
Dedicated alternative fuel vehicles, 266
Delco Electric, 60
Detergents, 65
Detonation, 160, 161, 194
  common causes of, 194
Diesel engines, 40–42
Differential, 166
Dispersants, 65
Displacement, 31–33
  calculating engine, 32–33
  -on-demand engines, 45
Distributor, 219, 220*f*
Distributor ignition (DI) systems, 26, 93
Dodge, 283
Downflow radiators, 77, 77*f*
Dual parallel system, 280
Duration, camshaft, 201
  methods used to measure, 201–202

E

E85, 97, 269–272
Eddy currents, 191
Efficiency, 37–39
  mechanical, 37
  thermal, 38–39
  volumetric, 37–38
Electrical energy, conversion of, 20, 59
Electric motors, 282
Electrolysis, 75
Electromagnets, 63
Electronic ignition (EI) systems, 93–94
Emission control system, 94–95
  devices in, 95
Emissions. *See also* Exhaust system
  credits, 277

reduction of, 28, 266
End gases, 189, 193
Energy
  conversion, 19–20
  heat, 19
  molecular, 21
  types of, 19
  work and, 19
Engine bearings, 15
Engine block, 10–11. *See also* Cylinder block
  components, 227–228, 229*f*. *See also specific components*
  construction, 229–234
  galleries, oil flow through, 230–231, 231*f*
Engine coolant temperature (ECT) sensor, 85
Engine cradle, 166, 167, 167*f*
Engine manifold vacuum, 134
Engine mounts, 4*f*, 168–169
  types of, 169–171, 170*f*
Engine oil, 65–67
  additives, 68
  breakdown, 68
  classifications, 65, 66
  organizations that rate, 65
  synthetic oils, 68–69
  viscosity, 67–68
Engine operation, 22–24
  basic, 2–5
  cooling system, 5–6, 74–87. *See also* Cooling systems
  fuel system, 91–92
  lubrication system, 64–74. *See also* Lubrication systems
  principles, 18–22
  starting system, 59–63
  systems, 58–98
  theory of, 14–47
Engine(s), 1, 2*f*. *See also* Cylinder block; Engine operation
  adhesives, sealants, and form-in-place gaskets, 121–122
  Atkinson cycle, 42, 284
  breathing, 7
  classifications, 24–27, 25*f*
  comparisons between gasoline and diesel, 41, 41*f*
  components, 15–18, 16*f*
  compression-ignition (CI), 26
  configurations, 166–168
  coolant flow through, 77*f*, 82*f*, 83–84
  crate, 166, 171, 171*f*, 172, 240
  cubic inch displacement (CID) of, 33, 43
  cylinder head, 8–9
  designs, 39–40
  diesel, 40–42
  displacement, 31–33
    calculating, 32–33
  displacement-on-demand, 45
  DOHC (Dual Overhead Cam) V8, 26
  efficiency, 37–39
  eight-cylinder, 7*f*, 10, 11*f*, 16*f*
    that can run on four, 42

engine block, 10–11
fasteners, 110–112
four-stroke, 10, 39, 40
free-wheeling, 183, 216
function of, 5
gaskets, 102, 115–120
gasoline, 283–284
"Hemi," 221
identification, 45–47
interference, 183, 216
internal combustion, 1, 10
location of, 3
longitudinally mounted, 3, 167, 168*f*
manufacturing processes for, 106–108
materials, 102–106
   treatment methods, 108
measurements, 33
mid-mounted, 168
Miller-cycle, 43
mounts. *See* Engine mounts
multivalve, 195
noises, 163
overhauling, 9*f*, 11, 157, 167*f*, 171
overhead camshaft, 8, 9*f*
overheated, 6, 90
oversquare, 31
performance, 8*f*
  factors affecting, 142–163
remanufactured, 166, 171–173
repair professional associations, 290
seals, 115, 119–121
spark-ignition (SI), 26
speed and spark delivery, 93
stratified charge, 43, 44*f*
supercharger. *See* Superchargers
timing mechanisms, 10
transverse-mounted, 3, 4*f*, 25, 166
turbocharger. *See* Turbochargers
two-stroke, 39–40
types of, 26
undersquare, 31
V6, power stroke of, 236
vacuum, 22, 155, 155*f*
vibration, 28–30
Environmental Protection Agency (EPA), 160, 265, 269, 277
Erosion, metal, 182–183
Ethylene glycol (EG) coolant, 75, 76
Evaporative emissions system, 95
Exhaust gas recirculation (EGR) system, 95, 115, 222
Exhaust manifold, 7, 144–145
Exhaust manifold gaskets, 118, 125
Exhaust pipe, 145
Exhaust system, 143–149
  catalytic converters, 145–147, 147*f*
  clamps, brackets, and hangers, 148
  components of, 143–144

exhaust manifold, 144–145
exhaust pipe and seal, 145
heat shields, 148
mufflers, 147–148
resonator, 148
tailpipe, 148
Expansion plugs, 231–232

F

Face, valve, 178, 179*f*
Fast-burn combustion chambers, 190, 191
Fasteners, 110–112. *See also* Gaskets
  metric, 110
  replacement concerns for threaded, 110
  torque-to-yield, 112–113
  types of, used throughout engine, 110
Ferrous metals, 103
Fillets, 111*f*, 180, 237, 237*f*
Fire rings, 117*f*, 122
Flexible fuel vehicles, 269–272
  manufacturers of, 270–272
Flywheel, 59, 73*f*, 229*f*, 239–240
Followers, 199
Ford Motor Company, 67
Forging, 107
Form-in-place gaskets, 121
Free-wheeling engines, 183, 216
Freeze plugs, 231
Friction, 21
  energy conversion created by, 20
  reducing, 64
Front main seal, 120, 121*f*
Fuel
  automotive, 95–99
  alternative, use of, 266–268
  being considered to power fuel cells, 287
  combustion and, 158–162
  consumption, reduction of, 28
  delivery, 96
  diesel, 98–99
  emissions, 147, 160
  octane rating of, 158–159
  source, nonreplenishable, 193
  volatility, 159–160
Fuel cell vehicles, 287
Fuel economy
  ensuring, 126
  increased, 68
  Pogue carburetor and, 148
  without increased exhaust emissions, 131
Fuel injection systems, 28, 91, 266
Fuel system, 91–95
  components of, 91*f*
  emission control system, 94–95
  ignition system, 93–94
Full-filtration system, 70

G

Gases, behavior of, 21
Gaskets, 102, 115–119. *See also specific types of gaskets*
  classifications, 115
  form-in-place, 121
  miscellaneous, 120
Gasoline, 19, 95. *See also* Automotive fuel, Fuel
Gasoline direct injection (GDI) systems, 28, 91
Gasoline engines, 283–284. *See also* Engine(s)
Gas-to-liquid (GTL), 98
Gauge sending units, 88
General Motors, 67, 140, 228
  electric car of, 278
  engine specifications, 228
  HEV truck, 283
Glow plugs, 41
Grade marks, 111
Grade, of bolt, 110
Gray cast, 103, 230, 237
Gross horsepower, 36

H

Hangers, 148, 149
Harmonic balancers, 239, 239*f*
HD-5 LPG, 266
Head gaskets, 52, 84
  damage to, 158*f*
  failures of, 176
Head, valve, 178
Heat energy, 18, 19
  released during combustion, 19
Heater core, 79
Heat shields, 148
Hemispherical chamber, 189
Honda, 222
  Civic GX, 272, 277
  Insight, 279
Horsepower, 2*f*, 5, 19
  curves, 36*f*
  gross, 36
  net, 36
  supercharger increase of, 140
  torque and, 35–37
Hoses, 80
  bypass, 82*f*
Hybrid cars, 195
Hybrid electric vehicles (HEVs), 279, 280
  gasoline engines in, 283–284
  manufacturers of, 279–280
  operation of, 280–284
  technology components, 282–283
  Toyota Prius, 284–286
Hydrocarbons (HC), 145, 146
Hydrogen, 287

I

Ignition system, 93–94. *See also* Spark plugs
  types of, 27–28
Indicated horsepower, 36
Inertia, 20–21
Inlet check valve, 71, 72
Intake air temperature (IAT) sensor, 126*f*, 129
Intake manifold, 7, 129–133
  tuning, 131
Intake manifold gasket, 115, 116*f*, 118*f*
Intake manifold tuning valve (IMTV), 132
Intake system, 125–149
  air cleaner/filter, 127–129
  components of, 125
  ductwork, 126–127
  functions performed by, 126
  intake manifold, 129–133
  turbochargers, 134–135
  vacuum controls, 134
Integral valve seats, 184
Intercoolers, 138
Interference engines, 183, 216
Internal combustion engines (ICEs), 1, 14
  creation of first workable, 10
International Lubricant Standardization and Approval Committee
  (ILSAC), 67
Inverter, 283
Iron, 103

J

Japanese Automobile Standards Organization (JASO), 67
Journals
  bearing, 30
  camshaft, 64*f*, 204*f*
  connecting rod bearing, 234
  crankshaft, 10, 15
  fillets on, 237
  main, 10, 15
    bearing, 235, 236
  rod, 15
  splayed, 236*f*

K

Kettering, Charles F., 60
Kinetic energy, 21
  wasted during braking, 281
Knock sensor (KS), 161

L

Lenoir, Jean Joseph, 10
Lexplate, 291
License plates, appearance of, 24
Lifter bores, 238
Lifters, 199, 204–206
  hydraulic, 202, 204–205, 205*f*
  pump up of, 206

roller, 200, 205, 205*f*
   solid, 204, 205, 205*f*, 206
Light trucks, 121
Lip seals, 120, 121
Liquefied petroleum gas (LPG), 266
Liquids, behavior of, 21
Lobes, camshaft, 200
Lobe separation angle, 203
Locating tab, 246
Long blocks, 227, 240
Longitudinally mounted engines, 3, 166, 167, 168*f*
L terminals, 61
Lubrication systems, 6, 64–74, 73*f*
   components of, 65
   engine oil, 65–68
   oil coolers, 72
   oil filters, 70–71, 72*f*
   oil flow, 72–74
   oil pumps, 70
   synthetic oils, 68–70
   warning indicators, 87–91

M

M85, 269–270
Machining, 107
Magnaflux, 109, 109*f*
Magnesium, 104
Magnetic fields, 63, 63*f*
Magnetic particle inspection (MPI), 109
Main bearing journals, 234, 234*f*
Main bearings, 244*f*
Main bores, 232–234, 233*f*
Main caps, 232–234, 234*f*
Main journals, 10, 15
Major thrust surface, 255
Malfunction indicator lamp (MIL), 147
Manifold absolute pressure (MAP) sensor, 91
Margin, 179
Mass airflow (MAF) sensor, 91, 92*f*, 126
Mechanical efficiency, 37
Mechanical energy, conversion of, 20
Message center, 90–91
Metallurgy, 102
   powder, 108
Metals, 102–104
   erosion of, 182–183
   expansion and contraction of, 115
   ferrous, 102
   nonferrous, 102
   tensile strength of, 104
Micron, 70
Mid-mounted engines, 168
Miller-cycle engine, 43
Minor thrust surface, 255

Misfire, 155, 161–162
   causes for, 177
   mechanical faults causing, 195
Modular blocks, 230
Molecular energy, 21
Momentum, 21
Motor mounts, 3, 4*f*
Mufflers, 147–148
   design changes for, 147–148
Multi-point fuel injection (MPFI) systems, 28, 91
Multivalve engines, 195

N

National Propane Gas Association, 268
Natural gas, 272–273
Necking, 183
Net horsepower, 36
Newton's Laws of Motion, 20
Nitriding, 238
Nodular iron, 103
Nonferrous metals, 103
Nucleate boiling, 77
Nuts, 110
   torque specifications for, 113
     metric, 114*f*

O

OBD-II regulations, 146–147
Octane rating, 158–159
   of CNG, 272
   MON (motor octane number), 95
   RON (research octane number), 95
Offset pins, 234, 236*f*, 255*f*
Off-square seating, 183
Oil coking, 138
Oil control rings, 260
Oil coolers, 72
Oil, engine, 65–68
Oil filters, 70–71
Oil flow, 72–74
Oil galleries, 70–71, 230–231
Oil grooves, 247
Oil holes, 236
Oil pan gaskets, 119
Oil pumps, 70
Open loop system, 268
Organic Additive Technology (OAT) corrosion inhibitors, 75
O-rings, 116, 118*f*, 186–187, 186*f*, 230
Otto, Nicholas, 10
Overhead camshaft engine (OHC), 8, 9*f*, 26, 175, 176*f*, 199, 200
   valvetrain designs, 206–207
Overhead valve (OHV), 26
Oversize bearings, 249
Oxidation, 65
Oxides of nitrogen (NOx), 146

P

Parallel system, 280
Pedestal-mounted rocker arms, 209, 210f
Pentroof combustion chambers, 189
Piezoresistive sensor, 88, *88*
Piston pins, 251, 255f. *See also* Wrist pins
   offset, 255, 255f
Piston rings, 16, 258–260
   materials used for, 260
   oversize, 261
   wear on, 256
Pistons, 2, 251–252
   cam ground, 252, 253
   designs and construction, 256–258
   formula for determining travel of, 35
   parts of, 251–252
   speed, 256
   tapered, 252, 253f
Piston skirts, 252f, 253
Pitch, 110, 111f
Plastics, 105
Plugs, 41, 231–232
Pogue carburetor, 148
Pogue, Charles Nelson, 148
Pole shoes, 63
Pollutants, toxic, 58, 95
Poppet valves, 178, 179f
Positive crankcase ventilation valve (PCV) system, 95
Positive displacement pumps, 70
Post terminals, 60
Potential energy, 19
Pour point depressants, 65
Powder metallurgy, 108
Power splitting device, 282, 283
Powertrain control module (PCM), 2
   activation of solenoid by, 45, 224, 276
   alternative fuel vehicles and, 268, 269
   atmospheric pressure and, 22
   control of cooling fans by, 85–86
   control of fuel delivery by, 14, 91–92, 126, 127f
   ignition system and, 93–94
   knock sensor, 96
   monitoring of catalytic converter by, 146
   sensitivity of, 154
   systems controlled by, 93–94
   warning messages displayed by, 90–91. *See also* Warning indicators
Powertrains, 1, 2
Preignition, 34, 160–161, 162f, 193–194
   common causes of, 194
Pressure
   compressibility and, 21
   vacuum and, 22
Pressure gauges, 133
Propane vehicles, 266–268
Propylene glycol coolant, 75

Prove-out circuit, 90, 90f
psia (pounds per square inch absolute), 133
psig (pounds per square inch gauge), 133
psi (pounds per square inch), 133
Pumping losses, 284
Pump up, lifter, 206
Pushrods, 8, 200, 205f

Q

Quenching, 188, 193f

R

Radiators, 77–78
   air flowing across, 5–6
   crossflow, 77
   downflow, 77
Rear main seal, 120, 121
Reciprocating, 24
Reed valve, 39
Regenerative braking, 281
Reid Vapor Pressure (RVP), 159
Remanufactured engines, 166, 171–173
   advantages of, 171
   disadvantages of, 172–173
Resonator, 148
Retarded camshafts, 203
Retrofitted alternative fuel vechicles, 266
Reverse flow cooling systems, 84–85, 84
Reversion, 203
Ring gear, 59
Rocker arm ratio, 208–209
Rocker arm(s), 9
   design of, 208, 209f
   geometry, 209–211
   materials used in construction of, 208
   mounting methods, 209
   pedestal-mounted, 209
   shaft-mounted, 209
   stud-mounted, 209
Rod journals, 15
Rolls Royce, 238
Room-temperature vulcanizing (RTV) compound, 122
Roots-type supercharger, 141
Rotary valve, 39
Rust inhibitors, 65

S

Saddles, 72
S-class oil, 65
Sealants, 115, 120, 121–122
Seals, 102, 119–121
   valve, 188
Secondary air injection system, 95
Sequential fuel injection (SFI) systems, 28
Series system, 280
Shaft-mounted rocker arms, 209, 210f
Shock absorbers, 170, 170f
Short blocks, 227, 240

Shot peening, 108
Side terminals, 61
Silencer, 148
Sintering, 108
Skin effect, 193
Smoke, 152, 153*f*, 176–177, 187
Society of Automotive Engineers (SAE)
  -measured camshaft, 202
  oil ratings, 65, 66*f*
  standard for engine rotation, 33
  standards for measuring horsepower, 36
Solenoids, 45, 224–225, 219
  vacuum, 134, 134*f*
Spark-ignition (SI) engine, 26
Spark plugs, 18, 94*f*, 152–154
Squish areas, 188
Stainless steel, 104
  alloys, 179
Stamping, 107–108
Starter motor, 63
Starting system, 59–63
  battery, 60–62
  starter motor, 63
Steel, 103–104, 179
  austenitic, 180
Stellite, 184
Stem necking, 183, 183*f*
Stratified-charge combustion chambers, 191, 192*f*
Stratified-charge engine, 43, 44*f*
Strip seals, 115
Strokes, 23, 40, 41*f*, 42
  crankshaft throw and, 235
  four, of automotive engine, 29
Stud-mounted rocker arms, 209
Superchargers, 139–141
  airflow through, 141, 142*f*
  disadvantage of, 138, 141
  operation of, 141–143
  Roots-type, 141
Super-Ultra-Low-Emission Vehicle (SULEV) standard, 277, 278
Surface-to-volume (S/V) ratio, 189
  emissions and high, 189
Swirl chambers, 190
Synthetic oils, 68–70

T

Tailpipe, 148
Tapered pistons, 252
Tappetts, 204
Temperature, 21, 38
  atmospheric pressure and, 22
Tempering, 108
Tensile strength, 104
Thermal efficiency, 38–39
Thermal energy, conversion of, 19
Thermistor, 88, 89*f*
Thermodynamics, 18

Thermostat, 5, 80, 81*f*
Thread depth, 111
Throttle, open, 22
Thrust bearings, 246*f*, 247
Timing belt, 219–221, 220*f*
Timing chain guides, 218, 219
Timing chains, 10
Timing chain tensioner, 215*f*, 218
Timing cover seal, 120
Timing mechanisms, 10, 215–224
  belt-driven systems, 219–221
  chain-driven systems, 218
  gear-driven systems, 221
  variable valve timing and lift, 222–225
Tipping, 183
Titanium, 105
Top dead center (TDC), 23, 201
Top terminals, 60
Torque, 5, 19, 28
  curves, 35*f*
  example of increased, 35*f*
  horsepower and, 35–36
  proper fastener, 112
Torque converter, 5, 240
Torque-to-yield fasteners, 113
Torsional vibrations, 239
Toyota, 224
  Prius, 279, 284, 285–286
Transaxle, 166
Transmission, 16
Transverse-mounted engines, 25, 166, 167*f*
Tumble port combustion chambers, 191
Turbine wheel, 135, 137
Turbo boost, 135
Turbochargers, 134–135
  basic operation of, 135–136
  boost pressure control, 136
  components of, 135*f*
  cooling, 137–138
  disadvantage of, 138
  history of, 134–135
  scheduled maintenance for, 138–139
  turbo lag, 136–137
Turbo lag, 136–137
Turbulence, 193, 193*f*
Two-stroke engines, 39–40

U

Ultra-Low-Emission Vehicle (ULEV) standard, 277
Undersize bearings, 249
Unified National Series, 110
U.S. Department of Transportation, 274

V

Vacuum
  basics, 133–134
  controls, 134

engine, 22, 155*f*

    pressure and, 22

Valley pans, 115

Valve cover gaskets, 118–119

Valve guides, 186

Valve lift, 200

Valve overlap, 24, 202

Valves, 24, 178–181. *See also* Valvetrains

    aluminized, 181

    breakage of, 183

    burned, 181–182

    butterfly, 132–133, 133*f*

    deposits on, 181

    determining malfunctions of, 181–184

    inlet check, 71, 72*f*

    materials used in construction of, 179–180

    poppet, 178–179, 179*f*

    servicing of, 157

    wear on intake and exhaust, 157

Valve seat inserts, 184

Valve seat recession, 185

Valve seats, 108, 184–186

    recession of, 186

Valve springs

    design of, 211–212, 211*f*

    functions of, 211

    retainers and rotators, 212

Valve stem, 199

    seals, 186–187

    wear, 184

Valve timing, 3. *See also* Valve timing mechanism

Valve timing mechanism, 10, 216–217

Valvetrains, 8, 9*f*, 175. *See also* Valves

    camshafts and, 199–213

    components of, 199–200. *See also specific components*

      reconditioning, 180–181

    designs, OHC, 206–207

    types of, 26

Variable-length intake manifold, 131, 132*f*

Variable valve lift systems, 222–225

Variable valve timing (VVT) systems, 28, 222–224

Vehicle identification number (VIN), 45–47, 45*f*, 46*f*

Vehicles

    alternative fuel and advanced technology, 265–287

    compressed natural gas, 272–273

    E85 and flexible fuel, 269–272

    electric, 278–279, 279*f*

    fuel cell, 287

    hybrid electric, 279–284

    propane, 266–268, 267*f*

Vibrations

    engine, 28–30

    torsional, 239, 239*f*

VIN. *See* Vehicle identification number (VIN)

Viscosity, 65–68

Viscosity index improver, 65

Volatility, fuel, 97, 159–160

Volumetric efficiency (VE), 37–38

W

Warmup converters, 146

Warning indicators, 87

    gauge sending units, 88

    lubrication system, 64–65

    message center, 90–91

    warning lights, 88–90

Warning lights, 88–90

Wastegate, 136–137, 137*f*

Water pump, 80

Wedge chamber, 188–189

Wrist pins, 16, 251, 251*f*

Z

Zero-emission vehicles (ZEVs), 279

Zero lash, 205